33

Topics in Heterocyclic Chemistry

For further volumes:
http://www.springer.com/series/7081

Aims and Scope

The series Topics in Heterocyclic Chemistry presents critical reviews on present and future trends in the research of heterocyclic compounds. Overall the scope is to cover topics dealing with all areas within heterocyclic chemistry, both experimental and theoretical, of interest to the general heterocyclic chemistry community.

The series consists of topic related volumes edited by renowned editors with contributions of experts in the field.

Roberto Paolesse
Editor

Synthesis and Modifications of Porphyrinoids

With contributions by
C.M.A. Alonso • J.F.B. Barata • J.A.S. Cavaleiro •
M.A.F. Faustino • D. Monti • S. Nardis • M.G.P.M.S. Neves •
S.M.G. Pires • C.I.M. Santos • K.M. Smith • A.C. Tomé •
V.I. Vaz Serra • J. Wojaczyński

Springer

Editor
Roberto Paolesse
Dept. of Chemical Science and Technologies
University of Rome Tor Vergata
Rome
Italy

ISSN 1861-9282 ISSN 1861-9290 (electronic)
ISBN 978-3-642-38532-2 ISBN 978-3-642-38533-9 (eBook)
DOI 10.1007/978-3-642-38533-9
Springer Heidelberg New York Dordrecht London

Library of Congress Control Number: 2013951146

Printed on acid-free paper

Springer is part of Springer Science+Business Media (www.springer.com)

Preface

Porphyrins and related macrocycles have fascinated and still continue to attract the attention of researchers belonging to different disciplines. It is impressive to note how these macrocycles are included in research articles spanning a huge number of different specialties, from medicine to material chemistry. The life is possible thanks to these macrocycles, but the richness of their properties makes porphyrins interesting for a wide range of fields, ranging from clinical applications to photovoltaic cells.

The first volume dedicated to porphyrinoids of Topics in Heterocyclic Chemistry is focused on some synthetic aspects of porphyrins and related macrocycles, because for all the studies and applications of the beautiful collections of porphyrinoids, the preparation and modification of these macrocycles is of course a necessary step.

In the first chapter titled "The McMurry Reaction in Porphyrinoid Chemistry" Kevin M. Smith, a leading researcher in the porphyrin area, describes the exploitation of the McMurry reaction for the macrocyclization route to the preparation of porphyrin analogues and for the preparation of bis-porphyrinoid systems.

The second chapter titled "Meso-tetraarylporphyrins: Synthetic Strategies and Reactivity Profiles Based on Nitro/Amino Substituents" presents an update on the preparation of nitro- and amino-derivatives of *meso*-tetraarylporphyrins and their exploitation for the further modification of the macroring to give other functionalized porphyrins; this chapter is authored by Maria G. P. M. S. Neves with Vanda I. Vaz Serra, Sónia M. G. Pires, Cristina M. A. Alonso, Augusto C. Tomé and José A. S. Cavaleiro.

In the third chapter titled "Functionalization of Corroles" the attention is focused on the functionalization of corrole, a contracted porphyrinoid that has object of intensive researches in the last decade, due to some peculiar characteristics of such a macrocycle. The chapter is surveyed by José A. S. Cavaleiro, Joana F. B. Barata, Carla I. M. Santos, M. Graça P. M. S. Neves, M. Amparo and F. Faustino and discusses the reactivity of corrole towards both electrophilic and nucleophilic reagents, describing the different functionalities that can be introduced in the corrole skeleton.

The fourth chapter titled "Degradation Pathways for Porphyrinoids" describes the degradation pathways of porphyrinoids, which are processes of interest for several aspects, ranging from catalysis to biochemistry. This chapter authored by Jacek Wojaczyński indicates that the degradation term in the case of porphyrinoids should not be considered negatively, because they can offer the opportunity to easily obtain linear oligopyrroles or to convert the parent porphyrinoid in a different macrocycle.

In the fifth chapter titled "Synthetic Routes to Porphyrinoids" Sara Nardis is back to describe porphyrin chemistry, presenting an update of the synthetic routes for the preparation of porphyrin bearing an unsymmetrical pattern of peripheral substituents, which is interesting for the exploitation of porphyrin in different application fields. This goal can be achieved by both the direct preparation of the macrocycle and the successive regioselective functionalization of a symmetrical porphyrin.

In the sixth chapter titled "Recent Developments of Non-covalent Porphyrin Assemblies" Donato Monti describes the recent advancements related to non-covalent porphyrin aggregates featuring supramolecular chirality, illustrating the preparation, the properties and the potential applications of these suprastructures.

Finally I would like to sincerely thank all the authors for their kind cooperation in the composition of their excellent contributions to this volume.

Contents

Top Heterocycl Chem (2014) 33:1–34
DOI: 10.1007/7081_2013_100

Published online: 30 March 2013

The McMurry Reaction in Porphyrinoid Chemistry

Kevin M. Smith

Abstract The McMurry reaction, first reported in 1974, is an organometallic coupling reaction of aldehydes and ketones, to give alkenes, that relies upon the use of low-valent titanium; the reactive titanium species is obtained from titanium (III) or titanium(IV) chloride and an in situ reducing agent. The first application of the McMurry reaction to polypyrrole chemistry took place in 1986 with the first synthesis of the (2,0,2,0)-porphyrin isomer, porphycene. Since then, numerous macrocyclization reactions of polypyrrole (and other heteroaromatic) compounds have been described, and many new molecules have been reported. Peripheral carbonyl substituents on metallo-porphyrins and -chlorins have been coupled to yield a plethora of bis-porphyrins, bis-chlorins, and their homo- and hetero-bimetallic complexes. Some porphyrin-chlorin systems have also been reported. Thus, this contemporary organometallic coupling reaction has been successfully applied to polypyrrole syntheses and substituent manipulations, and a large number of interesting new compounds have been obtained.

Keywords Alkene syntheses · Bisporphyrinoid alkenes · Expanded porphyrinoids · Low-valent titanium · Macrocyclization · McMurry reaction · Porphycene · Porphyrin isomers · Porphyrinoids

Contents

K.M. Smith (✉)
Department of Chemistry, Louisiana State University, Baton Rouge, LA 70803, USA
e-mail: kmsmith@lsu.edu

1 The McMurry Reaction

The McMurry reaction was first reported by McMurry and Fleming in 1974 [1]. In its simplest form it involves the self-coupling of an aldehyde or ketone, using titanium(III) or titanium(IV) chloride and a reducing agent, to give a symmetrical alkene (**1**) (Scheme 1).

Intramolecular McMurry reactions of dicarbonyl compounds can, of course, yield highly unsymmetrical products. Numerous solvent-dependent mechanisms have been proposed [2] for the reaction, which basically involves two steps. Firstly, a low-valent titanium species (produced by reduction of $TiCl_3$ or $TiCl_4$) transfers an electron to the carbonyl groups, and the resulting anion-radicals couple to give a pinacolate (**2**). Next, deoxygenation of the pinacolate takes place, to give the product alkene (**1**) (Scheme 2).

2 Macrocyclization Reactions

2.1 Porphycenes

Porphycene (**3**) is the (2.0.2.0) isomer of porphyrin [which is designated (1.1.1.1)] wherein the bracketed numbers indicate the number of interpyrrolic carbon atoms. The first synthesis of (**3**) by Vogel and coworkers in 1986 [3] was both a landmark in the soon-to-be-burgeoning field of porphyrin isomers and also the first example of the application of the McMurry reaction to cyclic tetrapyrrole synthesis. Scheme 3 shows the route followed by Vogel and coworkers [3] to the synthesis of (**3**). 5,5′-Diformyl-2,2′-bipyrrole (**4**) was self-condensed, using standard McMurry conditions, to give a 3% yield of (**3**), presumably by way of the initial macrocyclization product (**5**).

The analogous MacDonald (so-called “2 + 2”) synthesis of porphyrins from self-condensation of a 1-formyldipyrromethane (**6**) [4] also proceeds via a porphodimethene intermediate (**7**) which then oxidizes (with air) during the course of the reaction, to give porphyrin (**8**) (Scheme 4).

As is the case for unsubstituted porphyrin (so-called porphin), the unsubstituted porphycene (**3**) was shown to be fairly insoluble in organic solvents, and this caused Vogel and his coworkers to synthesize a host of more useful symmetrically and unsymmetrically substituted porphycenes, some in yields as high at 25% [5–9]. Since then, numerous research groups have used one or other version of the

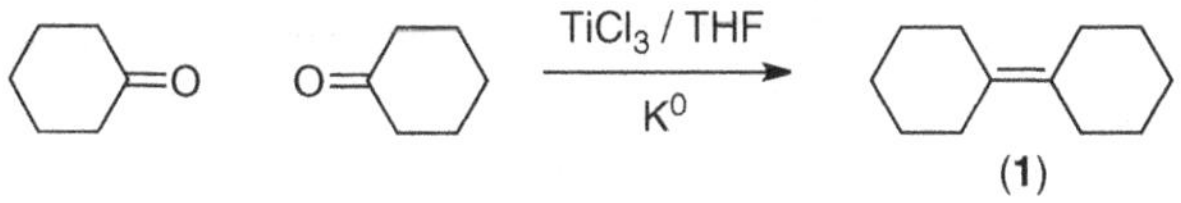

Scheme 1 A generic McMurry reaction

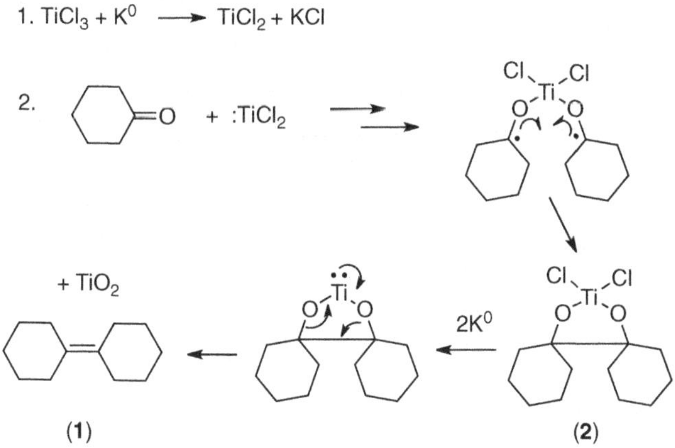

Scheme 2 Possible mechanism for the McMurry reaction

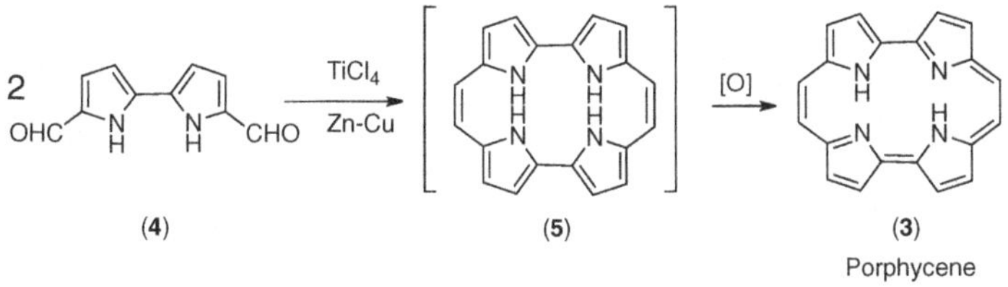

Scheme 3 The first synthesis of porphycene [3]

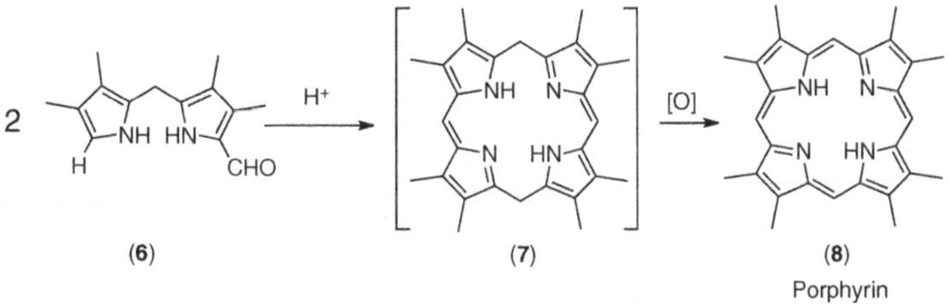

Scheme 4 The MacDonald "2 + 2" synthesis of a porphyrin [4]

McMurry reaction to prepare a multitude of variously substituted porphycenes from diformylbipyrroles (e.g. [10–14]).

If a McMurry reaction will work with aldehydes, the prospect is good that it will also work with ketones. Thus, for example, 5,5′-diacetyl-2,2′-bipyrrole (**9**) was self-condensed to give initially the dihydroporphycene (**10**) which in this case was

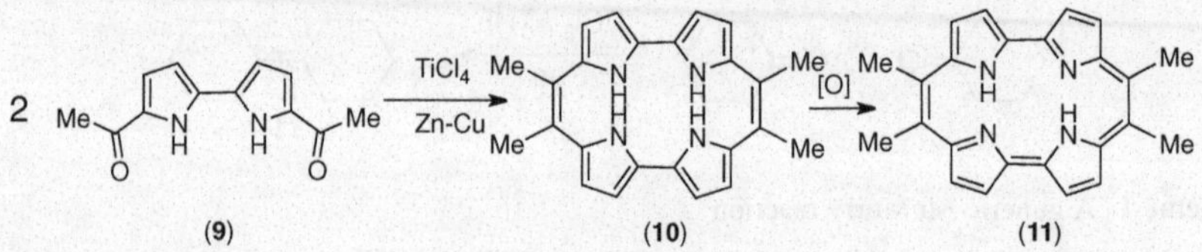

Scheme 5 Porphycene synthesis via bipyrrole diketones [15]

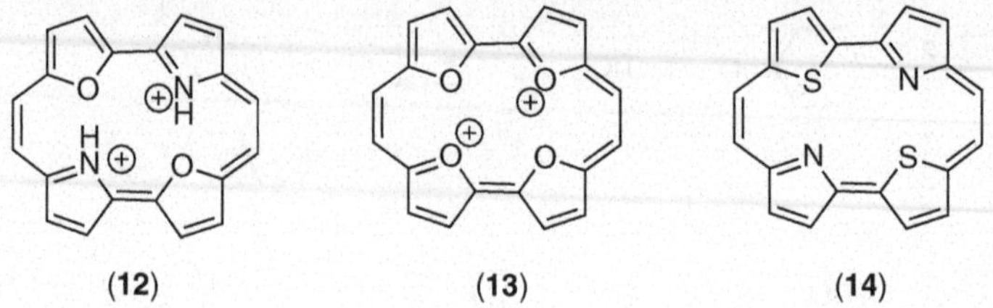

Chart 1 Typical oxa- and thia-analogues of porphycene prepared via the McMurry reaction

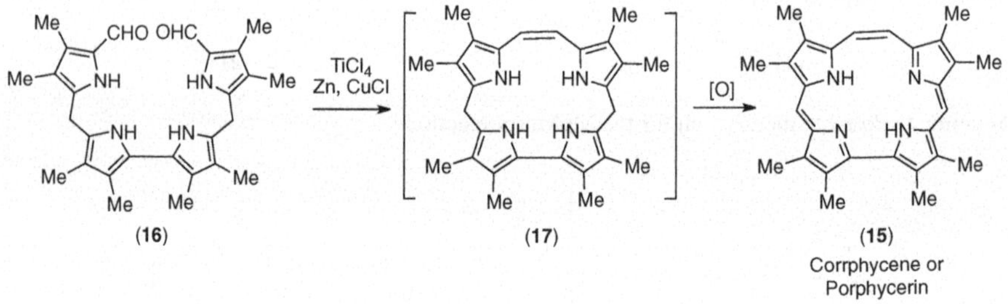

Scheme 6 Synthesis of corrphycene/porphycerin [22, 23]

isolated and characterized and was then oxidized to give the tetrasubstituted porphycene (**11**) [15] (Scheme 5).

Following the same basic self-condensation of 5,5′-diformyl compounds, but usually with a stronger final oxidant, di-oxa- (e.g. **12**) [16, 17], tetraoxa- (**13**) [18], dithia- (**14**) [19], and even tetrathia- [20, 21] analogues of porphycene have been synthesized (Chart 1).

2.2 Other Porphyrin Isomers

Two other porphyrin isomers [namely (2.1.0.1) and (2.1.1.0)] have been prepared via the McMurry reaction as the critical macrocyclization step.

Firstly, the (2.1.0.1) isomer (**15**), called corrphycene [22] or porphycerin [23], was synthesized by two different groups following the same fundamental approach. For example [22], the 5,5′-di(2-formyl-5-methylenepyrrolyl)-2,2′-bipyrrole (**16**) was subjected to an intramolecular McMurry reaction (Scheme 6) to give the

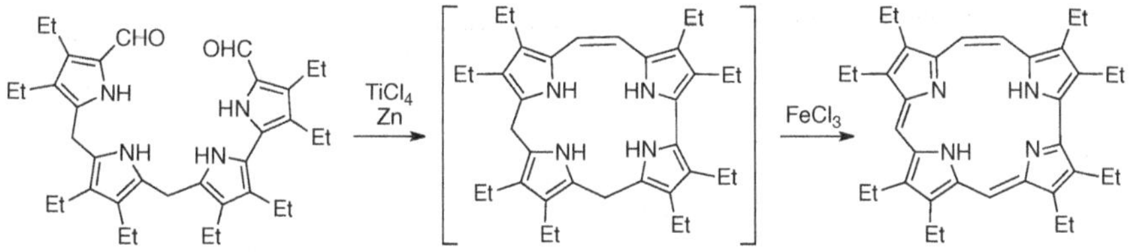

(19) (20) (18)

Hemiporphycene

Scheme 7 Synthesis of hemiporphycene (**18**) [24]

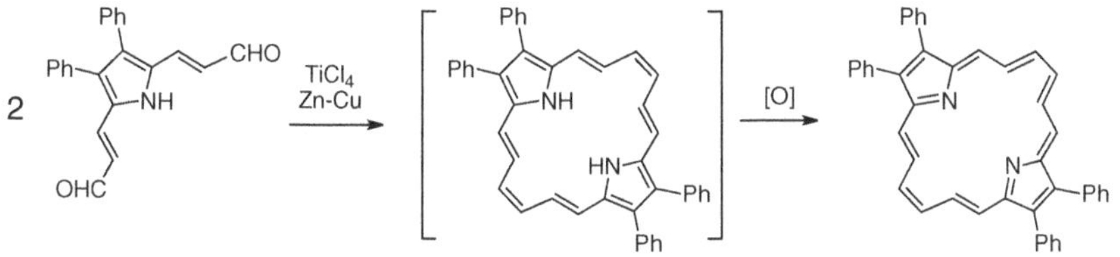

(22) (23) (21)

Scheme 8 Synthesis of 21,23-dideazoporphyrin (**21**) [25]

intermediate (**17**) which was not isolated, but was immediately oxidized with air or ferric chloride to give the corrphycene/porphycerin (**15**).

The second porphyrin isomer, (2.1.1.0) (**18**), called hemiporphycene was prepared [24] from the (5-formyl-5′-dipyrromethanyl)-2,2′-bipyrrole (**19**), using standard McMurry reaction conditions (Scheme 7). Once again, the reaction proceeded through a transient intermediate (presumably **20**), which was oxidized with ferric chloride to give the octaethylhemiporphycene (**18**).

2.3 *Dideazaporphyrin System*

The dideazaporphyrin (**21**) with absent opposite (21- and 23-) porphyrin-type nitrogens is a novel annulene which retains the overall aromatic character of the porphyrin system. It was prepared in 2010 by subjecting the bis-vinylogous-formylpyrrole (**22**) to standard McMurry reaction conditions and was obtained in 23% yield. Once again the anticipated intermediate (**23**) was not isolated (Scheme 8) [25].

2.4 *Stretched Porphycenes*

Vogel and coworkers [26] reported the first example of a "stretched" porphycene {i.e. [22]-tetradehydroporphyrin-(2.2.2.2) (**24**)}. Bis-pyrrolylalkyne (**25**) reacted under standard McMurry reaction conditions (Scheme 9) to give the ethyne-cumulene

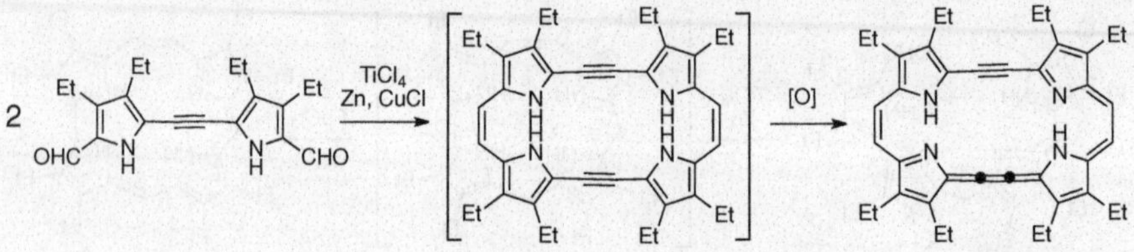

Scheme 9 Synthesis of a stretched porphycene (**24**) [26]

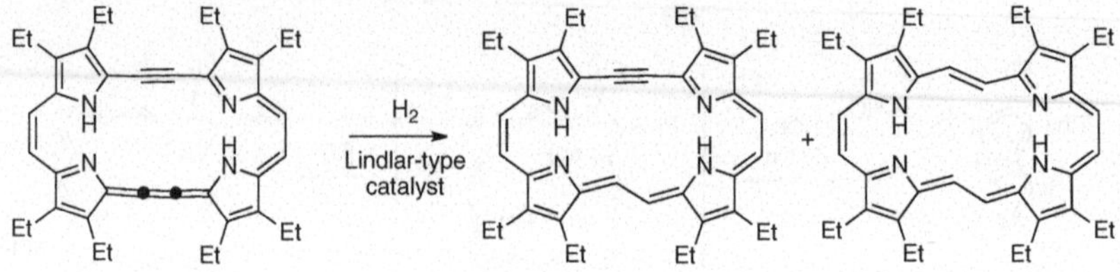

Scheme 10 Partial catalytic hydrogenation of stretched porphycene (**24**) [27]

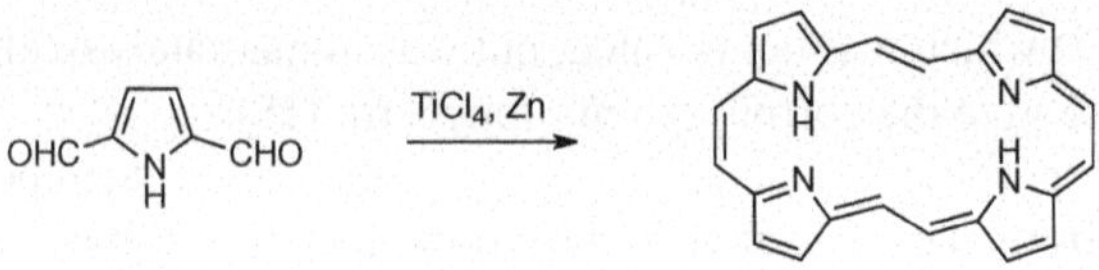

Scheme 11 Low yield synthesis of a stretched porphycene (**29**) [27]

macrocycle (**24**) in 18% yield after spontaneous in situ oxidation of the presumed intermediate (**26**).

The strongly reducing conditions used for the formation of (**24**) resulted in the formation of two other (2.2.2.2) systems, (**27**) and (**28**) [26]. Vogel and coworkers showed that (**27**) and (**28**) could be prepared independently by Lindlar-type partial catalytic hydrogenation of (**24**) (Scheme 10) [27].

Alternatively, the unsubstituted analogue (**29**) of (**28**) could be prepared in very low yield by tetra-macrocyclization of 2,5-diformylpyrrole (**30**) under McMurry conditions (Scheme 11) [27]. The tetra-*N*-methylated analogue (**31**) of (**28**) has also been prepared by McMurry macrocyclization of the *cis*-1,2-dipyrrolylethene (**32**) obtained by partial hydrogenation of the alkyne (**33**) [28, 29] (Scheme 12). Compound (**31**) was unstable but could be isolated and characterized as the oxidized bis-perchlorate salt (**34**).

A number of other stretched tetrapyrrole systems, as shown in Scheme 13 [30–35], have been synthesized, occasionally in low yield, via the McMurry reaction. Related oxa- and thia-analogues that have been prepared are not shown.

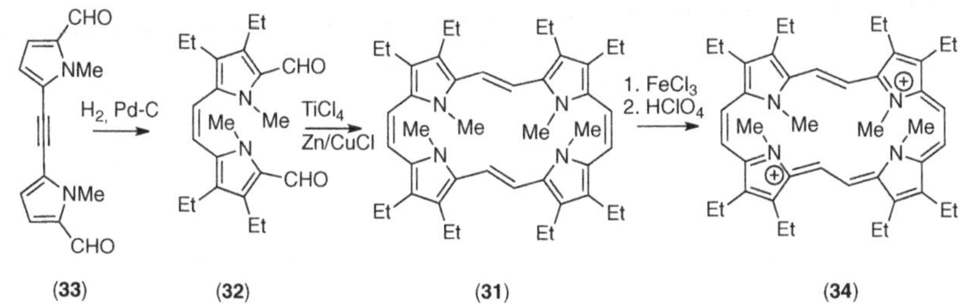

Scheme 12 McMurry synthesis of tetra-*N*-methylated stretched porphycene (**31**) [28, 29]

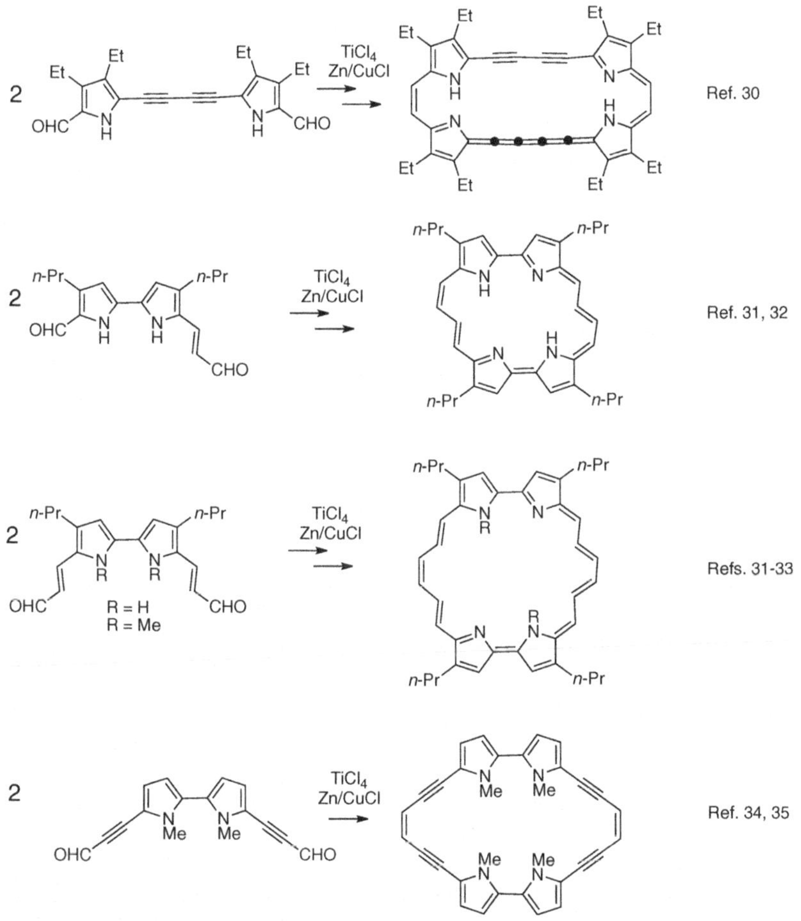

Scheme 13 Other examples of McMurry-derived stretched tetrapyrroles

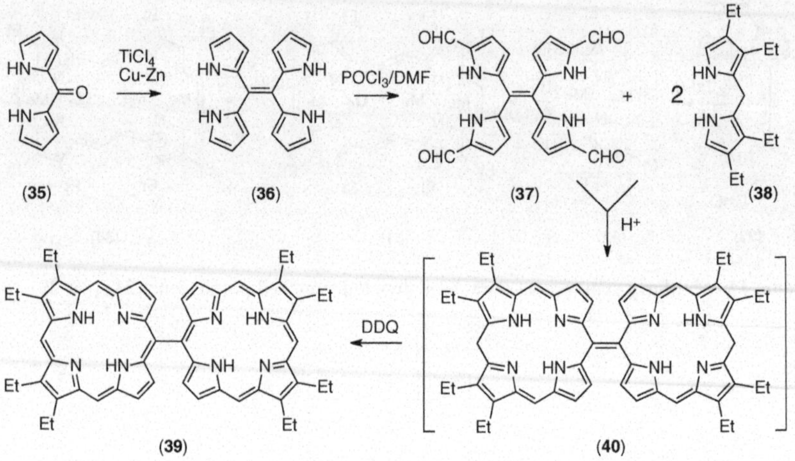

Scheme 14 Synthesis of a *meso–meso*-linked bisporphyrin (**39**) [36]

2.5 Meso–Meso *Linked Bisporphyrin*

Though the McMurry reaction was not used in the actual macrocyclization step, there does exist an example [36] wherein it was used for formation of a non-cyclic tetrapyrrole which eventually was transformed into a *meso–meso*-bisporphyrin. Thus, the dipyrroketone (**35**) was subjected to McMurry reaction conditions to give the 1,1,2,2-tetra(2-pyrrolyl)ethene (**36**) in 56% yield. This compound was then formylated using $POCl_3$/DMF and gave the tetraformyl compound (**37**) in 92% yield. Condensation of (**37**) with two equivalents of the 1,9-di-unsubstituted dipyrromethane (**38**) gave the bisporphyrin (**39**), in 7% yield after 2,3-dichloro-5,6-dicyanobenzoquinone (DDQ) treatment, presumably via the intermediate (**40**) (Scheme 14).

3 McMurry Reactions Around the Periphery of Tetrapyrrole Macrocycles

3.1 1,2-Diporphyrinylethenes

3.1.1 From β-Formylporphyrins

Zhilina and coworkers were the first to attempt reductive dimerization of a β-formylmetalloporphyrin (**41**) to give (**42**) [37]. These workers used a McMurry reaction with $TiCl_4$ and Zn(Hg) as the reductant, and obtained a 15% yield of (**42**) when employing a tenfold excess of the reagent at room temperature. At 12°C,

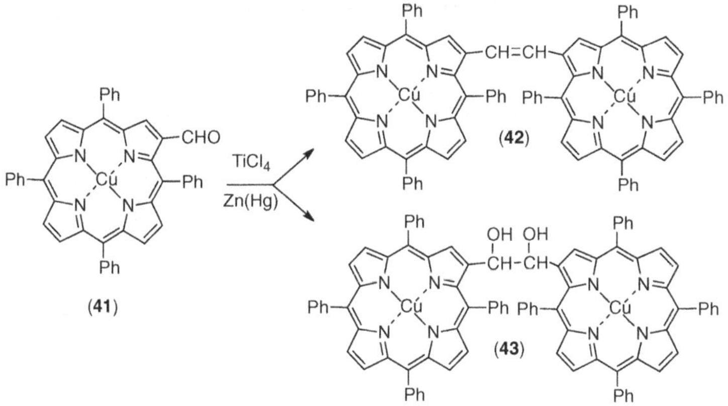

Scheme 15 McMurry reaction on copper(II) β-formyltetraphenylporphyrin (**41**) [37]

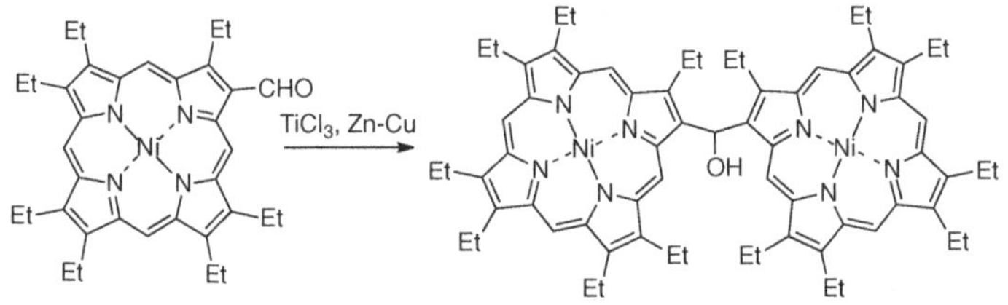

Scheme 16 Anomalous McMurry reaction yielding bis-Ni(II)-porphyrinylcarbinol (**45**) [38, 39]

the pinacol (**43**) was the major product, with only a small amount of (**42**) being observed (Scheme 15).

When nickel(II) 3,7,8,12,13,17,18-heptaethyl-2-formylporphyrin (**44**) was subjected to the McMurry reaction ($TiCl_3$, Zn–Cu), the unexpected bis-porphyrinyl-carbinol (**45**) was obtained in 61% yield [38] (Scheme 16). Figure 1 shows the X-ray structure of this unexpected product in which the two macrocycles form a skewed structure with a 77° angle between the two nitrogen planes [39].

A possible mechanism for the anomalous formation of the bis-porphyrinyl-carbinol (**45**) is shown in Scheme 17, with the main uncertainty being the dashed-line ligands on the titanium ion.

3.1.2 From *Meso*-Formyl- and *Meso*-Formylvinylporphyrins

Use of nickel(II) or copper(II) *meso*-formylporphyrins (**46**) yielded the expected 1,2-bis-metallporphyrinylethenes (**47**). Treatment of (**46** M = Ni) with $TiCl_3$ and Zn–Cu gave a 49% yield of the dimer (**47** M = Ni), while use of (**46** M = Cu) have a 64% yield of (**47** M = Cu) [38] (Scheme 18).

Based on analytical chromatography and proton-NMR spectroscopy of (**47** M = Ni), which is diamagnetic, the alkene products from these reactions were shown to be

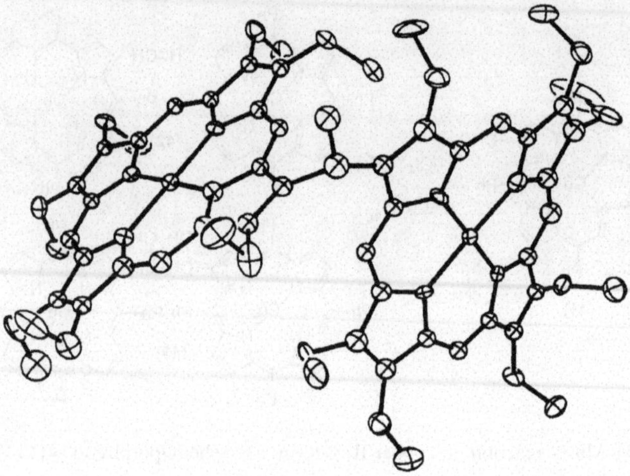

Fig. 1 X-ray structure of nickel(II) carbinol bisporphyrin (**45**). Adapted from [39]

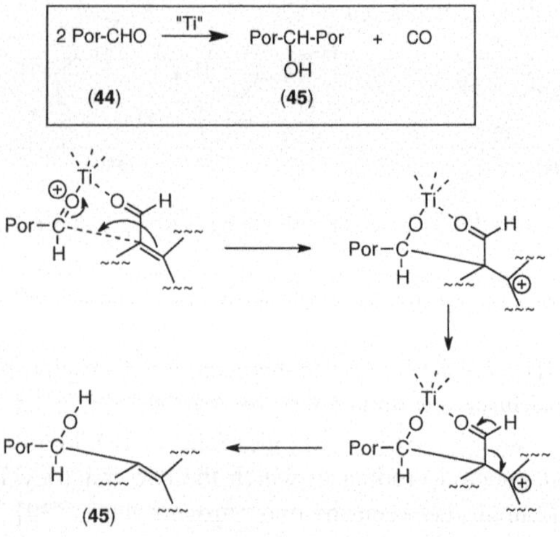

Scheme 17 Possible mechanism (with uncertainty regarding ligands on the titanium) for the anomalous formation of the bis-porphyrinyl-carbinol (**45**) from (**44**)

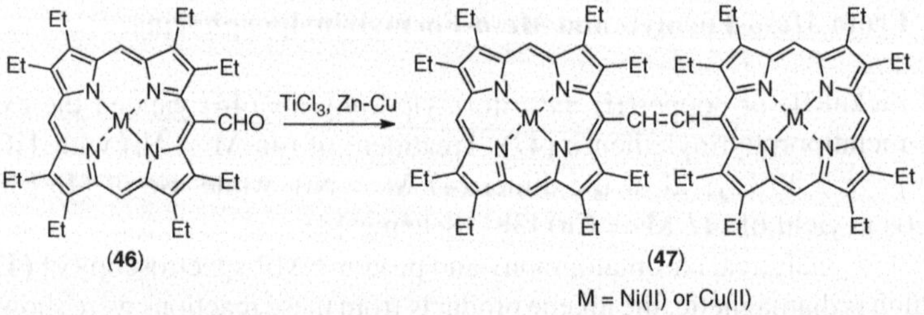

Scheme 18 McMurry reaction of Ni(II) and Cu(II) *meso*-formylporphyrins (**46**) [38]

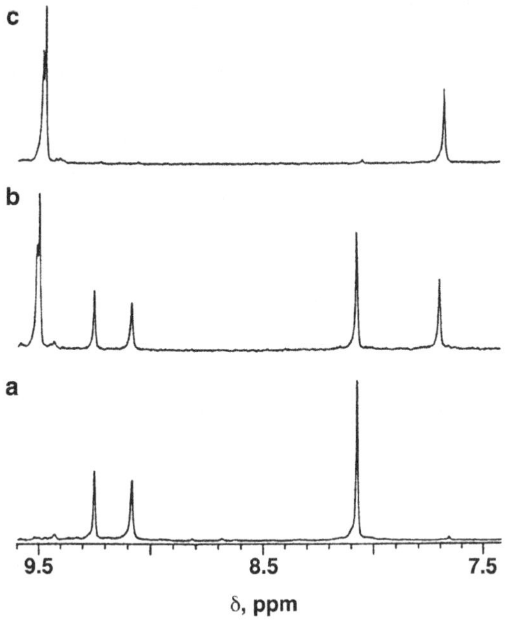

Fig. 2 Proton NMR spectra (9.5–7.5 ppm) in [D_2]-1,1,2,2-tetrachloroethane. (**a**) Pure *cis* compound (**48**); (**b**) mixture of *cis* and *trans* compounds; (**c**) after being kept overnight at 120°C in [D_2]-1,1,2,2-tetrachloroethane

a mixture of both *cis* and *trans* isomers. Surprisingly, the *cis* isomers predominated [39] for both the nickel(II) and copper(II) products; McMurry reactions usually provide *trans* alkene isomers [40]. It was rationalized that the *cis* isomers were preferred in the metalloporphyrin coupling reactions because of the π–π aggregation tendencies of metalloporphyrins [41–44]. During the coupling reaction, presumably at the pinacolate stage, there is an opportunity for the two porphyrin faces to π-stack prior to the final deoxygenation step (Scheme 2).

Chromatographic separation of the *cis* and *trans* isomers of (**47** M = Ni and Cu) was difficult, but the two isomers could be separated by fractional crystallization, wherein the *cis* isomers (in each case, M = Ni or Cu) were significantly more soluble in organic solvents that were the *trans* compounds [39].

Variable temperature proton NMR spectroscopy using a sample of approximately 50% *cis* and 50% *trans* (**47** M = Ni) between 25°C and 140°C in [D_2]-1,1,2,2-tetrachloroethane showed that with increased temperature, the *cis* form was transformed into the presumably thermodynamically stable *trans* form (see Fig. 2) [39]. As temperature was raised from 25°C to 100°C there was very little change in the proton NMR spectrum. However, as temperature was further increased the peaks corresponding to the *meso*-protons of the *trans* form (9.50, 9.48, and 7.66 ppm) increased at the expense of those from the *cis* form (9.30, 9.05 and 8.06). At 120°C overnight in the dark, the almost pure *cis* form was apparent (Fig. 2, trace c). These data imply that the *cis* compound (**48**) isomerizes to *trans* (**49**) in slightly acidic 1,1,2,2-tetrachloroethane (Scheme 19); however, refluxing acidic (H_2SO_4) tetrahydrofuran did not promote the isomerization from *cis* to *trans*, indicating a large thermal barrier for the process.

In the UV–visible spectra the *cis* compound (**48**) absorbs at 393 and 568 nm, bracketing the *trans* compound (**49**) at 414 and 566 nm (Fig. 3).

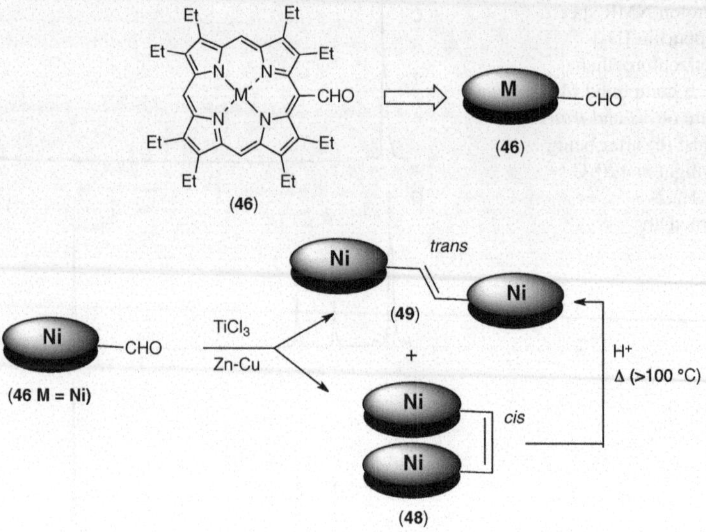

Scheme 19 Conversion of the *cis*-bis(Ni-porphyrin) (**48**) into *trans*-bis(Ni-porphyrin) (**49**) [39]

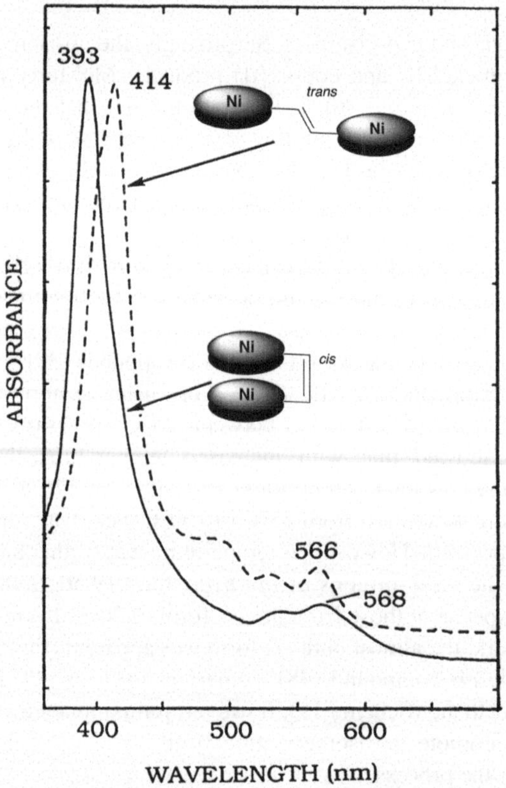

Fig. 3 UV–visible spectra (in CH_2Cl_2) of the *cis* (**48**) and *trans* (**49**) isomers of the bis(Ni-porphyrin)

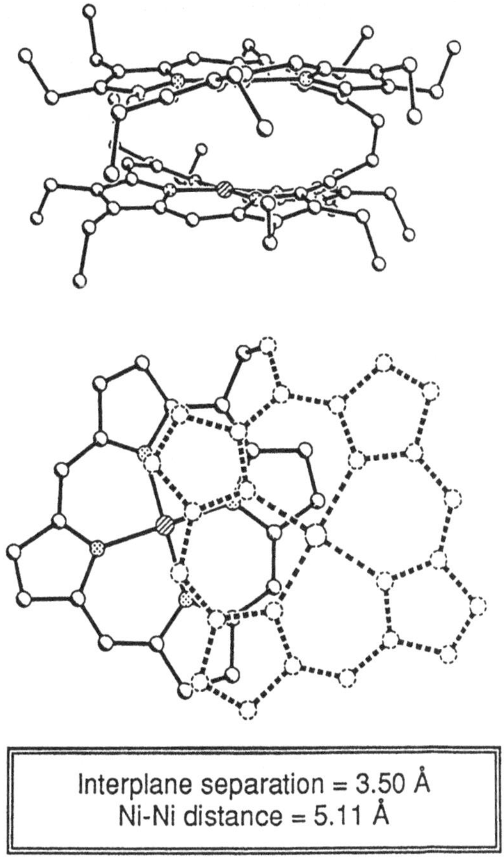

Fig. 4 X-ray structure of the *cis*-bis(Ni-porphyrin) (**48**); end-on view (*above*) and top view showing overlap (*below*) [39]. The top view has the ethyl substituents missing for clarity

All of the above structural conclusions were confirmed by X-ray diffraction analyses of compounds (**48**) (Fig. 4), *cis*-(**47** M = Cu) (Fig. 5) and (**49**) (Fig. 6).

Vinylogous systems also efficiently undergo the McMurry reaction. Thus, for example, nickel(II) *meso*-(2-formylvinyl)octaethylporphyrin (**50**) was treated with the standard McMurry reagent and gave a 96% yield of the bis-porphyrin (**51**) [38] (Scheme 20).

Chart 2 shows a number of other bis-porphyrinyl- and bis-chlorinyl-trienes that have been prepared from nickel(II) 2-formylvinyltetrapyrroles [38]; the arrow indicates the site of the new alkene bond formed during the McMurry reaction. Chelation with nickel(II) rather than copper(II) was preferred because the bis-Ni products are diamagnetic and can therefore be characterized by proton NMR spectroscopy.

Nickel(II) formylporphyrins with the aldehyde group even more distant from the porphyrin ring (e.g. **52** and **53**) can also be self-coupled using the McMurry reaction to give the corresponding bis-porphyrinylstilbenes (**54** and **55**) [45] (Scheme 21).

If the McMurry substrate has both carbonyls in the same molecule, there is a good chance that an intramolecular reaction can take place to give a single product. However, if the two carbonyl groups required for the McMurry coupling are in

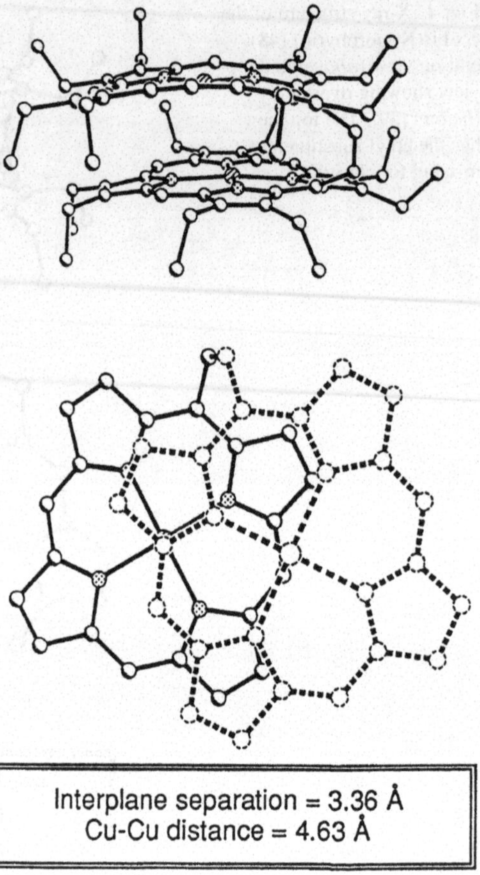

Fig. 5 X-ray structure of the *cis*-bis(Cu-porphyrin) (**47** M = Cu); end-on view (*above*) and top view showing overlap (*below*) [39]. The top view has the ethyl substituents missing for clarity

different molecules, then a mixture of compounds (statistically three, neglecting stereochemistry) will almost certainly result. Such an approach allows porphyrins and chlorins (dihydroporphyrins) to be linked together in one molecule, and such bis-tetrapyrroles have potential for investigation of electron transfer, and attendant applications. Two such examples are shown in Schemes 22 and 23. In Scheme 22 the required unsymmetrical system (**56**) was obtained by McMurry coupling of the nickel(II) porphyrin (**57**) and nickel(II) chlorin (**58**); along with the novel product (**56**) was also formed the bis-porphyrin and bis-chlorin nickel complexes (not shown). Chromatographic separation of the three products was facilitated by the gradient in polarities of the three components.

Likewise, when nickel(II) porphyrin (**59**) and nickel(II) chlorin (**60**) were cross-coupled (Scheme 23), the unsymmetrical product (**61**) was obtained, along with the corresponding bis-porphyrin (**62**) and bis-chlorin (**63**). Figure 7 shows the X-ray structure of compound (**61**), and Fig. 8 shows the optical spectra of the three products (**61**)–(**63**) obtained from the reaction between (**59**) and (**60**) which enables each component of the separable mixture to be clearly identified.

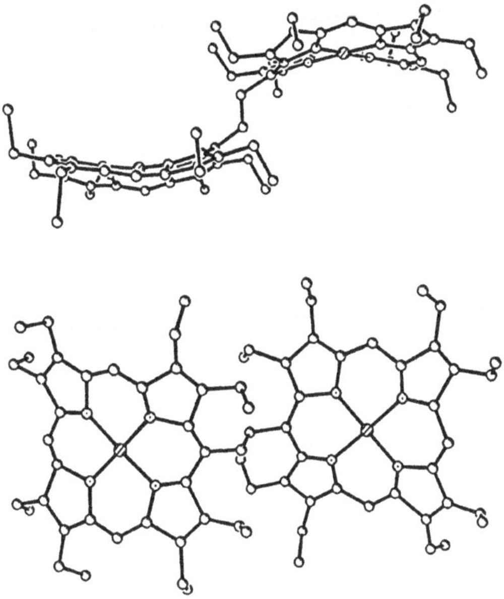

Fig. 6 X-ray structure of the *trans*-bis(Ni-porphyrin) (**49**); end-on view (*above*) and top view showing overlap (*below*) [39]

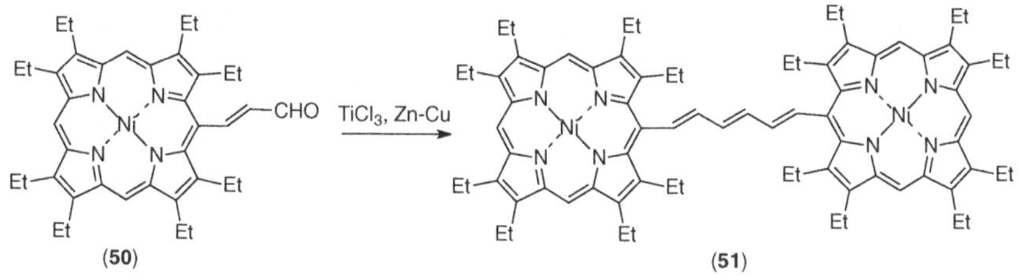

Scheme 20 bis-[Ni(II) porphyrin] (**51**) formation from Ni(II) *meso*-(2-formylvinyl)porphyrin (**50**) [38]

Interestingly, access to unsymmetrical heterobimetallic complexes was established when it was shown that the porphyrin component of (**61**) could be demetalated to give (**64**) without removal of the nickel from the chlorin portion. Subsequent metalation of the porphyrin in (**64**) will lead to heterobimetallic compounds. Such heterobimetallic (and particularly cofacial, see below) bisporphyrins are potentially good models for the "special-pair" in photosynthesis since it should be possible to design species that undergo facile charge-separation. In principle, heterobimetallic 1,2-bisporphyrinyl-alkenes should be accessible by McMurry cross-coupling of two different *meso*-formylmetalloporphyrins (along with the two homobimetallic products), but in numerous cases, partial remetalation of the unmetalated bisporphyrin has been shown to be a more effective approach. In a major study [46], the syntheses and

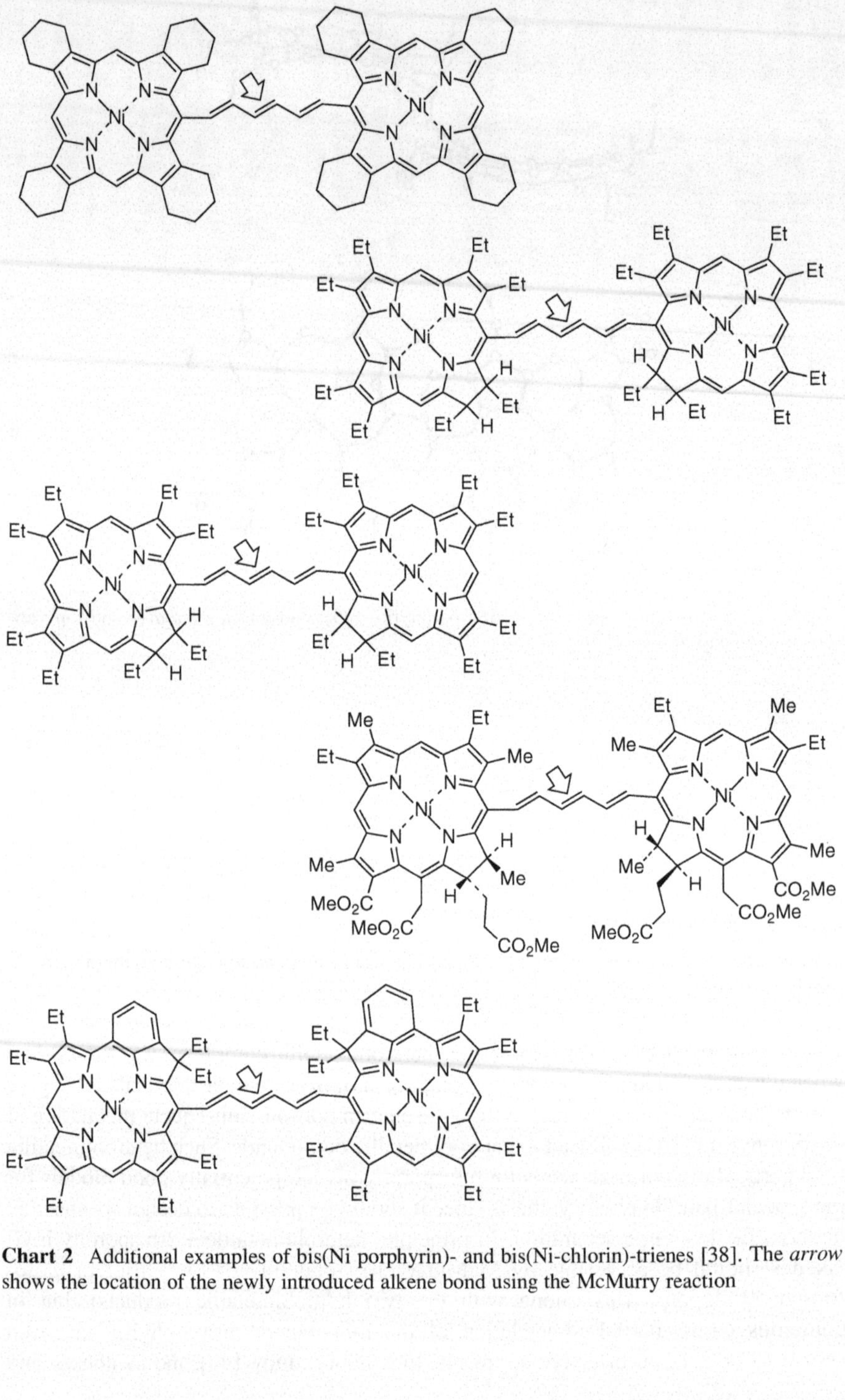

Chart 2 Additional examples of bis(Ni porphyrin)- and bis(Ni-chlorin)-trienes [38]. The *arrow* shows the location of the newly introduced alkene bond using the McMurry reaction

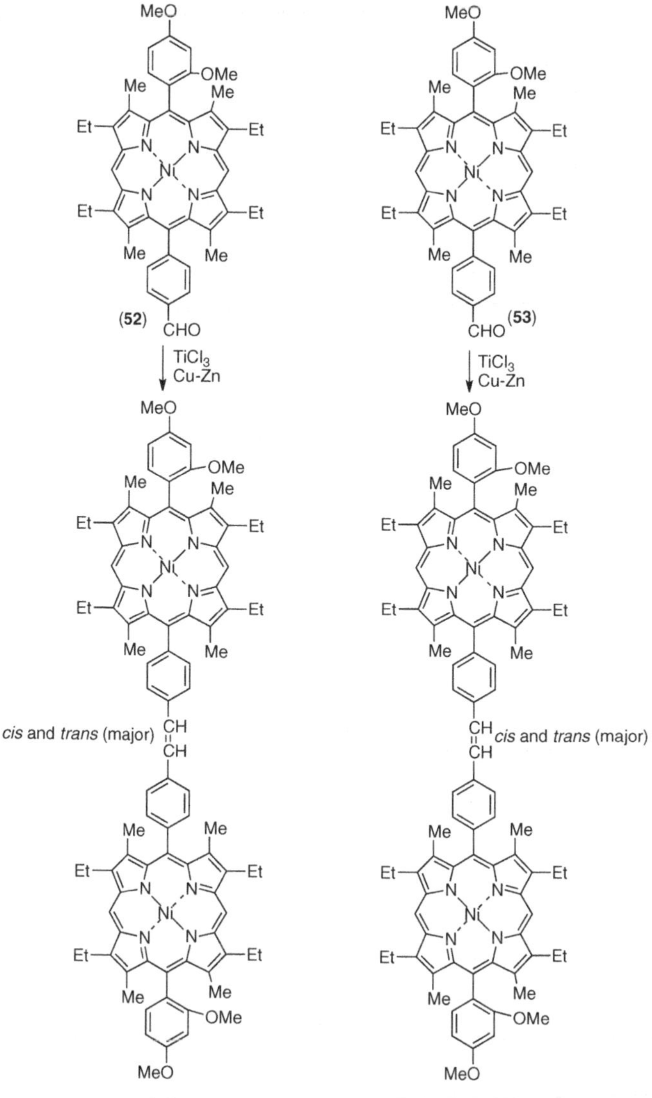

(54) (55)

Scheme 21 Syntheses of bis(Ni porphyrinyl)stilbenes (**54**) and (**55**) using the McMurry reaction [45]

characterization of a large number of 1,2-bisporphyrinylethenes (Chart 3) were reported. These included *cis*-1,2-biporphyrinylethenes [(**47** M = Cu-*cis*), (**48**), (**65**)–(**67**)], *trans*-1,2-bisporphyrinylethenes [(**47** M = Cu-*trans*), (**49**) **68**)], monometalated *cis*-[(**69**),(**70**)] and *trans*-1,2-bisporphyrinylethenes [(**71**), (**72**)] and heterobimetallic *cis*-[(**73**), (**74**)], and *trans*-1,2-bisporphyrinylethenes (**75**).

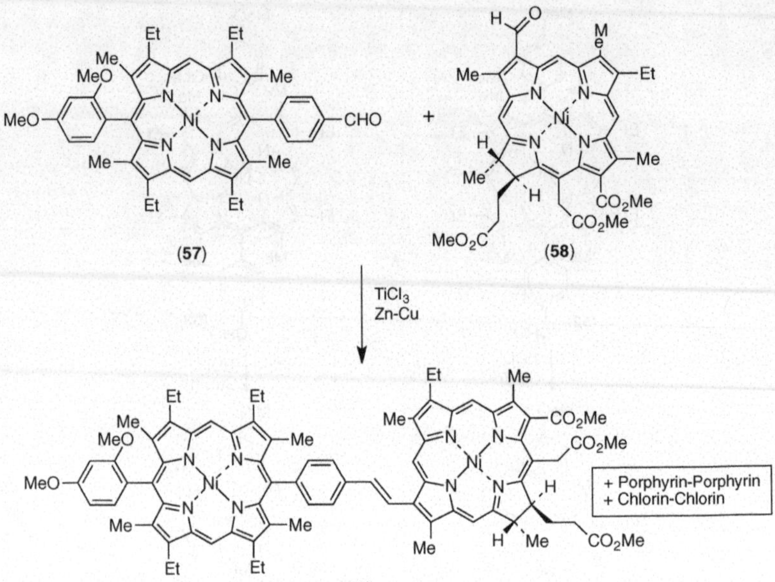

Scheme 22 Cross McMurry coupling of a nickel(II) porphyrin with a nickel(II) chlorin [45]

The free-base *trans* bis-porphyrin (**68**) was prepared in 56% yield by McMurry couping of the copper(II) *meso*-formyloctaethylporphyrin (**46**) to give (**47** M = Cu) followed by complete demetalation using 15% H_2SO_4/TFA [46]. The *cis*-isomer (**65**) was then obtained according to a process developed by Ponomarev and coworkers by heating in acetic acid to give [47]. Monometalated species were obtained without isomerization back to (**68**), after multiple trials reactions, by heating (**65**) in DMF with one equivalent of the corresponding copper(II) or nickel(II) acetate to give (**69**) or (**70**). The monometalated *trans* compounds (**71**) and (**72**) were likewise obtained from (**68**), but by refluxing with one equivalent of the corresponding metal acetate in dichloromethane/methanol. Metal insertion to give the heterobimetallic complexes (**73**)–(**75**) was accomplished using an excess of the appropriate metal acetate [46]; overmetalation or transmetalation was not an issue in these reactions. However, attempts to prepare the *cis*-Cu/Mn bisporphyrin by manganese insertion into the *cis*-mono-copper compound (**69**) were unsuccessful, the *trans*-compound (**75**) (Fig. 9) always being isolated.

Figures 10, 11, and 12 show typical X-ray structures for the copper-free base (**69**), nickel-zinc (**73**), and nickel-copper (**74**) *cis*-compounds. Figure 13 shows the typical crystal packing in the nickel-zinc case, with zinc porphyrin pairs and nickel porphyrin pairs in the system stacking with each other. Because the two porphyrins within these cofacial systems have enforced proximity and experience $\pi-\pi$ stacking, these

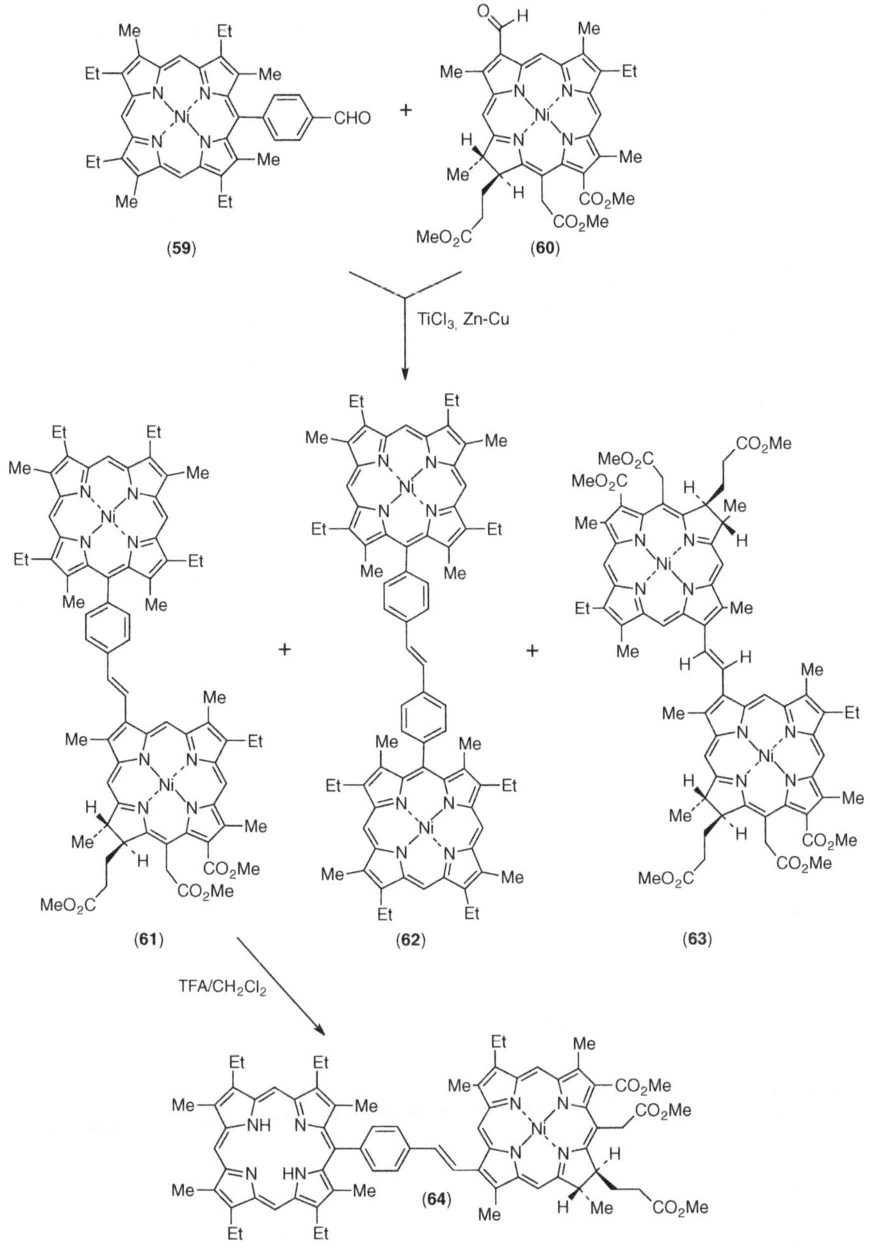

Scheme 23 Formation of porphyrin–chlorin (**61**), porphyrin–porphyrin (**62**) and chlorin–chlorin (**63**) from McMurry coupling of (**59**) and (**60**), and selective partial demetalation of (**63**) to give (**64**) [45]

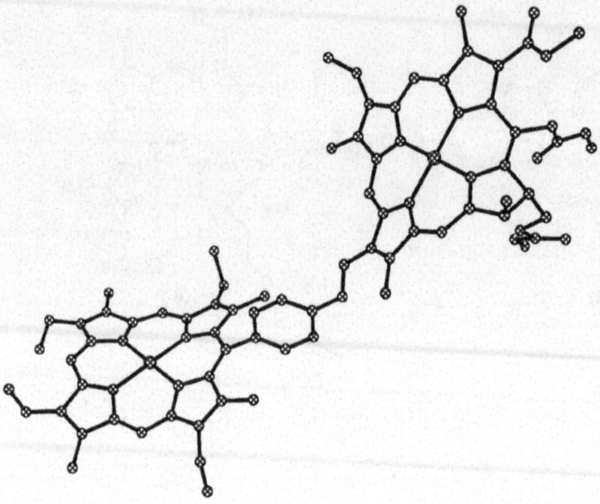

Fig. 7 X-ray structure of the unsymmetrical McMurry product (**61**) [45]

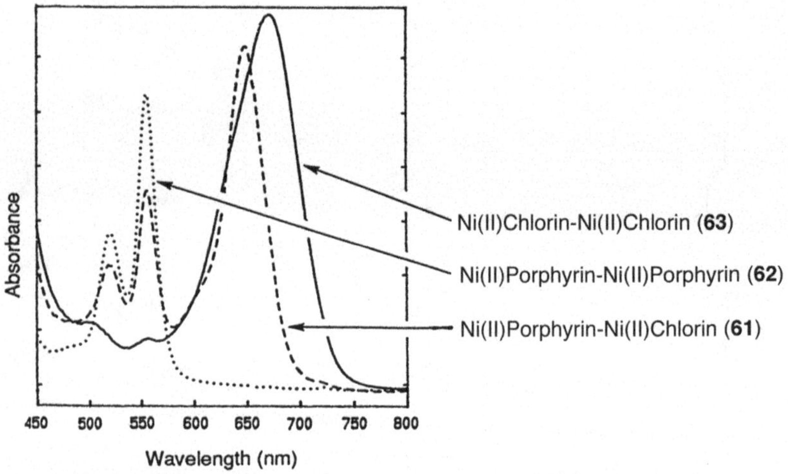

Fig. 8 UV–visible spectra, in CH_2Cl_2 of the three bis-nickel components (**61**)–(**63**) obtained from reaction of (**59**) with (**60**) [45]

are valuable for determining the relative capacities for aggregation is metalloporphyrins. Indeed, such *cis*-systems do not show equal intraplanar or intrametallic distances because of the strong effect of the metal ion upon the π–π-stacking characteristics [41, 42]. Table 1 shows the interplanar and intermetallic distances in the various compounds, including the homobimetallic copper and nickel systems.

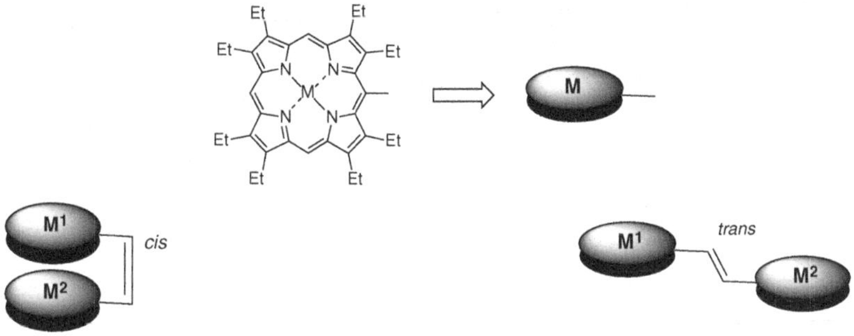

Compound	M^1	M^2
(47)	Cu(II)	Cu(II)
(48)	Ni(II)	Ni(II)
(65)	2H	2H
(66)	Co(III)	Co(III)
(67)	Fe(III)Cl	Fe(III)Cl
(69)	Cu(II)	2H
(70)	Ni(II)	2H
(73)	Ni(II)	Zn(II)
(74)	Ni(II)	Cu(II)

Compound	M^1	M^2
(47)	Cu(II)	Cu(II)
(49)	Ni(II)	Ni(II)
(68)	2H	2H
(71)	Zn(II)	2H
(72)	Ni(II)	2H
(75)	Cu(II)	Mn(III)Cl

Chart 3 1,2-Bisporphyrinylethenes prepared using the McMurry reaction [46]

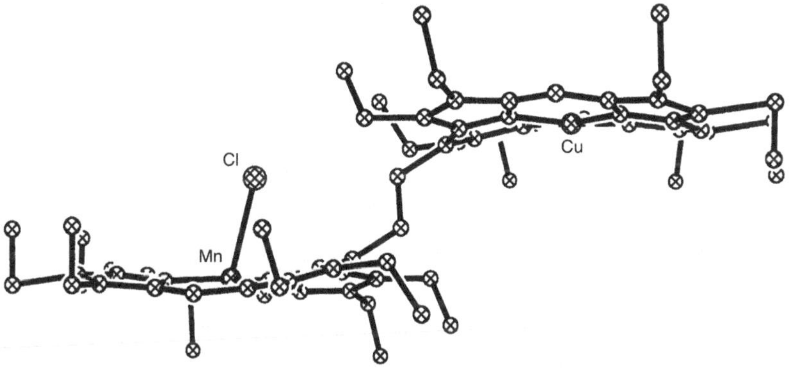

Fig. 9 X-ray structure for the *trans*-mono-copper mono-manganese product (**75**) obtained from reaction of (**69**) with manganese ions [46]

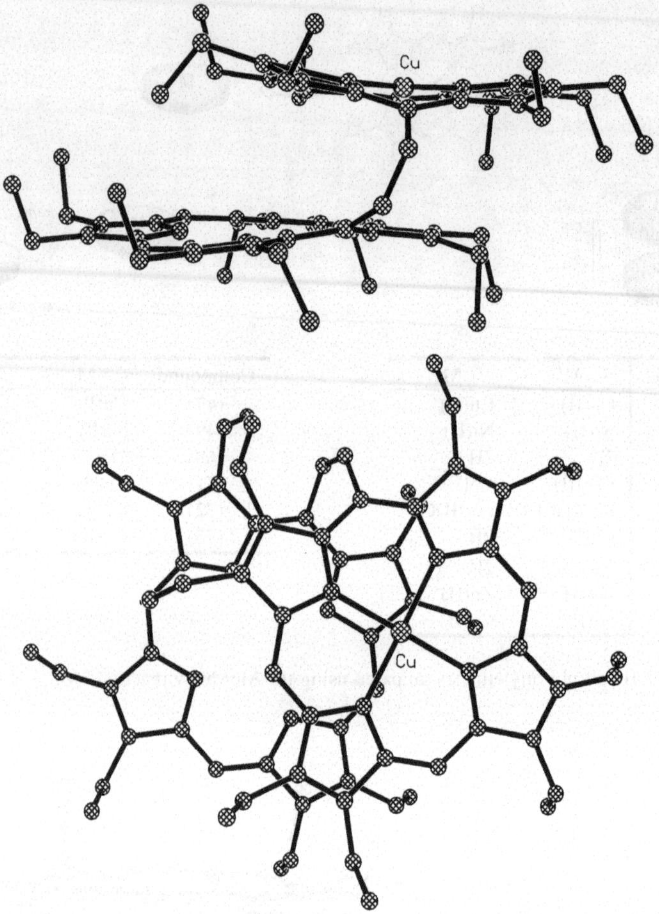

Fig. 10 X-ray structure for the *cis*-mono-copper bis-porphyrin (**69**) with 35% probability thermal ellipsoids [46]. *Top*, edge-on view; *bottom*, top view

3.2 1,2-Bis(Chlorinyl)Ethenes

3.2.1 From β-Formylchlorins

As mentioned earlier (Scheme 23) the bis-(nickel chlorin-e_6 trimethyl ester)ethene (**63**) was first obtained in 42% yield [along with (**62**)(6%) and (**61**) (41%)] from McMurry cross-coupling of the nickel(II) porphyrin (**59**) and nickel(II) 2-formylchlorin-e_6 trimethyl ester (**60**) (Scheme 23) [45]. It was obtained in higher yield (67%) when (**60**) was simply coupled on its own [48]. The formation of this product (**63**) rests in stark contrast to the bis-(porphyrinyl)-carbinol (**45**) obtained when the corresponding nickel(II) β-formylporphyrin

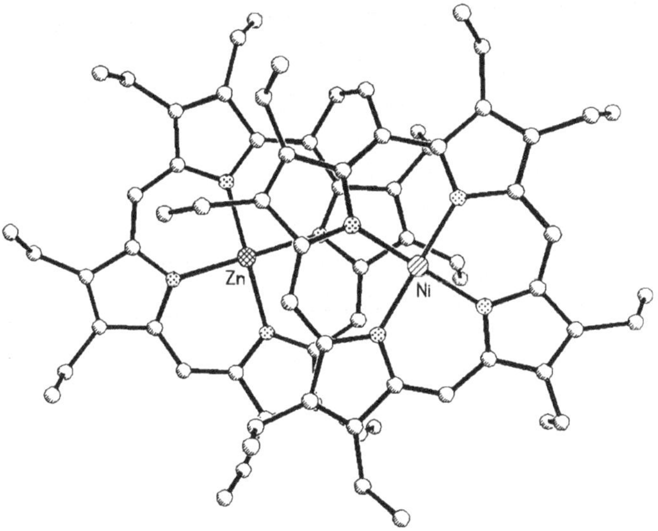

Fig. 11 X-ray structure (*top view* only) for the *cis*-mono-nickel mono-zinc 1,2-bis-porphyrinyl-ethene (**73**) with 35% probability thermal ellipsoids [46]

(**44**) was used [38, 39]; heating of (**63**) in toluene/acetic acid caused no double bond isomerization, and this, along with variable temperature NMR studies, indicated that the product (**63**) was indeed the thermodynamically most stable *trans* alkene.

3.2.2 From *Meso*-Formylchlorins

McMurry coupling of the nickel(II) *meso*-formylchlorin (**76**) gave a single bis-chlorin (**77**) in 52% yield (Scheme 24) [48]. The *trans*-arrangement of the alkene was suspected based on the fact that heating of (**76**) in boiling acetic acid did not accomplish any isomerization (NMR observation), but an X-ray structure (Fig. 14) was subsequently obtained that confirmed the initial supposition. Variable temperature NMR spectroscopy showed C_2 symmetry at room temperature, but cooling induced some broadening of peaks due to a dynamic process believed to involve restricted rotation about the alkene-chlorin bond.

Similar McMurry coupling of the nickel(II) 20-formyl-13^2-deoxomeso-pyropheophorbide (**78**) gave the bis-(nickel chlorin) (**79**) in 72% yield (Scheme 25) [48]. Variable temperature proton NMR studies also revealed dynamic processes at low temperature, but the *trans* stereochemistry of the alkene was again definitively established by X-ray crystallography (Fig. 15).

The virtually complete preponderance of the *trans*-geometry in compounds (**77**) and (**79**) is presumably due to the steric congestion at the two sp^3 hybridized bonds of the reduced macrocyclic ring. If the bis-nickel chlorin complex (**79**) was

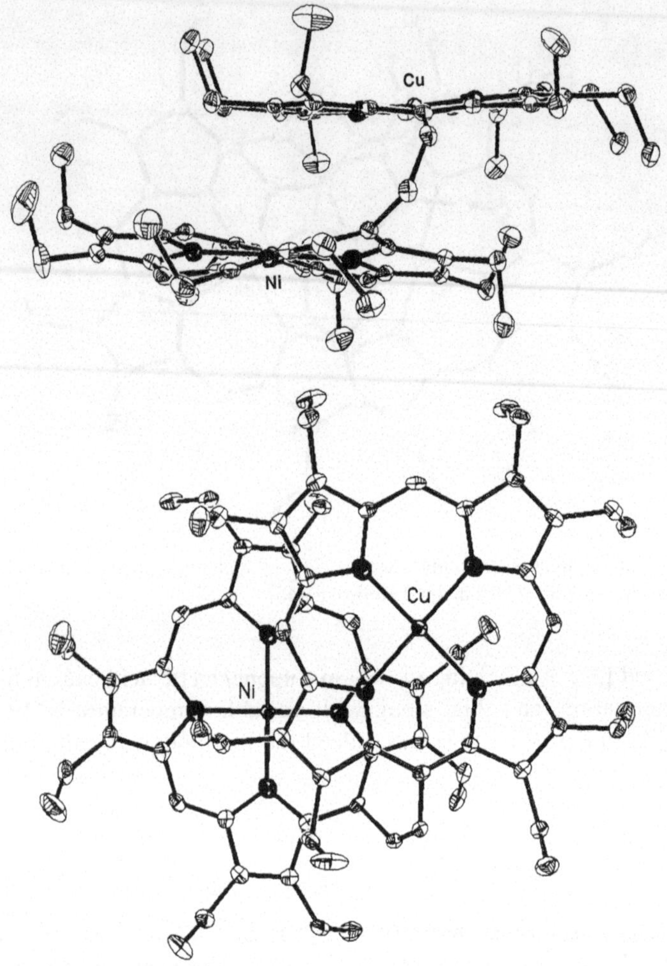

Fig. 12 X-ray structure for the *cis*-mono-copper mono-nickel 1,2-bis-porphyrinylethene (**74**) with 35% probability thermal ellipsoids [46]. *Top*, edge-on view; *bottom*, top view

oxidized by refluxing in toluene/acetic acid (3/1) in the presence of air, the resulting bis-nickel(II) porphyrin, obtained in 80% yield, was shown [by X-ray crystallography (Fig. 16) and NMR spectroscopy] to have the *cis*-geometry (**80**) (Scheme 25) [48].

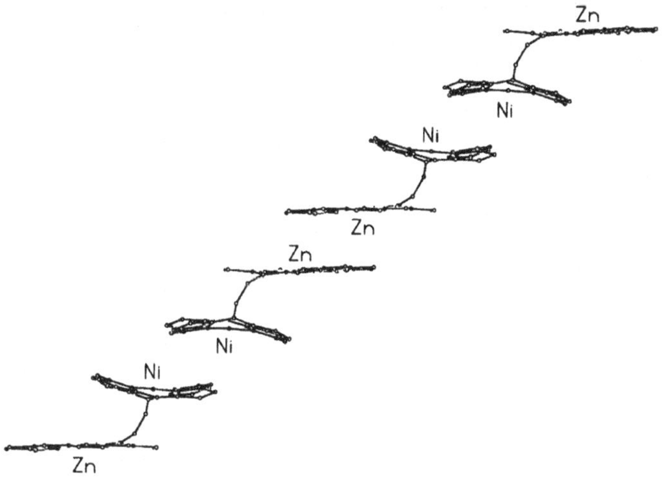

Fig. 13 Crystal packing in the *cis*-mono-nickel mono-zinc 1,2-bis-porphyrinylethene (**73**) [46]. Substituents have been omitted for clarity

Table 1 Average intramolecular 4N plane spacings and M^1–M^2-distances for *cis*-bis-metalloporphyrinylethenes [46]

Compound	M^1 and M^2	Average intramolecular 4N plane spacing (Å)	Intramolecular M^1–M^2 distances (Å)
(**47**) *cis*	Cu, Cu	3.36	4.63
(**48**)	Ni, Ni	3.53	5.11
(**73**)	Ni, Zn	3.43	5.37
(**74**)	Ni, Cu	3.46	5.33

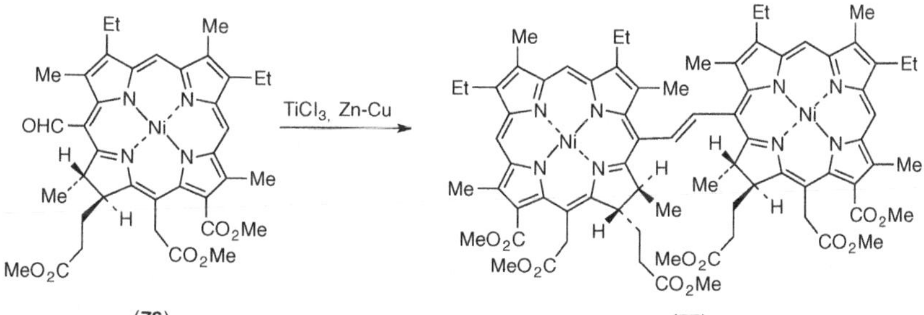

Scheme 24 Formation of nickel(II) bis-chlorin (**77**) from McMurry coupling of *meso*-formylchlorin (**76**) [48]

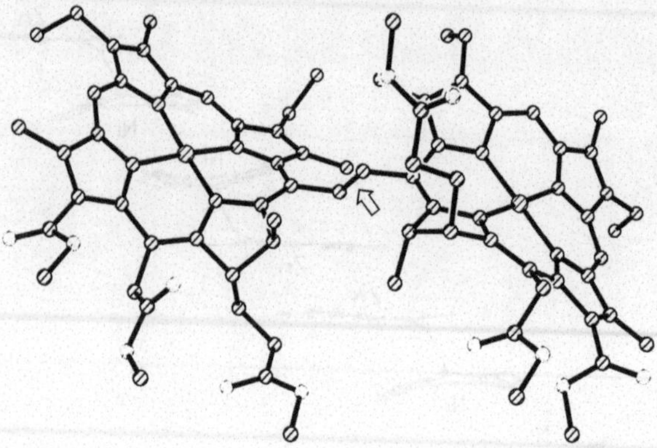

Fig. 14 Crystal structure of the bis-nickel chlorin (**77**) [48]. The *arrow* shows the *trans* alkene linkage

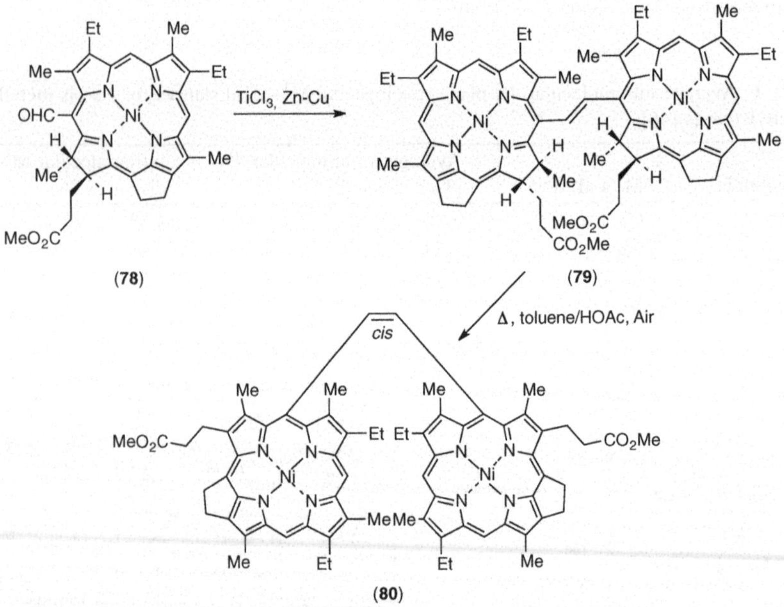

Scheme 25 McMurry synthesis of bis-(nickel deoxomesopyropheophorbide) (**79**) and its oxidation and isomerization to give (**80**) [48]

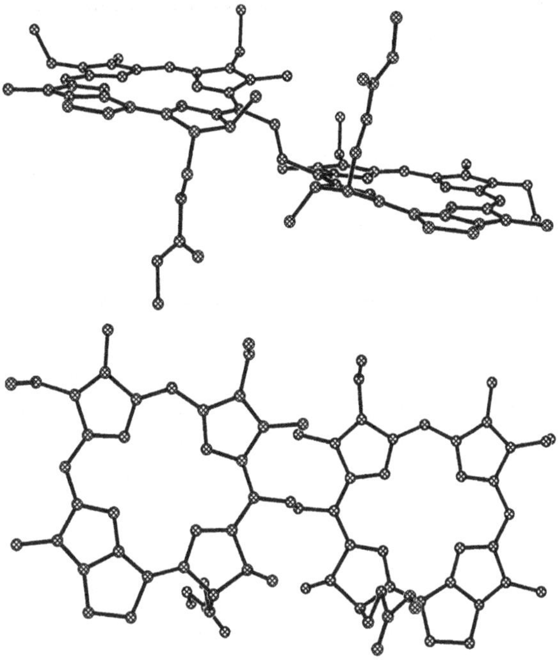

Fig. 15 Crystal structure of the bis-(nickel deoxomesopyropheophorbide) (**79**). *Above*, edge-on view; *below*, top view [48]

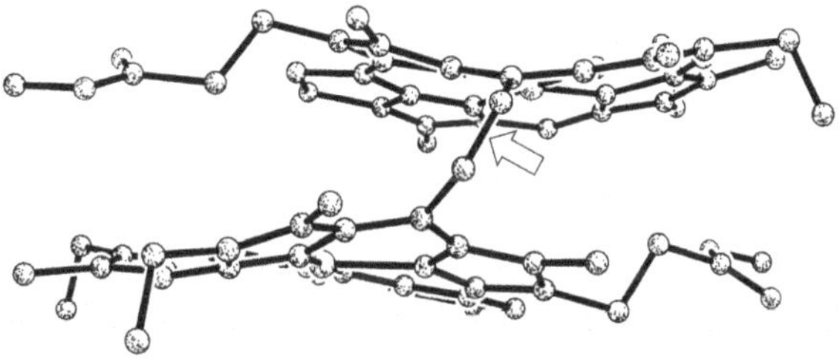

Fig. 16 Crystal structure of the *cis*-bis-(nickel deoxopyropheoporphyrin) (**80**) [48]

3.2.3 From β-Ketochlorins

A McMurry coupling reaction on the chlorophyll-a degradation product nickel(II) methyl mesopyropheophorbide-a (**81**) gave the nickel(II) bis-pyropheophorbide (**82**) in 61% yield (Scheme 26) [49]. Figure 17 shows the highly planar fully conjugated core of the molecule. The optical spectra of (**81**) and (**82**) are shown in Fig. 18 and show a very significant red-shift of the bis-chlorin compound (**82**) from 636 to 729 nm.

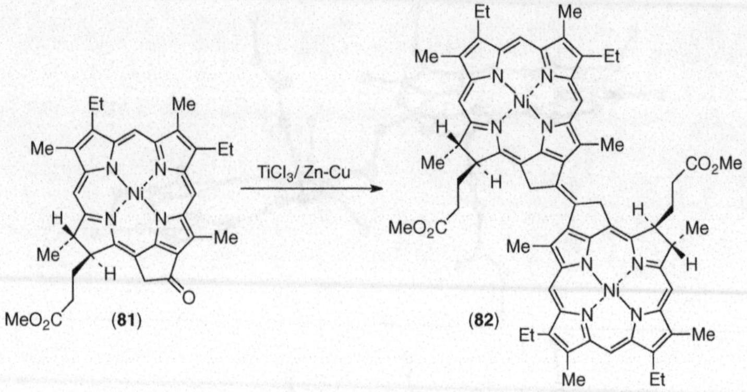

Scheme 26 McMurry coupling of nickel(II) methyl mesopyropheophorbide-a (**81**) to give the planar bis-nickel(II) complex (**82**) [49]

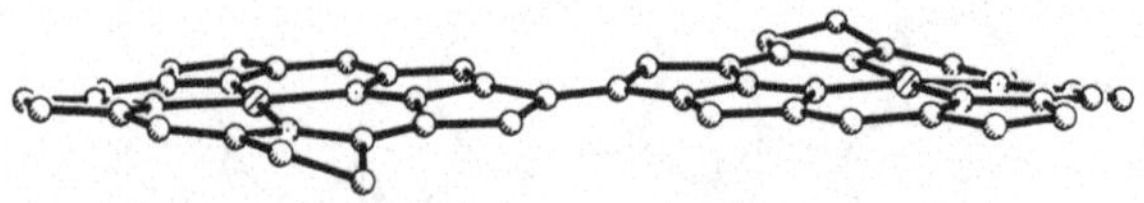

Fig. 17 Edge-on view of the crystal structure of the bis-(nickel methyl mesopyropheophorbide-a) (**82**) [49]. Substituents are removed for clarity

Attempts to demetalate (**82**) to give (**83**) (see Scheme 27) using sulfuric acid were unsuccessful due to decomposition. Thus, zinc(II) methyl pyropheophorbide-a (**84**) was obtained from methyl pyropheophorbide-a (**85**) by treatment with zinc(II) acetate in methanol/dichloromethane and was subjected to McMurry coupling in the presence of pyridine (to stabilize the acid-sensitive zinc ion) and gave a 46% yield of the bis-chlorin (**86**); from this, the demetalated product (**83**) was obtained by treatment with TFA, to give the metal-free bis-vinylchlorin (**87**), followed by catalytic hydrogenation of the two vinyl groups (Scheme 27). Figure 19 shows the X-ray structure of the metal-free bis-chlorin (**83**). The corresponding UV–visible spectra of (**85**) and (**87**) are shown in Fig. 20, with once again an almost 100 nm red-shift being caused by the McMurry coupling reaction (656–740 nm).

Finally, slow crystallization of (**83**) in dichloromethane/methanol in the presence of air resulted in the formation of the bis-methanol oxidation adduct (**88**) (Scheme 27), in which the two methanols are uniquely added, for steric reasons, so as to be *trans* to the reduced ring propionate functions. Figure 21 shows the X-ray structure of (**88**) [49].

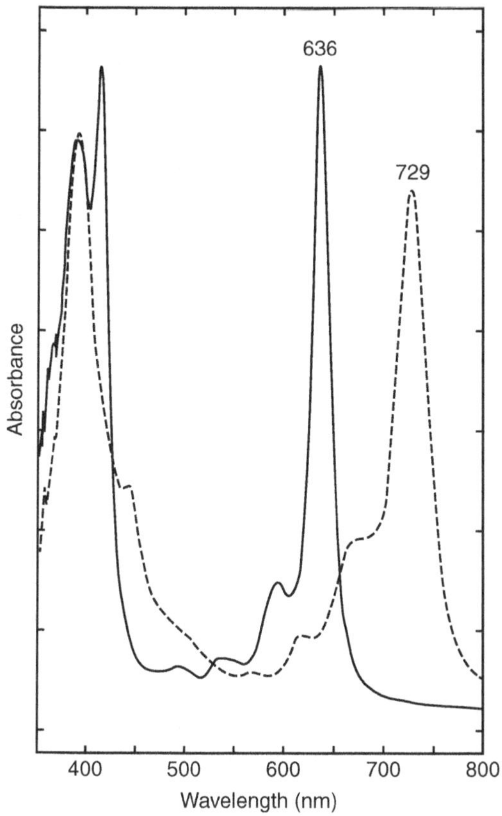

Fig. 18 UV–visible spectra, in dichloromethane, of the nickel(II) methyl mesopyropheophorbide-a (**81**) (*full line*) and the bis-(nickel methyl mesopyropheophorbide-a) (**82**) (*dashed line*) [49]. The two spectra are not normalized, the intensity of the peak at 729 nm being approximately three times as intense as that at 636 nm

4 Conclusion

For many decades of the last century porphyrinoid synthesis and chemistry was steeped in tradition and few moved outside of standard chemical techniques in their research projects. But in the past 30 years, new methodologies have been introduced into synthetic chemistry as a whole, and a number of Nobel prizes have signaled the success that these new inventions, and particularly organometallic cross-coupling reactions, have experienced. Aromatic/benzenoid chemistry in particular has benefitted from applications of such new procedures, and it is very gratifying that porphyrinoids have also enjoyed expanded horizons as aromatic chemistry itself has advanced. This chapter describes new synthetic and functionalization reactions that have been reported as a result of the invention of the McCurry coupling reaction.

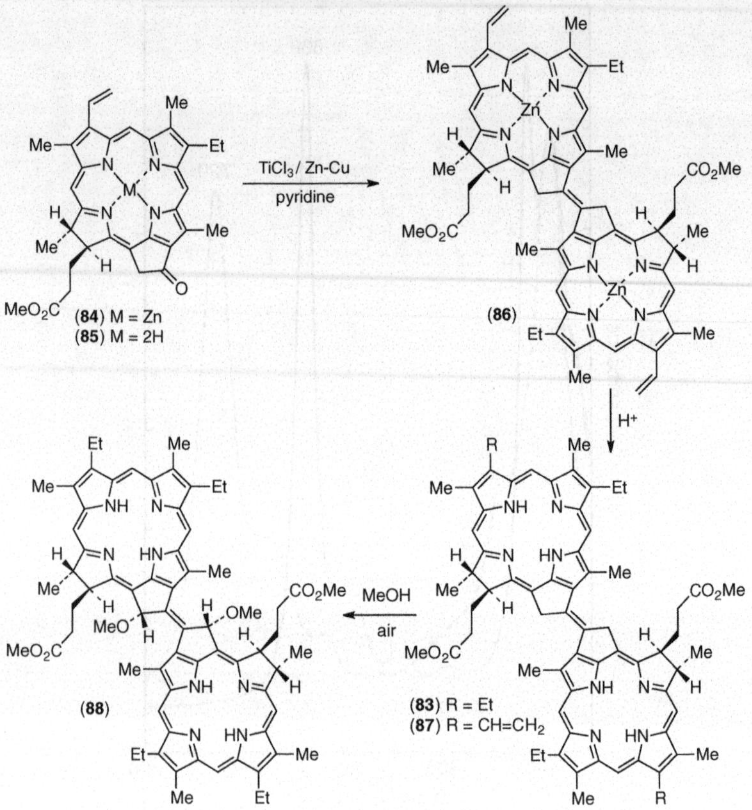

Scheme 27 Formation of planar bis-chlorins from McMurry coupling of methyl pheophorbides [49]

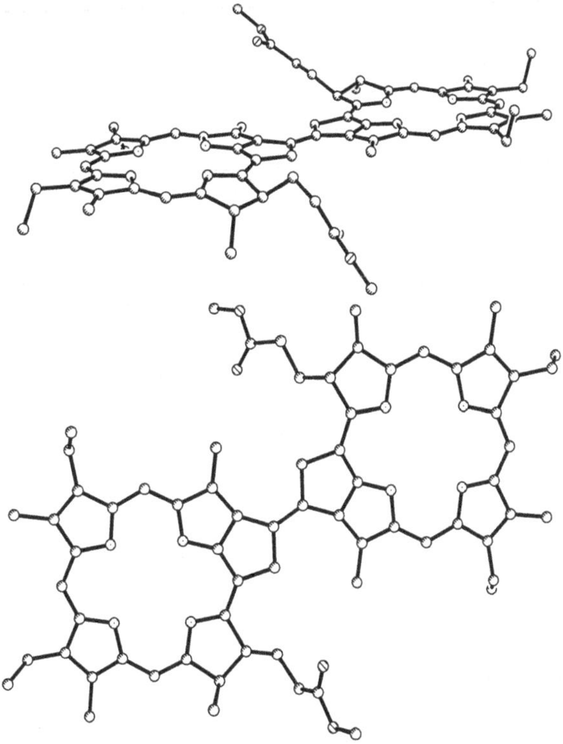

Fig. 19 Crystal structure of the bis-(methyl mesopyropheophorbide-a) (**83**) [49]. *Above*, edge-on view, and *below*, top view

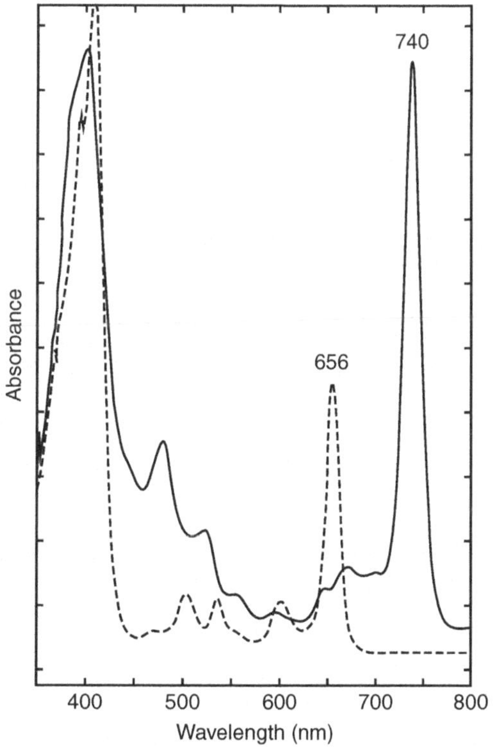

Fig. 20 UV–visible spectra, in dichloromethane, of methyl pyropheophorbide-a (**85**) (*dashed line*) and the bis-chlorin product (**87**) (*full line*) [49]

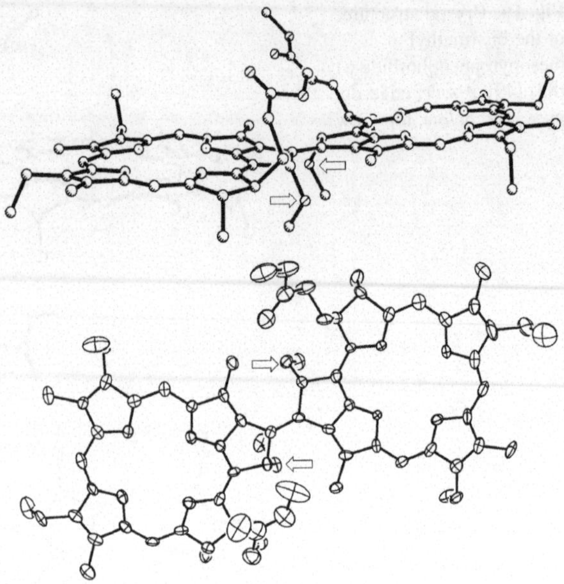

Fig. 21 Crystal structure of the bis-(methyl mesopyropheophorbide-a) methanol adduct (**88**) [49]. *Above*, edge-on view, and *below*, top view. *Arrows* show the locations of the methoxy additions

References

1. McMurry JE, Fleming MP (1974) New method for the reductive coupling of carbonyls to olefins. Synthesis of β-carotene. J Am Chem Soc 96:4708–4709
2. Ephritikhine M (1998) A new look at the McMurry reaction. Chem Commun 2549–2554
3. Vogel E, Kocher M, Schmickler H, Lex J (1986) Porphycene, a new type of porphine isomer. Angew Chem Int Ed Engl 25:257–259
4. Arsenault GP, Bullock E, MacDonald SF (1960) Pyrromethanes and porphyrins therefrom. J Am Chem Soc 82:4384–4389
5. Vogel E, Balci M, Pramod K, Koch P, Lex J, Ermer O (1987) 2,7,12,17-Tetrapropylporphycene – pendant of octaethylporphyrin in the porphycene series. Angew Chem Int Ed Engl 26:928–931
6. Vogel E, Koch P, Hou X-L, Lex J, Lausmann M, Kisters M, Aukauloo MA, Richard P, Guilard R (1993) New porphycene ligands: octaethyl- and etioporphycene (OEPc and EtioPc): four and five-fold coordinated zinc complexes from OEPc. Angew Chem Int Ed Engl 32:1600–1604
7. Barbe J-M, Richard P, Aukauloo MA, Lecomte C, Petit P, Guilard R (1994) Electrocrystallization and X-ray structure of new porphycene-based material, [Ni(OMPc)]2.5(BF_4) 2-$C_{10}H_7Cl$. Chem Commun 2757–2758
8. Richert C, Wessels JM, Muller M, Kisters M, Benninghaus T, Goetz AE (1994) Photodynamic antitumor agents: β-methoxyethyl groups give access to functionalized porphycenes and enhance cellular uptake and activity. J Med Chem 37:2797–2807
9. Lausmann M, Zimmer I, Lex J, Lueken H, Wieghardt K, Vogel E (1994) μ-Oxodiiron(III) complexes of porphycenes. Angew Chem Int Ed Engl 33:736–739
10. Stepien M, Donnio B, Sessler JL (2007) Discotic liquid-crystalline materials based on porphycenes: a mesogenic metalloporphycene-tetracyanoquinodimethane (TCNQ) adduct. Chemistry 13:6853–6863

11. Anju KS, Ramakrishnan S, Thomas AP, Suresh E, Srinivasan A (2008) 9,10,19,20-Tetraaryl-porphycenes. Org Lett 10:5545–5548
12. Sanchez-Garcia D, Borrell JI, Nonell S (2009) One-pot synthesis of substituted 2,2′-bipyrroles. A straightforward route to aryl porphycenes. Org Lett 11:77079
13. Kuzuhara D, Mack J, Yamada H, Okujima T, Ono N, Kobayashi N (2009) Synthesis, structures and optical and electrochemical properties of benzoporphycenes. Chemistry 15:10060–10069
14. Vogel E (1990) Novel porphyrinoids. Pure Appl Chem 62:557–564
15. Vogel E, Kocher M, Lex J, Ermer O (1989) Steric modulation of the porphycene system by alkyl substituents: 9,10,19,20-tetraalkylporphycenes. Israel J Chem 29:257–266
16. Sessler JL, Weghorn SJ (1997) Expanded, contracted and isomeric porphyrins. Pergamon, Oxford, p 137
17. Coln D (1991) PhD Dissertation, University of Cologne
18. Vogel E, Sicken M, Rohrig P, Schmickler H, Lex J, Ermer O (1988) Tetraoxaporphycene dication. Angew Chem Int Ed Engl 27:411–414
19. De Munno G, Lucchesini F, Neidlein R (1993) 21,23-Dithiaporphycene: the first aromatic sulfur-containing system with porphycene structure. Tetrahedron 49:6863–6872
20. Ellinger F, Gieren A, Hunner T, Lex J, Luchesini F, Merz A, Neidlein A, Salbeck J (1993) The tetrathiaporphycene redox system: electrochemical reduction of a 20π cyclophane to its diatropic dianion. Monatsh Chem 124:931–943
21. Hu Z, Atwood JL, Cava MP (1994) A simple route to sulfur-bridged annulenes. J Org Chem 59:8071–8075
22. Sessler JL, Brucker EA, Weghorn SJ, Kisters M, Schafer M, Lex J, Vogel E (1994) Corrphycene: a new porphyrin isomer. Angew Chem Int Ed Engl 33:2308–2312
23. Aukauloo MA, Guilard R (1994) Etioporphycerin: synthesis and characterization of a new porphyrin isomer. New J Chem 18:1205–1207
24. Vogel E, Broering M, Scholz P, Deponte R, Lex J, Schmickler H, Schaffner K, Braslavsky SE, Mueller M, Poerting S, Weghorn SJ, Fowler CJ, Sessler JL (1997) Octaethylhemiporphycene: synthesis, molecular structure, and photophysics. Angew Chem Int Ed Engl 36:1651–1654
25. Lash TD, Jones SA, Ferrence GM (2010) Synthesis and characterization of tetraphenyl-21,23-dideazoporphyrin. The best evidence yet that porphyrins really are the [18]annulenes of naure. J Am Chem Soc 132:12786–12787
26. Jux N, Koch P, Schmickler H, Lex J, Vogel E (1990) Acetylene-cumulated-porphyrinoid. Angew Chem Int Ed Engl 29:1385–1387
27. Vogel E, Jux N, Rodriguez-Val E, Lex J, Schmickler H (1990) Porphyrin homologs. [22]-Porphyrin(2.2.2.2), an extended porphycene. Angew Chem Int Ed Engl 29:1387–1390
28. Sessler JL, Weghorn SJ (1997) Expanded, contracted and isomeric porphyrins. Pergamon, Oxford, p 220
29. Rodriguez-Val R (1994) PhD dissertation, University of Cologne
30. Martire DO, Jux N, Aramendia PF, Negri RM, Lex J, Braslavsky SE, Schaffner K, Vogel E (1992) Photophysics and photochemistry of 22π and 26π acetylene-cumulene porphyrinoids. J Am Chem Soc 114:9969–9978
31. Sessler JL, Weghorn SJ (1997) Expanded, contracted and isomeric porphyrins. Pergamon, Oxford, p 225
32. Dietrich H-J (1994) PhD dissertation, University of Cologne
33. Vogel E (1993) The porphyrins from the 'annulene chemist's' perspective. Pure Appl Chem 65:143–152
34. Sessler JL, Weghorn SJ (1997) Expanded, contracted and isomeric porphyrins. Pergamon, Oxford, p 230
35. Sauer F (1993) PhD Dissertation, University of Cologne
36. Khoury RG, Laquinod L, Smith KM (1997) Rational approach to the synthesis of meso-meso (5,5′) linked bis-porphyrins. Chem Commun 1057–1058

37. Zhilina ZI, Ishkov YV, Voloshanovskii IS, Andronati SA (1988) Reductive dimerization of the cuprous complex of 2-formyl-5,10,15,20-tetraphenylporphyrin. Dokl Akad Nauk SSSR 303:377–380
38. Vicente MGH, Smith KM (1991) Vilsmeier reactions of porphyrins and chlorins with 3-(dimethlamino)acrolein to give meso-(2-formylvinyl)porphyrins: new syntheses of benzochlorins, benzoisobacteriochlorins and benzobacteriochlorins and reductive coupling of porphyrins and chlorins using low-valent titanium complexes. J Org Chem 56:4407–4418
39. Senge MO, Vicente MGH, Gerzevske KR, Forsyth TP, Smith KM (1994) Models for the photosynthetic reaction center: preparation, spectroscopy, and crystal and molecular structures of cofacial bisporphyrins linked by cis-1,2- and trans 1,2-ethene bridges and of 1,1-carbinol-bridged bisporphyrins. Inorg Chem 33:5625–5638
40. McMurry JE (1989) Carbonyl coupling reactions using low-valent titanium. Chem Rev 89:1513–1524
41. Abraham RJ, Eivazi F, Pearson H, Smith KM (1976) Mechanisms of aggregation in metalloporphyrins. Demonstration of a mechanistic dichotomy. J Chem Soc Chem Commun 698–699
42. Abraham RJ, Eivazi F, Pearson H, Smith KM (1976). π-π Aggregation in metalloporphyrins. Causative factors. J Chem Soc Chem Commun 699–700
43. Abraham RJ, Fell SCM, Pearson H, Smith KM (1979) The NMR spectra of porphyrins 15. Self-aggregation in zinc(II) protoporphyrin-IX dimethyl ester. Tetrahedron 35:1759–1766
44. Abraham RJ, Smith KM (1983) The NMR spectra of porphyrins 21. Applications of the ring current model to porphyrin and chlorophyll aggregation. J Amer Chem Soc 105:5734–5741
45. Paolesse R, Pandey RK, Forsyth TP, Jaquinod J, Gerzeveske KR, Nurco DJ, Senge MO, Liccocia S, Boschi T, Smith KM (1996) Stepwise syntheses of bisporphyrins, bischlorins, and biscorroles, and of porphyrin-chlorin and porphyrin-corrole heterodimers. J Am Chem Soc 118:3869–3882
46. Clement TE, Nurco DJ, Smith KM (1998) Synthesis and characterization of a series of monometallo-, bimetallo-, and heterobimetallo-1,2-ethene-linked cofacial bisporphyrins. Inorg Chem 37:1150–1160
47. Ponomarev GV, Borovkov VV, Sugiura K, Yoshiteru S, Shul'ga AM (1993) Synthesis and properties of *cis*-1,2-bis(octaethylporphyrinyl)ethylene. Tetrahedron Lett 34:2153–2156
48. Jaquinod L, Nurco DJ, Medforth CJ, Pandey RK, Forsyth TP, Olmstead MM, Smith KM (1996) Synthesis and characterization of bis(chlorin)s from the McMurry reaction of formylchlorins. Angew Chem Int Ed Engl 35:1013–1016
49. Jaquinod L, Senge MO, Pandey RK, Forsyth TP, Smith KM (1996) Planar bischlorophyll derivatives with a completely conjugated π-system: model compounds for the special pair in photosynthesis. Angew Chem Int Ed Engl 35:1840–1842

Top Heterocycl Chem (2014) 33: 35–78
DOI: 10.1007/7081_2013_101

Published online: 17 May 2013

Meso-Tetraarylporphyrins Bearing Nitro or Amino Groups: Synthetic Strategies and Reactivity Profiles

Vanda I. Vaz Serra, Sónia M.G. Pires, Cristina M.A. Alonso, Maria G.P.M.S. Neves, Augusto C. Tomé, and José A.S. Cavaleiro

Abstract Porphyrins bearing nitro and amino substituents have been used as excellent synthons for further functionalization in order to obtain new compounds with adequate features for a wide range of applications. This chapter brings an update on the effort of several research groups to study the synthesis and reactivity features of *meso*-tetraarylporphyrins bearing those functionalities.

Keywords Aminoporphyrins · Nitration · Nitroporphyrins · Porphyrins

Contents

V.I.V. Serra, S.M.G. Pires, C.M.A. Alonso, M.G.P.M.S. Neves (✉), A.C. Tomé and J.A.S. Cavaleiro
Department of Chemistry and QOPNA, University of Aveiro, 3810-193 Aveiro, Portugal
e-mail: gneves@ua.pt

1 Introduction

The extraordinary development observed in the porphyrin field after the structure elucidation of protoporphyrin IX by Fisher in 1929 [1] and the synthesis of chlorophyll *a* by Woodward in 1960 [2] shows that the scientific community has been having a great interest in the potentiality of these unique compounds. Today, it is accepted that porphyrin derivatives, besides their central role in respiration, photosynthesis, and other vital functions, have a promising future in several fields such as medicine [3], catalysis [4], and electronic materials [5]. Knowing that all those applications are strongly dependent on the structure of the macrocycle, there has been a considerable research directed towards the development of synthetic strategies to functionalize readily available porphyrins, especially *meso*-tetraarylporphyrins. Part of that work has been related to the functionalization of a primary group inserted in *meso*- or in β-pyrrolic positions of *meso*-tetraarylporphyrins. In this chapter, we highlight the most relevant and recent synthetic strategies concerning the functionalization of *meso*-tetraarylporphyrins through nitro or amino groups located at β-pyrrolic positions or in *meso*-phenyl groups. Occasionally, other porphyrin derivatives may also be discussed. The interest in nitro- and aminoporphyrins is mainly due to the very attractive reactivity and versatility of these two functional groups [6, 7].

The nitro group can improve the ability of porphyrin systems to act as radiosensitizers [8] and their usefulness in porphyrin functionalization has been demonstrated and is well documented. Nitroporphyrins themselves are widely used as starting materials: they can undergo direct nucleophilic addition and substitution reactions with a wide range of nucleophiles and displacement of the nitro group. The reduction of the nitro group to the amino group is a very useful reaction that extends the porphyrin potentialities for further functionalization via the amino group. In fact, a wide range of porphyrin derivatives with improved properties have been prepared via amide linkage, *N*-alkylation, nucleophilic substitution, diazotization, cycloaddition, palladium-catalyzed reactions, etc.

Besides the clear differences in reactivity of the β- and *meso*-aryl positions, the selective nitration of such positions can be controlled by the choice of the nitrating agent and the metal ion coordinated with the porphyrin macrocycle. The following sections cover useful nitration procedures and the exploitation of the nitro group in further functionalization of the porphyrin macrocycle. In the last topic it is highlighted the recent synthetic strategies concerning the functionalization of *meso*-tetraarylporphyrins through amino groups.

2 Synthesis and Reactivity of Nitroporphyrins

2.1 *Synthesis of Meso-(Nitrophenyl)porphyrins*

Meso-(Nitrophenyl)porphyrins can be obtained by the condensation of pyrrole with nitrobenzaldehydes (Scheme 1) or by nitration of *meso*-tetraarylporphyrins.

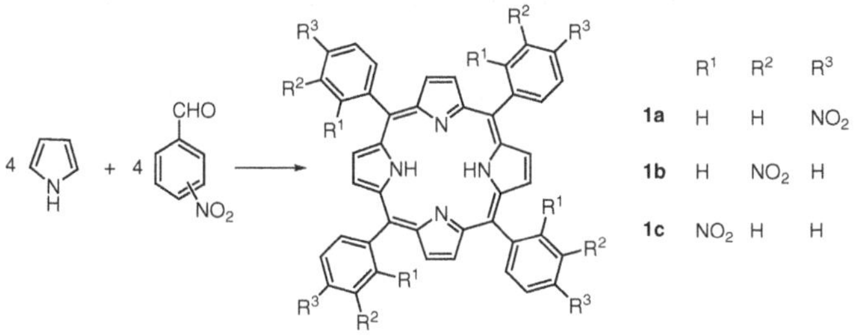

Scheme 1 Synthesis of *meso*-(nitrophenyl)porphyrins

The first strategy was used by Martell and coworkers [9, 10] to synthesize *meso*-tetrakis(4-nitrophenyl)porphyrin (**1a**). This synthesis was based on the Rothemund's landmark conditions [11] to prepare *meso*-substituted porphyrins. The authors referred that the best yield (2.6%) was obtained when equimolar amounts of pyrrole and 4-nitrobenzaldehyde were heated at 120°C in a mixture of pyridine and methanol for 24 h. The synthetic improvements that appeared afterwards to obtain *meso*-tetraarylporphyrins were also considered in the synthesis of porphyrin **1a** and its isomers **1b** and **1c**, respectively, with the nitro groups at *para*, *meta*, or *ortho* positions of the phenyl substituents [12, 13]. Porphyrin **1a**, for instance, can be obtained in 19–22% by refluxing a solution of pyrrole and 4-nitrobenzaldehyde in propanoic acid containing acetic anhydride [14]. The same porphyrin can be obtained in 28% yield if the cyclocondensation is mediated by microwave irradiation in the presence of small amounts of propanoic acid [15]. Under the same microwave conditions, the cyclocondensation of equimolar amounts of pyrrole and 2-nitrobenzaldehyde afforded porphyrin **1c** in 25% after ca. 5 min of reaction [15]. Under classical heating, porphyrin **1c**, frequently used in the development of synthetic models for oxygen-binding hemoproteins, was obtained in 13% yield after refluxing pyrrole and 2-nitrobenzaldehyde in acetic acid for 20 min [16]. Much poor yields (4%) were reported for porphyrin **1b** when 3-nitrobenzaldehyde and pyrrole were heated at reflux in propanoic acid containing acetic anhydride [14].

The cyclocondensation of pyrrole with a mixture of two aldehydes gives access to a wide variety of multifunctional porphyrins. Although not being considered an efficient and elegant strategy, due to the low yields (<5%) and the purification procedures required to separate the products mixture, this mixed-aldehyde approach is expeditious and is being largely exploited for the preparation of porphyrins bearing one or more *meso*-nitrophenyl groups. For instance, Tsuchida and coworkers [17–19] used that strategy for the preparation of the mono-(4-nitrophenyl)porphyrin **2** (Fig. 1). Using a 3:1 ratio of benzaldehyde and 4-nitrobenzaldehyde the desired porphyrin **2** was obtained in 2.7% yield; the related bis- (**3** and **4**) and tris(4-nitrophenyl) (**5**) substituted porphyrins were also isolated.

The same strategy was considered by Collman [20] to obtain the mono-(2-nitrophenyl) analogue and by Martell and coworkers [21] to obtain unsymmetrical 3-nitrophenyl substituted porphyrins. Little has also reported the synthesis of unsymmetrical porphyrins containing a 2,6-dinitrophenyl group, or a hydroxy-nitrophenyl group, as potential intermediates in the synthesis of difunctional

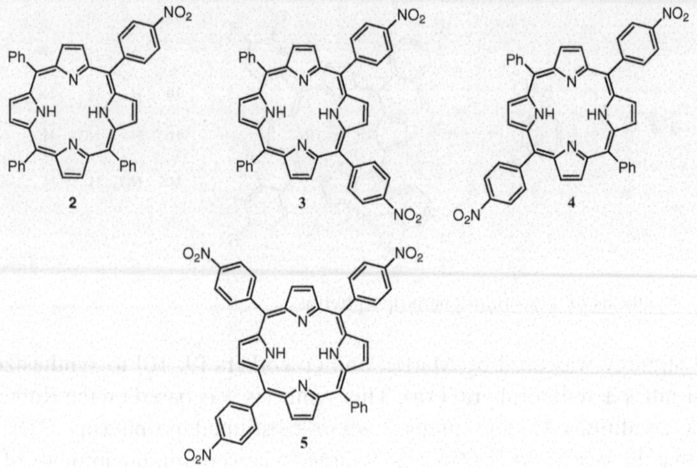

Fig. 1 Structures of *meso*-(4-nitrophenyl)porphyrins

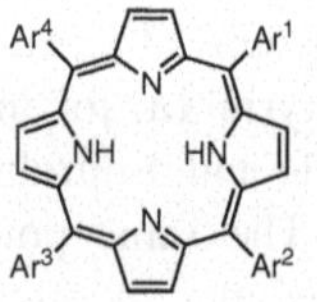

6a, $Ar^1 = o\text{-}C_6H_4NO_2$, $Ar^2 = Ar^3 = Ar^4 = p\text{-}C_6H_4CO_2H$
6b, $Ar^1 = Ar^2 = o\text{-}C_6H_4NO_2$, $Ar^3 = Ar^4 = p\text{-}C_6H_4CO_2H$
6c, $Ar^1 = Ar^3 = o\text{-}C_6H_4NO_2$, $Ar^2 = Ar^4 = p\text{-}C_6H_4CO_2H$

Fig. 2 Unsymmetrical porphyrins bearing *o*-nitrophenyl and *p*-carboxyphenyl groups

"tailed-porphyrins" [22]. The mixed-aldehyde approach was also followed to prepare unsymmetrical porphyrins bearing *o*-nitrophenyl and *p*-carboxyphenyl substituents (**6a–c**, Fig. 2) [23]. The authors were able to optimize the benzaldehyde derivatives molar ratio in order to obtain the desired porphyrins in high yields. Other examples of unsymmetrical porphyrins prepared by this approach can be found in a review by Lindsey [12].

As already mentioned, the direct nitration of *meso*-tetraarylporphyrins is another strategy to obtain porphyrins, mainly unsymmetrical ones, containing nitroaryl groups. This approach was considered for the first time by Kruper's group in the nitration of *meso*-tetraarylporphyrins, namely TPP [24, 25]. The authors used an excess of red or yellow fuming nitric acid in different solvents ($CHCl_3$, CH_2Cl_2, and AcOH) and referred higher yields for the mono-nitro derivative (ca 56%) in $CHCl_3$; under these conditions the dinitro derivatives **3** and **4** (as a mixture of 2–3:1, respectively) were also isolated but the yields rarely exceed 5%. Better yields for the dinitro derivatives (28%) were obtained under conditions forcing the conversion of the mono-substituted derivative (use of 29 equivalents of red fuming nitric acid in $CHCl_3$). The use of acetic acid as solvent allowed to obtain the tris(4-nitrophenyl) porphyrin **5** in 10% yield. The *para* regioselectivity was also observed in the nitration of porphyrins bearing 3-methyl- or 3-methoxyphenyl substituents.

Meng and coworkers [8] also studied the nitration of TPP but under slightly different conditions relatively to those reported by Kruper. The reactions were carried out by using a mixture of nitric acid and acetic acid and the degree of nitration was controlled by the reaction time. The 5-(4-nitrophenyl)-10,15,20-triphenylporphyrin **2** was obtained in 74% yield after 1 h of reaction. Significant improvements were also reported for the dinitro derivatives **3** and **4** (70% yield after 5 h of reaction) and for the trinitro derivative **5** (30% yield after 2 days of reaction). Attempts to obtain the tetrakis(4-nitrophenyl)porphyrin **1a** via nitration of TPP failed and this was due to the degradation of the macrocycle during the time required for this lengthy reaction. Porphyrin **1a** was obtained as a by-product (2%) during the tri-nitration conditions. The same authors studied the nitration of unsymmetrical porphyrins bearing phenyl and pyridyl groups. The nitration of these porphyrins requires higher reaction times than those used for TPP due to the protonation of the pyridyl groups; in certain cases the use of a mixture of acetic and sulfuric acids was required.

The selective nitration of TPP and other *meso*-tetraarylporphyrins at two neighboring aryl rings was described by Ostrowski and Lopuszynska [26]. Following a new protocol, which is based on the use of fuming yellow HNO_3 in $CHCl_3$, accompanied by a careful control of the reaction temperature, nitration of TPP afforded compound **3** in 42% yield. *meso*-Tetraarylporphyrins bearing 3-chlorophenyl, 3-methoxyphenyl or 3-methylphenyl groups afforded the corresponding 5,10-bis(4-nitroaryl)porphyrins in 30%, 37%, and 83% yields, respectively. The same group reported that under exhaustive nitration conditions tri-substituted derivatives can also be obtained in reasonable yields [27]. For instance, nitration of TPP afforded derivative **5** in 35% yield.

Smith and coworkers reported the selective nitration of the phenyl groups of TPP with sodium nitrite and trifluoroacetic acid [28]. The authors found that the degree of nitration can be efficiently controlled just by varying the amount of $NaNO_2$/TFA used and the reaction time. Compound **2** was obtained in excellent yield (80–90%) after 3 min of reaction at room temperature in the presence of 1.8 equivalents of $NaNO_2$ in TFA. The two isomeric bis(4-nitrophenyl)porphyrins **3** and **4** were obtained after 1.5 min of reaction in the presence of 8.1 equivalents of $NaNO_2$ in a total yield of 63%, while the tris(4-nitrophenyl)porphyrin **5** required 1 h of reaction and 36.7 equivalents of $NaNO_2$ to be isolated in 60% yield.

The nitration of the *meso*-phenyl groups of TPP with NO_2BF_4 has been also reported [29]. The authors found that the mode of addition is the key step for the success of this methodology. Mono-, bis-, and tris(4-nitrophenyl) derivatives were obtained in excellent yields (>90%) by dropwise addition of a sulfolane solution of 1.0, 2.9, or 4.6 equivalents, respectively, of that nitrating agent to a dichloromethane solution of TPP at room temperature. The authors have also reported the selectivity of the dinitration procedure for the 5,10-bis(4-nitrophenyl) isomer.

2.2 Synthesis of β-Nitro-Meso-Tetraarylporphyrins

Efficient nitrating procedures, based on electrophilic or radical conditions, are now well established for giving access to β-nitroporphyrins in excellent yields [30]. Most of the protocols are based on the use of metalloporphyrins. In fact, attempts to nitrate

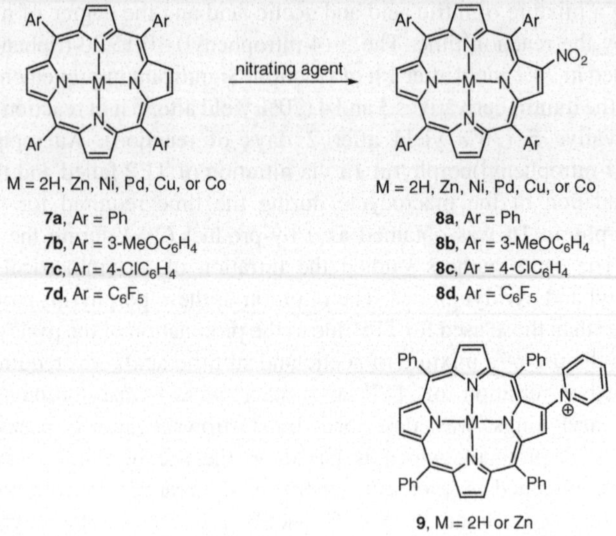

Scheme 2 Synthesis of β-nitro-*meso*-tetraarylporphyrins **8** and pyridinium salt **9**

the free-base TPP using a mixture of nitric acid and sulfuric acid afforded only 2-nitro-*meso*-tetraphenylporphyrin **8a** (M = 2H, Scheme 2) in low yield; this is due to the conversion of the starting porphyrin into the unreactive dication [31, 32]. The nitration of TPP (**7a**, M = 2H) under essentially neutral conditions was considered by Jackson and coworkers. Using nitronium tetrafluoroborate in a mixture of pyridine/chloroform at 140°C, the β-nitroporphyrin **8a** (M = 2H) was isolated in 15% yield being accompanied by the pyridinium salt **9** (M = 2H) (18% yield). Attempted nitration of the zinc complex of TPP with nitronium tetrafluoroborate in pyridine afforded the pyridinium derivative **9** (M = Zn) in 80% yield [32].

In contrast with the previous results, nitration of the copper, nickel, and palladium complexes of TPP with N_2O_4 occurs selectively at the β-pyrrolic positions, affording the corresponding complexes **8a** (M = Cu, Ni or Pd) in quantitative yields [33]. The extension of this protocol to *meso*-tetraarylporphyrins **7b** and **7c**, (M = Cu), afforded the corresponding derivatives **8b** and **8c** in high yields (>90%), thus confirming the generality of the process. The presence of extra substituents at the β-pyrrolic position does not affect the site of nitration. For instance, the nitration of the copper complex of β-nitroporphyrin afforded an isomeric mixture of β,β′-disubstituted derivatives **10a–e** (Fig. 3) in 85% total yield [34].

A procedure giving access to β-dinitro- and β-trinitro-*meso*-tetraphenylporphyrins using the controlled addition of fuming nitric acid to CuTPP was also reported [35]. The 2,12-dinitro- and 2,13-dinitro derivatives were obtained by the controlled addition of 0.7 mL of HNO_3 to 100 mg of CuTPP in $CHCl_3$ over a period of 1.2 min, while the 2,7-dinitro-, 2,8-dinitro-, and 2,18-dinitro derivatives were obtained by the addition of 1.0 mL of HNO_3 during 1.0 min to the same amount of porphyrin. Yields of 20% for the pure products were reported. The trinitroporphyrins **11a–c** (2,7,13-, 2,7,18-, and 2,8,12-trinitro) were obtained

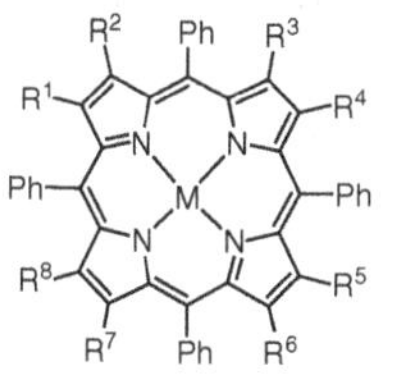

10a, R^1, R^3 = NO_2, R^2, R^4-R^8 = H
10b, R^1, R^4 = NO_2, R^2, R^3, R^5-R^8 = H
10c, R^1, R^5 = NO_2, R^2-R^4, R^6-R^8 = H
10d, R^1, R^6 = NO_2, R^2-R^5, R^7-R^8 = H
10e, R^1, R^8 = NO_2, R^2-R^7 = H

11a, R^1, R^3, R^6 = NO_2, R^2, R^4, R^5, R^7, R^8 = H
11b, R^1, R^3, R^8 = NO_2, R^1, R^2, R^4-R^7 = H
11c, R^1, R^4, R^5 = NO_2, R^2, R^3, R^6, R^7, R^8 = H

Fig. 3 Structures of β-nitro-*meso*-tetraarylporphyrins

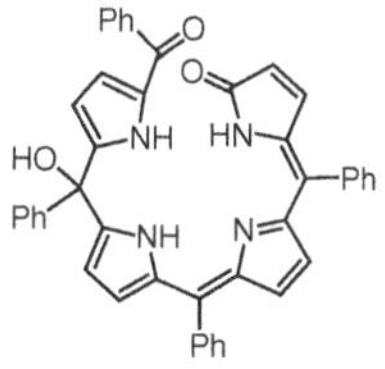

12

Fig. 4 Structure of the bilinone obtained by nitration of MgTPP or ZnTPP by N_2O_4

by increasing the amount of HNO_3 up to 2.0 mL and maintaining the addition during the period of 1 min. The corresponding free-bases were obtained by demetallation with sulfuric acid. Electrochemical studies revealed that successive insertion of nitro groups at the β-positions shifts the one-electron ring oxidations anodically while the ring reduction occurs at a less cathodic potential relatively to the unsubstituted porphyrin free-bases.

Nitration of porphyrins coordinated with less electronegative metal ions, such as magnesium(II), zinc(II), chloroiron(III) and cobalt(II), and the β-nitro derivatives are accompanied by products resulting from reactions at the *meso*-position. In fact, the nitration of magnesium and zinc chelates of TPP by N_2O_4 afforded the corresponding β-nitro derivatives in low yields (ca. 25%) giving mainly the ring-opened bilinone **12** (Fig. 4) [36] and other non-porphyrin products resulting from reactions at the *meso*-positions. This metal ion dependent selectivity was justified by considering that the metalloporphyrin π-cation radicals obtained via oxidation by $NO_2\cdot$ have different electron spin distributions (a_{1u} or a_{2u}) and, as a result, being the position of attack by further $NO_2\cdot$ dependent on its spin density. The preferential attack at the *meso*-positions was also reported when the zinc complex of TPP was treated with thallium(III) nitrate or cerium(IV) ammonium nitrate followed by acid treatment [37]. Under these conditions the β-nitro derivative **8a** (M = 2H) was isolated in low yields (15–28%) accompanied by porphodimethenes and the ring-opened bilinone **12**.

Callot and coworkers reported an excellent protocol to nitrate the copper complex of TPP based on the use of copper(I) nitrate in a mixture of chloroform, acetic acid, and acetic anhydride [38]. Using that nitrating mixture, Cavaleiro and coworkers prepared **8a** (M = Cu) in 86% yield directly from TPP without the previous preparation of the copper complex [39]. The extension of this nitrating procedure

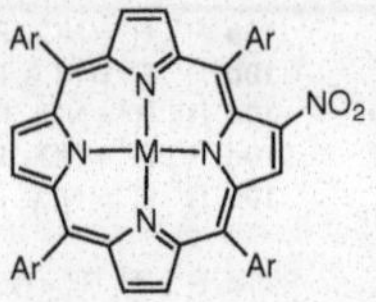

M = 2H, Zn, Cu

8e, Ar = 4-$Me_2NC_6F_4$

Fig. 5 Structure of porphyrin **8e**

to *meso*-tetrakis(pentafluorophenyl)porphyrin (**7d**, M = 2H) has provided access to the mono-nitroporphyrin **8d** (M = 2H) and to a mixture of dinitro and trinitro isomers in quantitative yield [40]. Krishnan and coworkers also used copper(I) nitrate in chloroform to obtain the donor–acceptor porphyrins **8e** (Fig. 5) for studies concerning quadratic nonlinear optics [41]. Callot and coworkers reported a mild procedure for the nitration of the nickel or copper complexes of *meso*-tetraarylporphyrins using lithium nitrate in $CHCl_3/Ac_2O/AcOH$, for 1.5 h at 40–45°C, affording the 2-nitro derivatives in 90–95% yield (aryl = phenyl, *p*-tolyl, and 3,5-di-*tert*-butylphenyl) [42].

The selective β-mononitration of *meso*-tetraphenylporphyrin complexes can also be achieved using aqueous HNO_3. Ostrowski et al. [43] found that several TPP complexes (**7a**, M = Zn, Cu, Ni, and Co) can be nitrated with adequate concentrations of HNO_3 (ca. 25%) to afford the corresponding complexes **8a** in very good yields (77% for M = Cu and 81% for M = Ni). A mixture of dinitro compounds (2–20%) is also detected in all cases. An extension of this work to other *meso*-tetraarylporphyrins (Ar = 3-$NO_2C_6H_4$, 3-$CH_3C_6H_4$, 3-ClC_6H_4, 2,6-$Cl_2C_6H_3$, C_6F_5) afforded the corresponding mono-β-nitrated products in yields ranging from 74% to 93% [44]. These results show that the type of *meso*-aryl substituent does not change the site of nitration; it only affects the reaction yield. In fact, the systems less prone to electrophilic substitution require slightly drastic conditions (higher concentration of nitric acid and longer reaction times) to ensure high yields. Again, some dinitro compounds were also formed.

The nitration of the free-base *meso*-tetrakis(2,6-dichlorophenyl)porphyrin with red fuming nitric acid, at room temperature, afforded a 1:9 mixture of β-pentanitro- and β-hexanitroporphyrins in 70% yield [45]. Nitration of *meso*-tetrakis(pentafluorophenyl) porphyrin under similar conditions led to a mixture of regioisomers containing one nitro group on each pyrrole ring (55% yield). All attempts to obtain *meso*-tetrakis (2,6-dichlorophenyl)porphyrin substituted by more than six β-nitro groups or *meso*-tetrakis(pentafluorophenyl)porphyrin substituted by more than four β-nitro groups by using more HNO_3, higher temperatures or longer reaction times in reactions between HNO_3 and those porphyrins or their Zn(II) or Fe(III) complexes were unsuccessful [45]. Later, it was reported that the nitration of the zinc complex of *meso*-tetrakis(2,6-dichlorophenyl)porphyrin with red fuming nitric acid in the presence of nitromethane, acetic anhydride, and montmorillonite K-10, for 2 h at room temperature, affords the β-heptanitro derivative **13b** (Fig. 6) in 50%

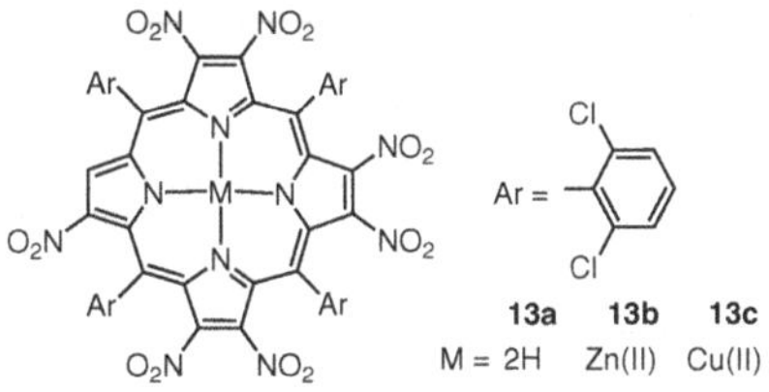

Fig. 6 β-Heptanitroporphyrin obtained by nitration of *meso*-tetrakis(2,6-dichlorophenyl)porphyrin with montmorillonite K10-HNO_3 under microwave irradiation

yield [46]. An expeditious β-polynitration of *meso*-tetrakis(2,6-dichlorophenyl)porphyrin with montmorillonite K10-HNO_3 using microwave irradiation has been described [47]. The microwave irradiation of that porphyrin with montmorillonite K10-HNO_3 for 1.5 min selectively gave β-heptanitro derivative **13a** in 72% yield. Similarly, the microwave irradiation of the Zn(II) and Cu(II) complexes with K10-HNO_3 gave the β-heptanitro derivatives **13b** and **13c** in 81% and 75% yield, respectively.

3 Functionalization of Porphyrins via Nitro Groups

3.1 Functionalization of Meso-(Nitrophenyl)porphyrins

The nucleophilic aromatic substitution methodology has been used to functionalize *meso*-(nitrophenyl)porphyrins. Ostrowski and coworkers [48], for instance, reported that the copper and the zinc complexes of the *meso*-(4-nitrophenyl)porphyrin **2** react with carbanions **14** affording the corresponding products **15** in yields ranging from 50% to 67% (Scheme 3). These vicarious nucleophilic substitutions take place selectively at the *ortho* position to NO_2 group; bulky carbanions of lower nucleophilicity do not react. Further studies showed that these reactions can also be performed using free-base porphyrins, with improved yields, if low temperatures are considered [49].

The potentiality of *meso*-(nitrophenyl)porphyrins for further functionalization was shown in the synthesis of the porphyrin-fullerene dyad **16** (Fig. 7), a new artificial photosynthetic model [50].

Ostrowski and coworkers also explored the activation of the nitro group towards the attack by nucleophiles to introduce the amino functionality in *meso*-tetraarylporphyrins bearing one or two nitrophenyl groups. For instance, the amino-functionalized porphyrins **17** (Fig. 8) were obtained from the reaction of the zinc, copper, and nickel complexes of porphyrin **2** with 1,1,1-trimethylhydrazinium iodide in the presence of KOH in DMSO [51, 52].

The reaction of 5-(4-nitrophenyl)-10,15,20-triphenylporphyrin (**2**) and the corresponding zinc and copper complexes with other nucleophiles was also studied [53]. The authors reported that in the reaction of **2** with NaCN the substitution

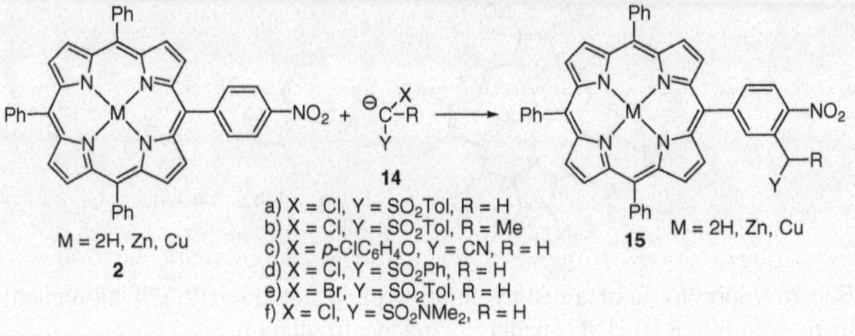

Scheme 3 Vicarious nucleophilic substitution in *meso*-(4-nitrophenyl)porphyrins

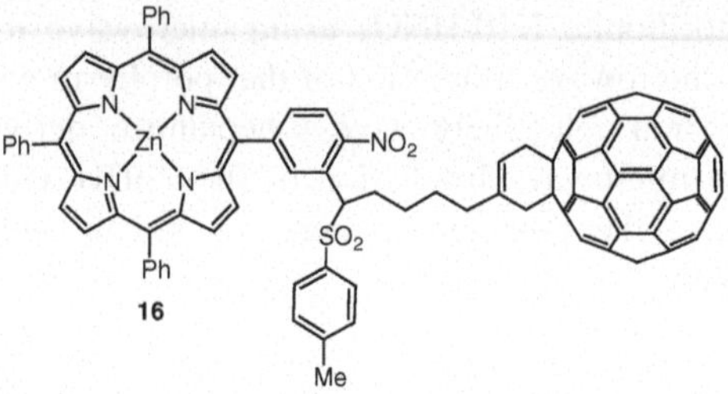

Fig. 7 Structure of a porphyrin-fullerene dyad

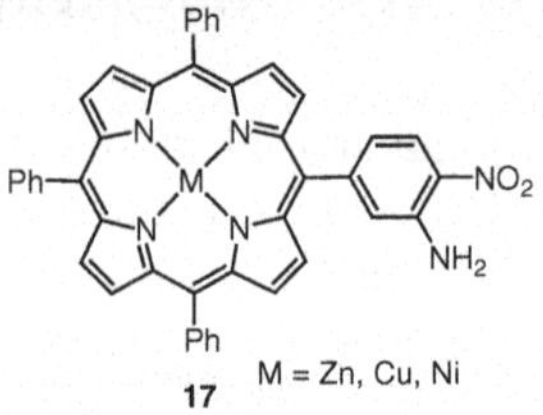

Fig. 8 A porphyrin functionalized with a *meso*-(3-amino-4-nitrophenyl) group

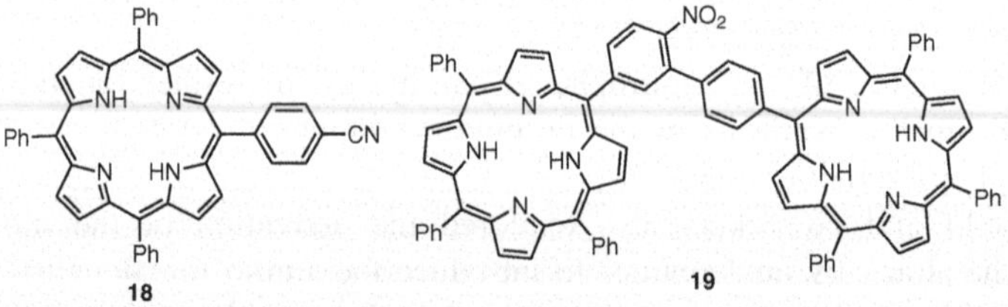

Fig. 9 Structures of porphyrins **18** and **19**

of the nitro group occurs and porphyrin **18** (Fig. 9) is obtained in reasonable yield. On the other hand, the reaction of **2** with phenoxides affords the diporphyrin derivative **19**, while under the same conditions, but using the metal complexes, the nitro group is reduced to the amino group. A similar reduction occurs in the reaction with thiolates [53].

3.2 *Functionalization of β-Nitroporphyrins*

As already mentioned, nitroporphyrins are excellent starting materials to prepare new derivatives with improved features for specific applications. In fact, a nitro group is particularly useful to activate the pyrrole unit where it is inserted towards the attack by nucleophiles, dienes or 1,3-dipoles and also to direct electrophilic substitutions to the antipodal pyrrole ring. The alkene-type reactivity of β-nitro-*meso*-tetraarylporphyrins can be justified by the preferential localization of the double bond adjacent to the electron-withdrawing group; that double bond is not involved in the major aromatic delocalization pathway.

The possibility of using β-nitro-*meso*-tetraarylporphyrins for further functionalization at the β-pyrrolic positions was firstly considered by Crossley and coworkers [54, 55] who found that β-nitro-*meso*-tetraphenylporphyrin (**8a**, M = 2H) reacts with benzenethiolate and ethanethiolate to afford the corresponding β-thioethers. This pioneering work was followed by other publications exploring the reaction of β-nitro-*meso*-tetraarylporphyrins with nucleophiles to insert a variety of substituents at the β-pyrrolic positions. This topic has been comprehensively reviewed by Jaquinod [30].

The attack by nucleophiles to a β-nitro-*meso*-tetraarylporphyrin can occur either at the carbon atom containing the nitro substituent (ipso-attack) or at the adjacent β-position (α-attack) (Scheme 4). In fact, based on deuterium labeling experiments, Crossley and coworkers found that "soft" nucleophiles such as thiolates and the anion of benzaldoxime lead to products of type **C**, resulting from an ipso-attack [54, 56]. They also found that, in general, "hard" nucleophiles such as oxyanions [57], hydride [58], acylamide ions [59], Grignard and organolithium reagents [60] attack the β-pyrrolic position next to the nitro group. The outcome of these reactions is dependent on the nucleophile, coordinated metal ion, and temperature.

The direct displacement of the nitro group was also observed when phenoxide ion and other phenols were used in reactions with the free-base 2-NO_2TPP and with the corresponding Cu(II) and Ni(II) complexes [61, 62]. It was found that the type of product (2-aryloxy- or 2-hydroxyaryl-) can be controlled by the choice of solvent. This methodology was recently explored to build the ditopic chemosensor **22** (Scheme 5) [63]. This compound interacts selectively with histamine when compared with L-histidine and nicotine. Following the same protocol, Chen et al. [64] found that **8a** and its Ni(II), Cu(II), and Zn(II) complexes react with 2-naphthoxide in protic solvents (2-naphthol, diglycol, and diglycol monomethyl ether), at 150°C, to afford only the *C*-coupling products **23** with yields varying from 50% to 81% (Scheme 6). In aprotic solvents (DMF or DMSO), at 150°C, the *O*-coupling products **24** are also obtained but at room temperature only compounds **23** are formed.

Following a synthetic methodology similar to the one developed by Crossley, Pan and coworkers [65] reported the synthesis of the β-(2,5-dihydroxyphenyl) porphyrins **25** (Scheme 7). The preliminary biological activity studies showed that

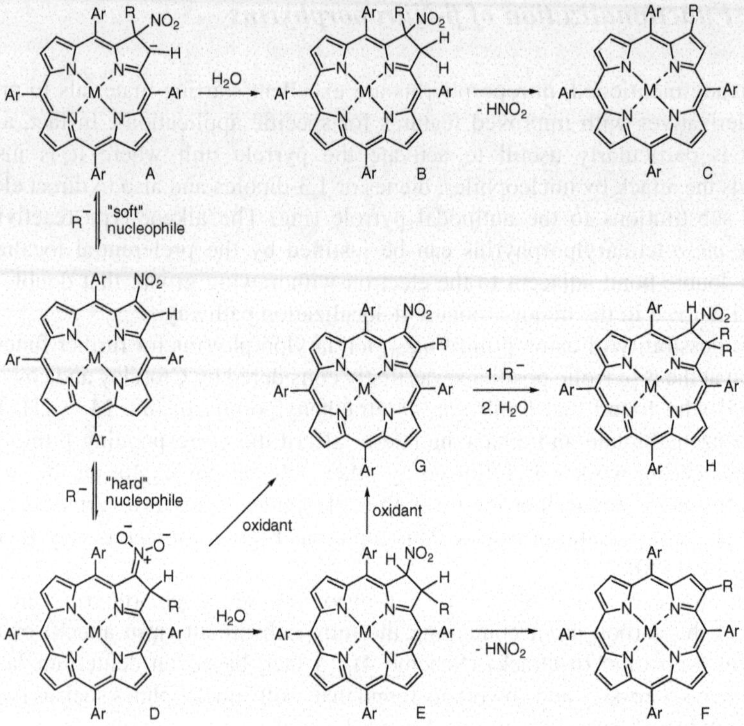

Scheme 4 Representative reactivity of β-nitro-*meso*-tetraarylporphyrins with nucleophiles

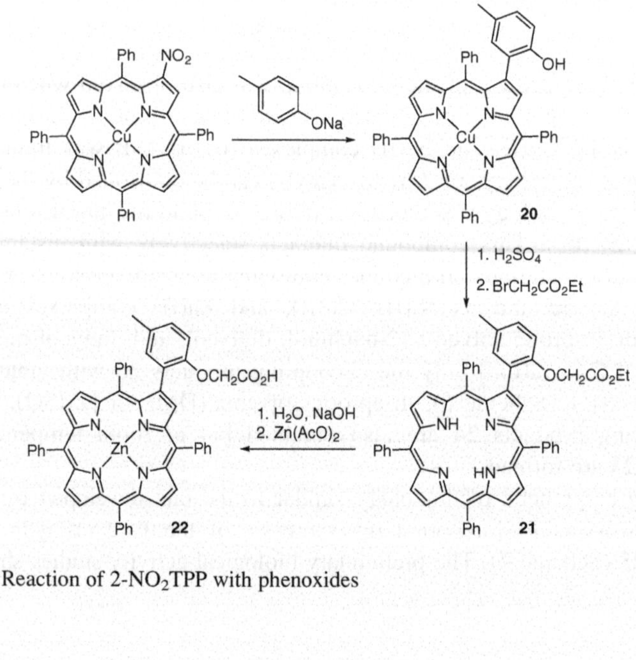

Scheme 5 Reaction of 2-NO_2TPP with phenoxides

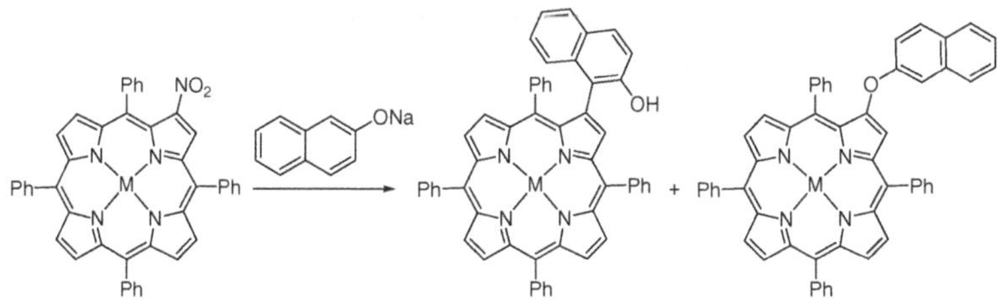

8a
M = 2H, Ni(II), Cu(II), Zn(II)
23
24

Scheme 6 Reaction of 2-NO_2TPP with 2-naphthoxide

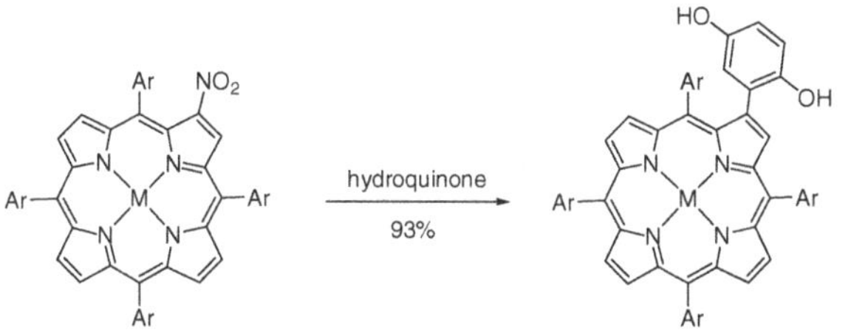

M = Cu(II), Zn(II)
8f, Ar = 4-HOC_6H_4
25a, Ar = 4-HOC_6H_4, M = Cu
25b, Ar = 4-HOC_6H_4, M = Zn

Scheme 7 Reaction of β-nitroporphyrins with hydroquinone

the zinc(II) derivative has photo-toxicity on human chronic myelogenous leukemia cell and is able to cleave supercoiled DNA (pBR 322 DNA) while the copper(II) complex has lower biological activity.

The use of anilines as nucleophiles in the reaction with β-nitro-*meso*-tetraarylporphyrins was considered by Cavaleiro and coworkers [66]. For instance, the reaction of **8a** with aniline, at reflux temperature, gives the 2-(phenylamino) porphyrin **26a** as the major product (53% yield), but the novel *N*-phenylquinolino [2,3,4-*at*]porphyrin **27a** (6% yield) and the chlorin **28** are also formed (Scheme 8). When this reaction is performed in refluxing *o*-dichlorobenzene, porphyrin **27a** is the main product (26% yield). This solvent was also adequate to obtain derivative **27b** when *p*-toluidine was selected as the nucleophile. The oxidative cyclization of 2-arylaminoporphyrins **26** to the corresponding *N*-arylquinolino[2,3,4-*at*] porphyrins **27** can be done in excellent yields in nitrobenzene. This synthetic strategy is not efficient with anilines with electron-withdrawing substituents.

Smith and coworkers explored the nitroalkene character of β-nitro-*meso*-tetraarylporphyrins to prepare β-fused pyrroloporphyrins **29** (Fig. 10) via the Barton–Zard condensation of the nickel complex of 2-NO_2TPP with α-isocyanoacetic esters in the presence of DBU [67]. Interestingly, when the zinc complex of 2-NO_2TPP was used, the cyclopropyl-annulated chlorin **30** was obtained.

Based on the conjugate addition of active methylene compounds, such as malonates or malononitrile, to β-nitro-*meso*-tetraarylporphyrins in the presence of a

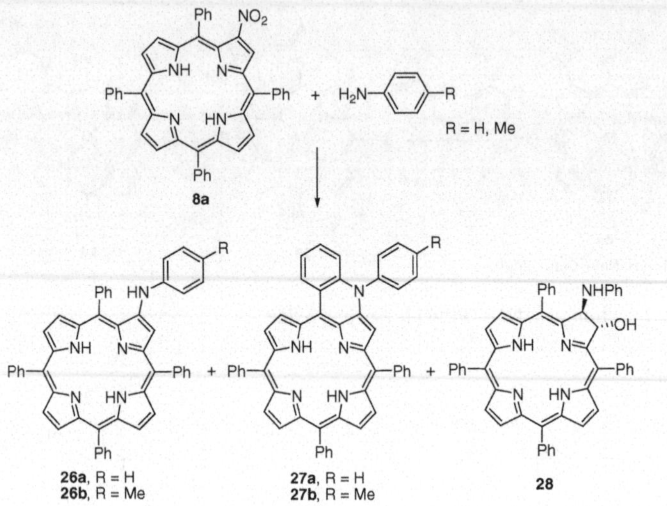

Scheme 8 Reaction of 2-NO_2TPP with anilines

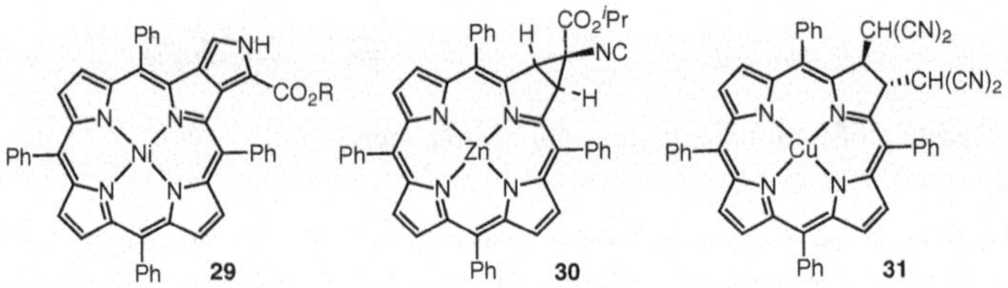

Fig. 10 Structures of porphyrins **29–31**

base, the same group was able to prepare a wide range of reduced porphyrins such as *trans*-nitrochlorins, cyclopropachlorins, or disubstituted *trans*-chlorins such as **31** [68, 69]. The product distribution can be controlled by the size of the carbanion, reaction time, and/or temperature as well as the use of free-bases or chelates.

Cavaleiro and coworkers reported that 1,3-diketones and 3-ketoesters such as acetylacetone and ethyl acetoacetate can act as efficient nucleophiles in Michael additions with β-nitro-*meso*-tetraarylporphyrins affording the corresponding derivatives **32** as the only products (Scheme 9) [70]. The central metal ion or the *meso*-aryl-substituents do not affect significantly the reactivity of the system, nor in terms of yields (72–88%) nor in terms of reaction times (40–50 min). The 1,3-dicarbonyl derivatives **32a,b** showed to be excellent C3 synthons for the synthesis of porphyrins bearing an heteroaromatic group at the β-position [70].

Cavaleiro and coworkers also considered the use of β-nitro-*meso*-tetraarylporphyrins as precursors to the novel [1,2,3]triazolo[4,5-*b*]porphyrins **33** (Scheme 10) [71]. Knowing that *N*-unsubstituted 1*H*-1,2,3-triazoles can be obtained from the reaction of sodium azide with alkenes bearing strongly electron-withdrawing, it was anticipated that the reaction β-nitroporphyrins with sodium azide could afford

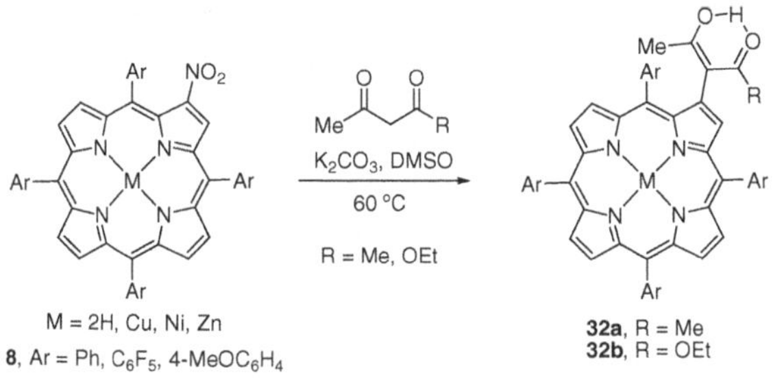

Scheme 9 Reaction of β-nitro-*meso*-tetraarylporphyrins with 1,3-diketones and 3-ketoesters

Scheme 10 Synthesis of [1,2,3]triazolo[4,5-*b*]porphyrins

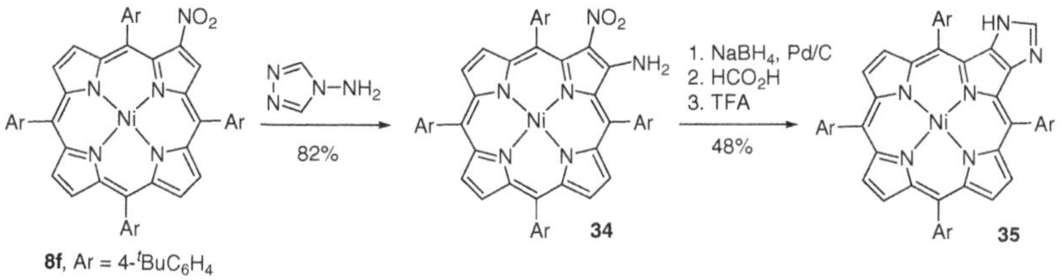

Scheme 11 Synthesis of imidazo[4,5-*b*]porphyrins

such type of compounds. As expected, the best yield (80%) was obtained with the porphyrin bearing electron-withdrawing groups at the *meso* positions.

A β-nitro-*meso*-tetraarylporphyrin was also used by Richeter et al. [72] to synthesize a porphyrin with an additional imidazole ring fused to a β,β′-pyrrolic bond (Scheme 11). The powerful amination reagent 4-amino-4*H*-1,2,4-triazole described by Callot and coworkers [73] was used to prepare intermediate **34** which, after reduction of the nitro group followed by reaction with formic acid and cyclization with trifluoroacetic acid, afforded the imidazo[4,5-*b*]porphyrin **35**. The same authors found that compound **35** can be obtained in better yield (70%) if trimethyl orthoformate is used as an alternative to formic acid [74].

Treatment of 2-nitro-*meso*-tetraarylporphyrins with excess triethyl phosphite at 155°C in 1,2-dichlorobenzene affords the corresponding cyclic enamines **36** in 70–75% yield (Scheme 12) [42].

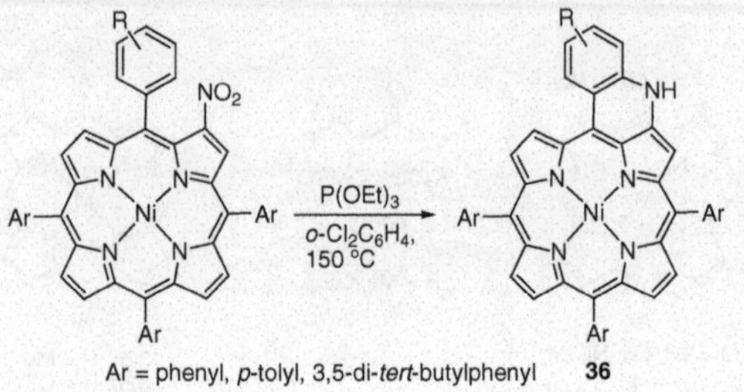

Scheme 12 Reaction of 2-nitro-*meso*-tetraarylporphyrins with triethyl phosphite

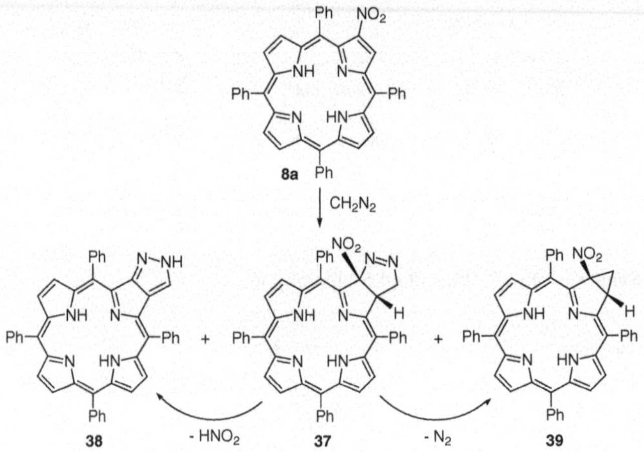

Scheme 13 Reaction of 2-NO_2TPP with diazomethane

Cavaleiro and coworkers demonstrated that *meso*-tetraarylporphyrins can participate as dienophiles in Diels–Alder reactions affording adducts with important biological significance [75]. The same group found that they also participate as dipolarophiles in 1,3-dipolar cycloaddition reactions [76, 77]. As observed in other types of reactions, the presence of a β-nitro group also activates the porphyrin macrocycle towards cycloaddition reactions. This effect is evident in the reaction of β-NO_2TPP **8a** with diazomethane (Scheme 13) [78]. In fact, the cycloaddition occurs selectively at the substituted pyrrolic unit affording the pyrazoline-fused chlorin **37** (in 41% yield) accompanied by two minor compounds (**38** and **39**). It was shown that chlorin **37** is the precursor of the two minor products.

The benefit of NO_2 as a substituent to activate the β,β′-double bond where it is inserted was also considered in Diels–Alder reactions. Ostrowski et al. [79] revisited the reaction of porphyrins with the highly reactive diene *ortho*-benzoquinodimethane but now using β-NO_2TPP as the dienophile (Scheme 14). The expected chlorin **40** was isolated as the main product in 54% yield accompanied by the naphthoporphyrin **41** and the dinaphthoporphyrin **42**. When the non-functionalized TPP was used the expected chlorin was isolated in only 26% yield [75].

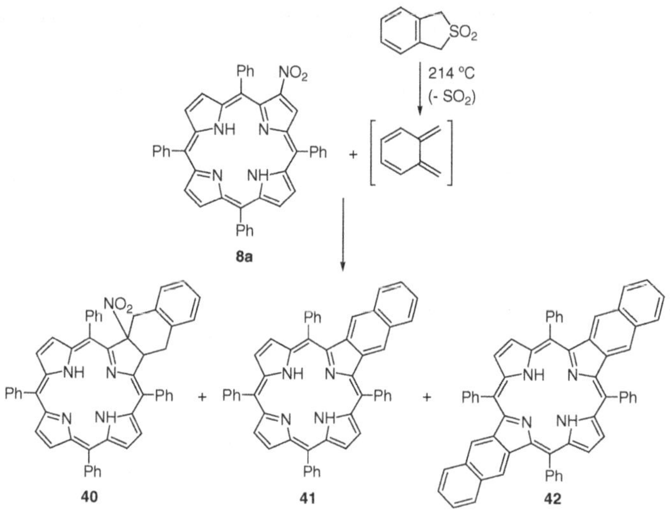

Scheme 14 Diels-Alder reaction of 2-NO_2TPP with *ortho*-benzoquinodimethane

4 Synthesis of Aminoporphyrins

In the last thirty years, porphyrins functionalized with amino groups have become popular starting materials for further functionalization. The possibility of using these versatile intermediates has been facilitated by their easy access in multigram scale from *meso*-tetraarylporphyrins, namely through well-known reduction procedures starting from adequate nitroporphyrins. Most of the protocols are based on the use of Sn/HCl or Sn/HCl /ultrasound [39], $SnCl_2$/HCl [34, 80], $NaBH_4$–Pd/C [54], $HCOONH_4$–Pd/C [39], $HCOONH_4$–Zn [81], or H_2–Pd/C [82]. Other strategies consider the acid hydrolysis of adequate acetamidoporphyrins, prepared by condensation of pyrrole with acetamidobenzaldehydes under acidic conditions [83].

Nitrogen nucleophiles bearing a potential leaving group such as hydroxylamine, hydrazine, tosylhydrazine, and hydroxylamine *O*-sulfonic acid, 4-amino-4*H*-1,2,4-triazole are also being used to introduce the amino functionality directly in electrophilic centers of the porphyrinic core according to Callot procedures [73]. This approach was already mentioned for the amination of porphyrins bearing nitro groups and it is an important alternative to a previous approach involving also the reaction of 2-nitroporphyrins but with acylamide ions at the 3-position followed by hydrolysis of the amide bond [59].

The nucleophilic aromatic substitution of the *para*-F atoms of 5,10,15, 20-tetrakis(pentafluorophenyl)porphyrin (TF_5PP) by amines, discovered by Kadish in 1990 [84], is also considered an efficient strategy to introduce different amino functionalities on that versatile platform. A comprehensive mini-review dedicated to this topic was recently published [85].

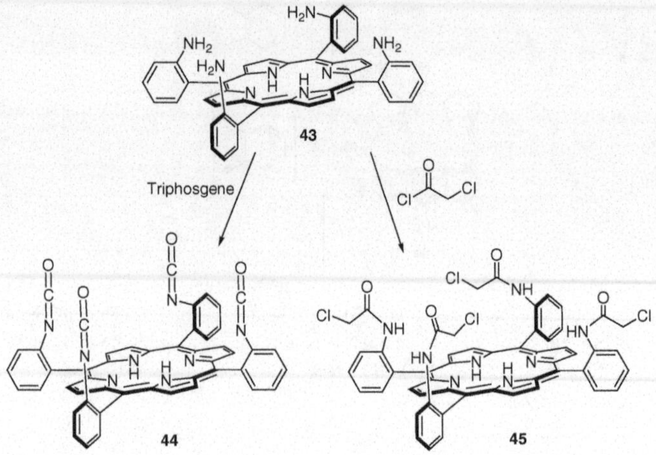

Scheme 15 Synthesis of "picket-fence" precursors

5 Functionalization of Aminoporphyrins

A wide range of porphyrin derivatives with improved properties have been prepared by functionalization of amino porphyrins, namely via amide linkage, alkylation, nucleophilic substitution, diazotization, cycloaddition reactions, and palladium-catalyzed transformations.

5.1 *Via Amide Bonds*

In the 1970 decade, *meso*-tetraarylporphyrins bearing aminophenyl groups, especially *ortho*-aminophenyl groups, were largely explored as excellent synthons in the construction of biomimetic models of heme proteins. The original work described by Collman in 1973 [86] was followed by the synthesis of a series of porphyrins named with fancy names such as "picket-fence", "strapped", or "pocket" porphyrins. Most of these compounds were constructed via amide bonds, and this topic has been the subject of exhaustive reviews [30, 87, 88].

The α,α,α,α-isomer of 5,10,15,20-tetrakis(*o*-aminophenyl)porphyrin **43** (Scheme 15) considered by Collman, also became very popular for the development of chiral catalysts and receptors for specific binding [89]. Most of the synthetic strategies reported on the development of receptors are based on the structural modification of porphyrin **43** using standard strategies or via its previous conversion into synthons **44** and **45** [90, 91].

Porphyrins **46–54** described below are examples of picket-fence porphyrin-type receptors prepared from the α,α,α,α-isomer **43** [92]. The α,α,α,α-tetrakis (*o*-isocyanatophenyl)porphyrin **44** was considered in the synthesis of the anion

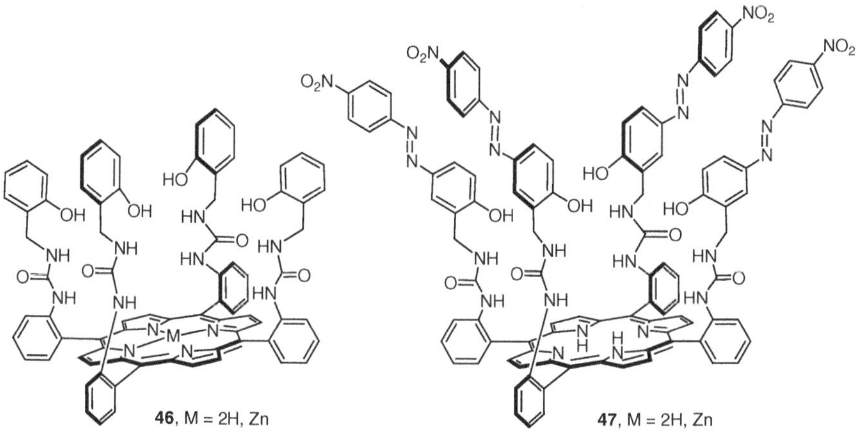

Fig. 11 "Picket-fence" porphyrins

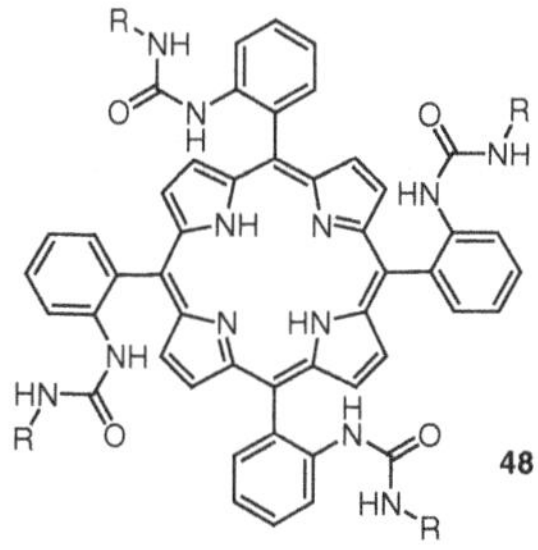

48

a) R = $C_{10}H_{21}$ (decyl)
b) R = L-Valine methyl ester
c) R = L-Threonine methyl ester
d) R = L-Tryptophan methyl ester

Fig. 12 "Picket-fence" porphyrins bearing amino acid residues

sensors **46** and **47** using the adequate 2-(aminomethyl)phenol derivatives (Fig. 11) [93]. Compounds **46** were described as exhibiting good selectivity for AcO^- and $H_2PO_4^-$ while the *p*-nitrophenylazo derivatives **47** showed a selective coloration for F^-, $H_2PO_4^-$ and AcO^-. Porphyrin derivatives **48** (Fig. 12), bearing different amino acid residues, were prepared by a similar approach [94]. These compounds showed promising features in sugar recognition.

The reaction of α,α,α,α-tetrakis(*o*-chloroacetamidophenyl)porphyrin **45** with sodium imidazolate afforded porphyrin **49** (Fig. 13) containing imidazolium subunits [95]. UV/visible spectroscopic studies revealed that this receptor is selective for sulfate anions. Cyclic and square wave voltammetry studies demonstrate the receptor's ability of compounds **45** and **49** to sense a variety of anions electrochemically via significant cathodic perturbations of the respective porphyrin's first oxidation wave.

The reaction of **43** with 3,4-dimethoxybenzoyl chloride afforded the picket-fence porphyrin **50** and the corresponding complexes **51** were prepared using standard literature methods (Scheme 16). The anion binding ability of these compounds was

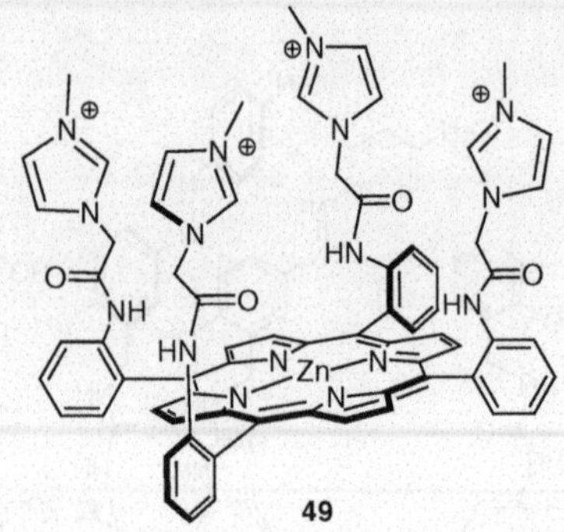

Fig. 13 "Picket-fence" porphyrins containing imidazolium subunits

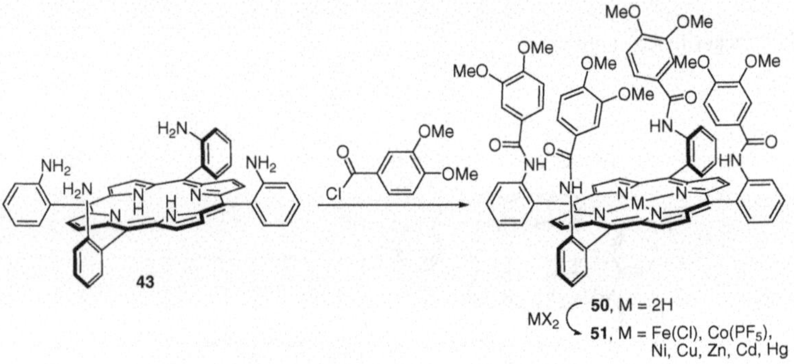

Scheme 16 Synthesis of "picket-fence" porphyrin **50** and complexes **51**

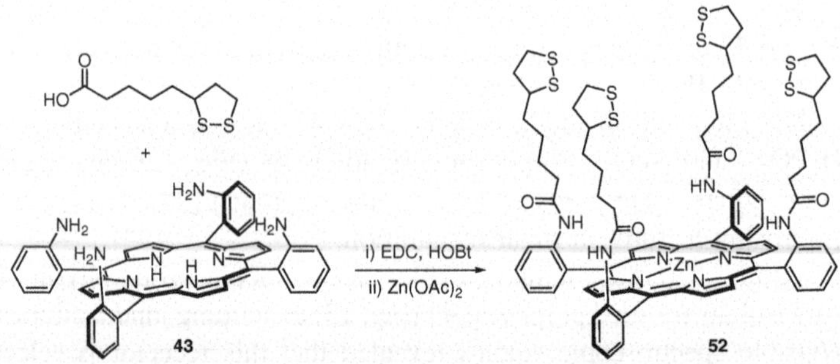

Scheme 17 Synthesis of a "picket-fence" porphyrin bearing disulfide groups

evaluated and the best results were obtained with the cadmium and mercury complexes that showed to bind anions strongly in highly competitive solvent mixtures [96].

The disulfide and dithiocarbamate functionalized porphyrins **52** (Scheme 17) and **54** (Scheme 18) were considered in the synthesis of gold nanoparticles. The nanoparticles show to be more efficient to recognize anions than the free receptors

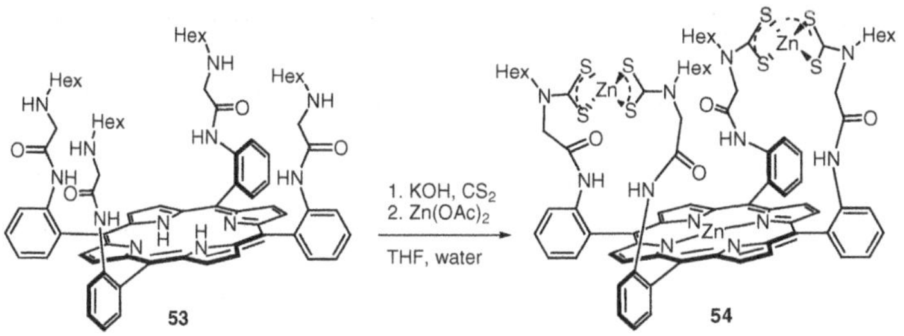

Scheme 18 Synthesis of a "picket-fence" porphyrin bearing dithiocarbamate groups

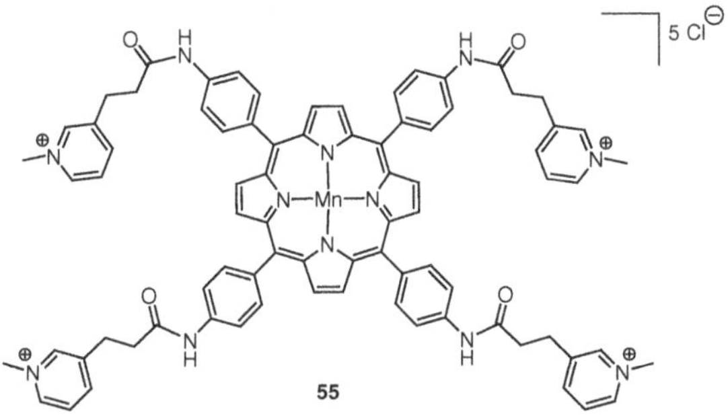

Fig. 14 Structure of the tetracationic porphyrin **55**

in solution. The tetraamide **52** was prepared by reaction of **43** with thioctic acid, in the presence of EDC and HOBt (1-hydroxybenzotriazole), followed by the addition of $Zn(AcO)_2$ (Scheme 17). The synthesis of the dithiocarbamateporphyrin **54** involved the reaction of synthon **45** with hexylamine, followed by reaction with carbon disulfide and $Zn(AcO)_2$ (Scheme 18) [97, 98].

meso-Tetraarylporphyrins bearing amino groups in *para* positions of the phenyl substituents were also considered in the design of receptors via amide bond. The cationic porphyrin **55** (Fig. 14), obtained by reacting 5,10,15,20-tetrakis (*p*-aminophenyl)porphyrin with 3-(pyridin-3-yl)propanoic acid in the presence of HOBt and HBTU (*O*-benzotriazol-1-yl-*N,N,N′,N′*-tetramethyluronium hexafluorophosphate) followed by quaternization of the pyridyl nitrogens with methyl iodide and metallation with manganese(III), was described as showing a much higher preference for G-quadruplexes as opposed to duplex DNA [99].

The asymmetric 5-(4-aminophenyl)-10,15,20-triphenylporphyrin was used to prepare the porphyrin-functionalized [2]rotaxane host molecule **59** (Scheme 19). The synthesis involved the condensation of two equivalents of the aminoporphyrin with pyridine-3,5-dicarboxylic acid using adequate coupling agents, followed by cationization with methyl iodide. Then, a ring-closing metathesis mediated cyclization of the adequate bis-vinyl-functionalized benzene-1,3-dicarboxamide **58** in the presence of Grubbs 2nd generation catalyst afforded **59**. This rotaxane exhibits a high binding affinity and general selectivity for chloride anions [100].

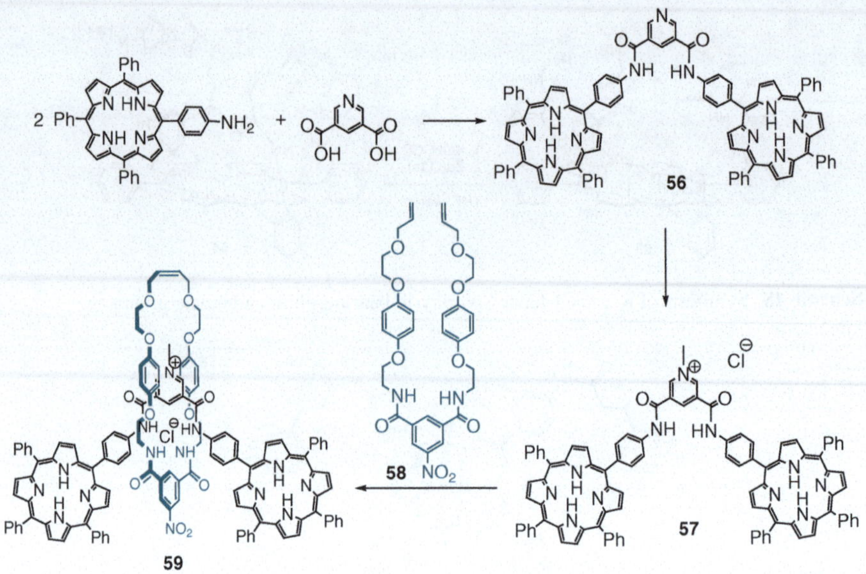

Scheme 19 Synthesis of a porphyrin-functionalized [2]rotaxane

meso-Tetraarylporphyrins bearing aminophenyl groups have been used in the preparation of porphyrin derivatives conjugated to other bioactive compounds via amide bonds. A recent review covering this type of covalent attachment of porphyrins to peptides and proteins was recently published [101].

5.2 *Via Alkylation of the Amino Group*

The cationization of the amino groups was explored by several research groups to improve porphyrin DNA binding, photodynamic effect, solubility in physiological fluids, and selectivity to cancer cells. For instance, the cationic β-tetra- and β-octasubstituted porphyrins **64–67** were prepared by cationization of the corresponding aminoporphyrins **60–63** (Scheme 20) [102]. The synthetic approach to β-tetrasubstituted porphyrins **60** and **61** involved the mononitration of the 2,3,12,13-tetrabromoTPP with fuming HNO_3, followed by Suzuki coupling with the adequate boronic acids and then reduction of the nitro group with $SnCl_2$. A similar strategy was used to obtain the octasubstituted porphyrins **62** and **63** although the nitration of the phenyl groups preceded the bromination step.

Alkylation of 5-(4-aminophenyl)-10,15,20-triphenylporphyrin with 6-iodo-1,2:3,4-di-*O*-isopropylidene-α-D-galactopyranose, followed by methylation with methyl iodide, afforded the cationic glycoporphyrin derivative **68** (Fig. 15) [103]. Removal of the isopropylidene groups from **68** by acid treatment afforded glycoporphyrin **69**.

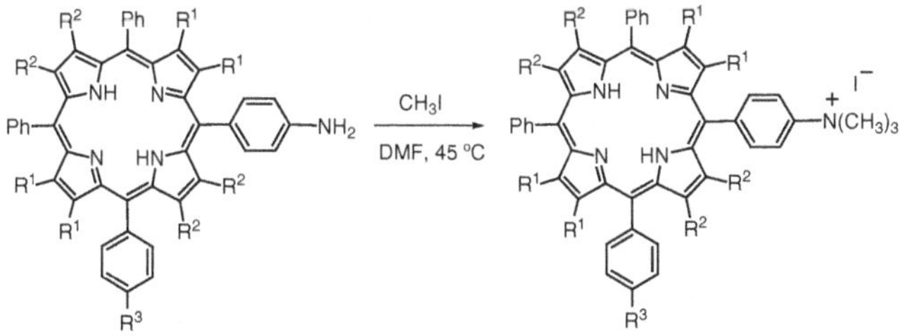

60, R^1 = Ph, R^2 = R^3 = H
61, R^1 = Me, R^2 = R^3 = H
62, R^1 = R^2 = Ph, R^3 = H
63, R^1 = R^2 = Ph, R^3 = NH_2

64, R^1 = Ph, R^2 = R^3 = H
65, R^1 = Me, R^2 = R^3 = H
66, R^1 = R^2 = Ph, R^3 = H
67, R^1 = R^2 = Ph, R^3 = $N^+(CH_3)_3$

Scheme 20 Cationization of *meso*-(4-aminophenyl)porphyrins

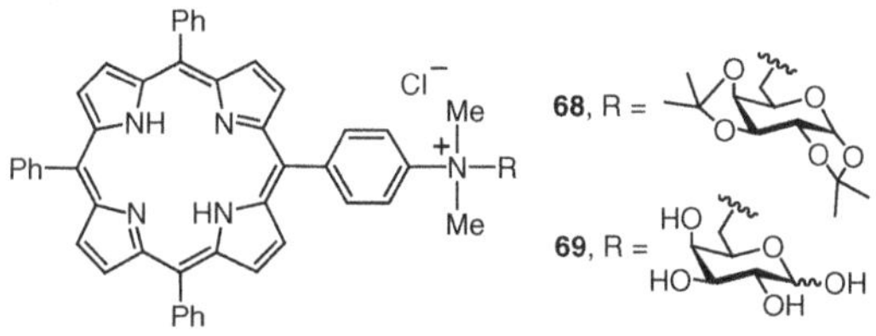

Fig. 15 Cationic glycoporphyrins

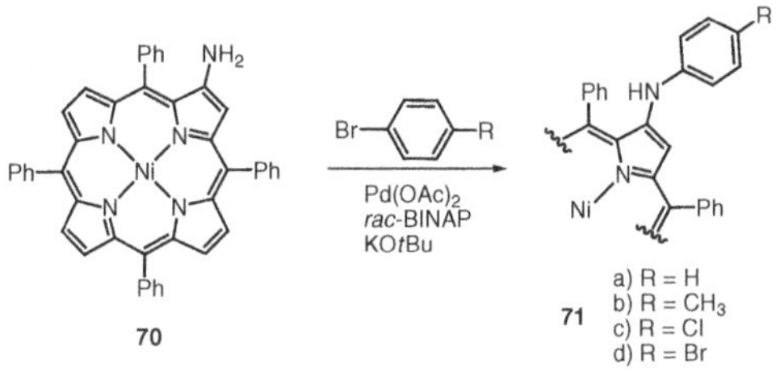

Scheme 21 Synthesis of (2-arylamino)porphyrins via Buchwald–Hartwig amination

5.3 *Via Transition Metal Catalysis*

Modification of amino groups mediated by transition metal complexes, such as palladium(0), is an interesting alternative to bromoporphyrins for carbon–nitrogen bond formation [13]. Van Lier and coworkers reported for the first time, but without experimental details, that 2-aminoporphyrins react with aryl halides affording 2-(arylamino)porphyrins [104]. Based on that methodology, usually known as Buchwald–Hartwig amination, Cavaleiro and coworkers were able to synthesize 2-arylaminoporphyrins **71** in excellent yields by reacting 2-NH_2-NiTPP (**70**) with bromobenzene derivatives, even with electron-withdrawing substituents (Scheme 21) [66].

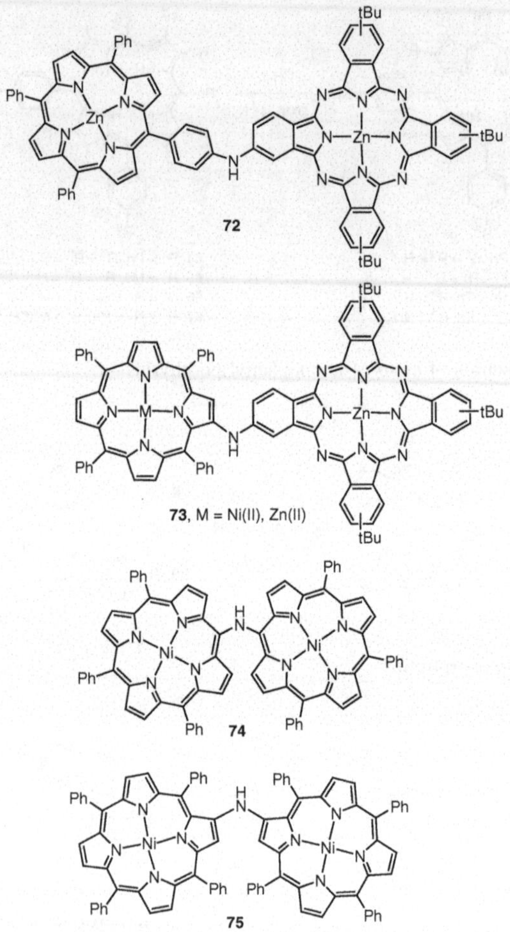

Fig. 16 Structures of porphyrin-phthalocyanine dyads and porphyrin dyads obtained using Buchwald-Hartwing amination reactions

Cavaleiro and coworkers also used the Buchwald–Hartwig amination conditions to synthesize the porphyrin–phthalocyanine dyads **72** and **73** [105] and the porphyrin–porphyrin dyads **74** and **75** [106] (Fig. 16) where the two chromophores are linked by a nitrogen atom.

Dyad **73** was prepared by two complementary routes (Scheme 22). One of them involved the direct coupling of 2-NH_2-NiTPP and the iodophthalocyanine **77** in the presence of $Pd(OAc)_2$, *rac*-BINAP and KO^tBu. The other approach involved the statistical cross-condensation of porphyrin-2-ylaminophthalonitrile **78** with 4-*t*-butylphthalonitrile. Phthalonitrile **78** was obtained from 2-aminoTPP and 4-iodophthalonitrile, using the same coupling conditions [$Pd(OAc)_2$, *rac*-BINAP and KO^tBu].

Similar catalytical conditions were used to couple 2-NH_2-NiTPP with 2-Br-NiTPP (**79**) and 5-I-NiTPP (**80**) (Scheme 23). The electronic spectra of the dimers **75** and

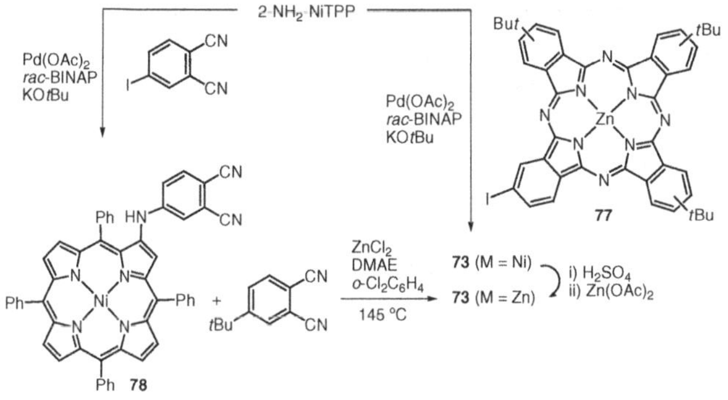

Scheme 22 Synthetic routes to a porphyrin-phthalocyanine dyad

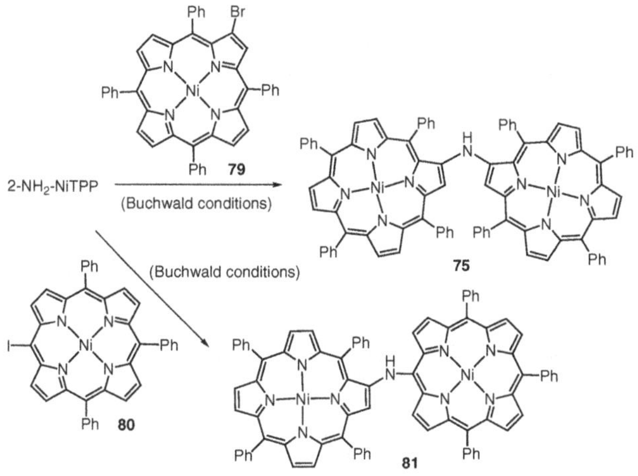

Scheme 23 Synthesis of porphyrin dyads

81 are typical of highly delocalized systems and electrochemistry studies have shown that the first oxidation step occurs on the connecting amine function.

The cyclic enamine **82** [42] (structurally related to 2-NH_2-NiTPP) reacts with iodobenzene under Ullmann amination conditions (copper iodide, L-proline, potassium carbonate) to give access to the *N*-phenylquinolino[2,3,4-*at*]porphyrin **83** in good yield (Scheme 24) [107]. Electrochemical studies with the free-base and the corresponding Ni, Cu, and Pd complexes have shown that the presence of the N-phenyl group is responsible for the formation of stable radical cations.

5.4 *Via Diazonium Salts*

The synthetic value of diazonium salts was considered in several synthetic approaches for further functionalization of the porphyrin core. These important synthons

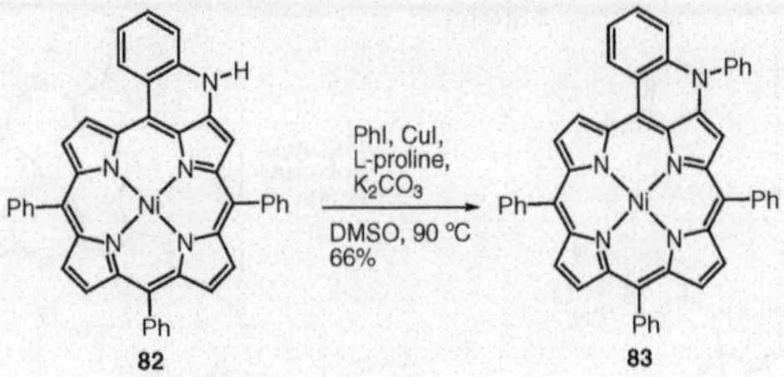

Scheme 24 *N*-Arylation of a quinolino[2,3,4-*at*]porphyrin

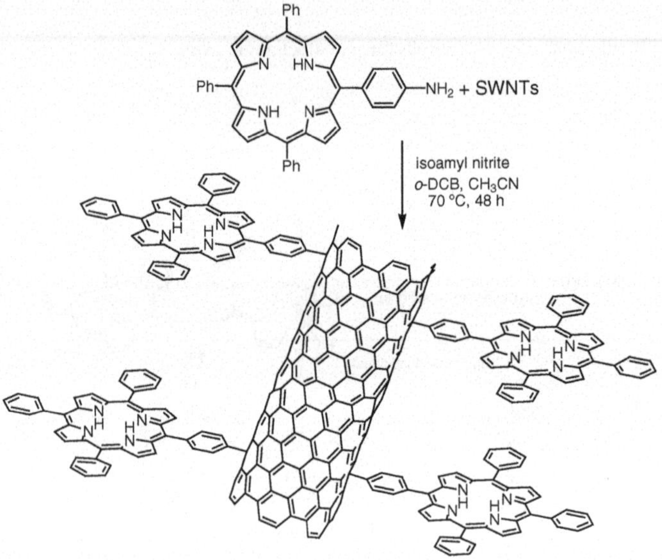

Scheme 25 Synthesis of pophyrin-SWNT nanohybrids

can be accessible from adequate aminoporphyrins using different diazotization conditions such as sodium nitrite and tetrafluoroboric acid at −5°C [108], sodium nitrite, and sulfuric acid [39] or with isoamyl nitrite [109].

The in situ decomposition of the diazonium salt obtained from 5-(4-aminophenyl)-10,15,20-triphenylporphyrin with isoamyl nitrite in the presence of single-walled carbon nanotubes (SWNTs) was considered in the synthesis of nanohybrids where the porphyrin was covalently attached to the nanotube (Scheme 25) [109]. The new materials showed better solubility and dispersion stability in organic solvents and superior optical limiting effects than SWNTs and C_{60}.

The diazonium salt of the nickel complex of 5-(4-aminophenyl)-10,15,20-triphenylporphyrin was used to graft the corresponding complex to glassy carbon and gold and indium tin oxide surfaces via reduction of the diazonium moiety. Nitrosium tetrafluoroborate ($NOBF_4$) was selected as the diazotizing agent. The characterization of the resulting materials confirms that the metallated

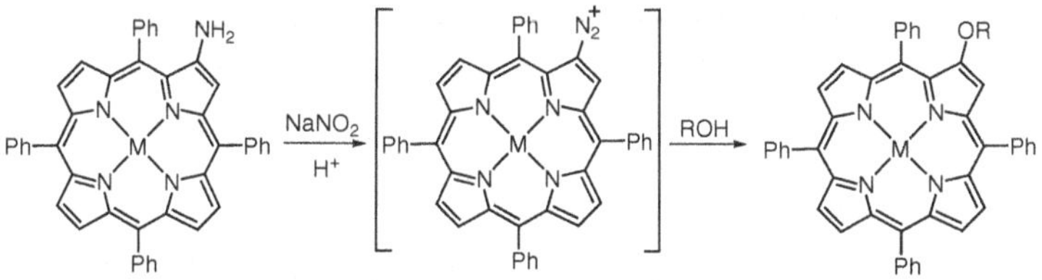

72, M = Ni, Cu
83
84
R = Me, Et, Pr, *t*-Bu, C_6H_{11}

Scheme 26 Synthesis of β-alkyloxy substituted porphyrins via diazonium salts

porphyrin is intact, stably attached to the surface but with highly solvent-dependent electrochemistry [110].

The diazonium salts of the nickel or copper complexes of 2-amino-*meso*-tetraphenylporphyrin **72** are efficient intermediates to new 2-substituted porphyrins. Several β-alkyloxy substituted porphyrins **84** were obtained from the reaction of the in situ generated diazonium salt **83** (M = Cu) with alcohols or alkoxides (Scheme 26) [39, 111].

The diazotization of the 2-aminoporphyrins **72a–c** with $NaNO_2$ and sulfuric acid in tetrahydrofuran containing hydroperoxide gave rise to 2-diazo-3-oxo-tetraphenylchlorins **85** (Scheme 27) [112]. The photochemical induced dediazoniation of metallo 2-diazo-3-oxo-tetraphenylchlorins **85** in the presence of alcohols afforded the corresponding 2-alkyloxy derivatives **86** and other compounds that were justified by the existence of different reaction pathways after the formation of ketocarbenes by dediazoniation [112].

The reaction of porphyrin diazonium salts with sodium azide afforded porphyrins bearing azido substituents in excellent yields (ex: **87** and **88**, Fig. 17). These compounds are important synthons for click chemistry [113–115].

The porphyrin diazonium salt **83** (M = Ni) was also used as a pseudo-halide in Heck reactions [116]. The reactions were performed in the presence of methyl acrylate, propenal, and methyl vinyl ketone and afforded the expected unsaturated 2-substituted porphyrins **89a–c** (Fig. 18). Depending on the α,β-unsaturated carbonyl compound used, the minor products **89d–f** were also obtained. The formation of pyridoporphyrins **90** was justified by the reaction of the unchanged 2-aminoporphyrin with the α,β-unsaturated carbonyl compounds [117].

The extension of the previous studies to 3-sulfolene gave access, after isomerization and thermal extrusion of sulfur dioxide, to porphyrin **92** bearing a buta-1,3-dien-2-yl group in the β-pyrrolic position (Scheme 28) [118].

The β-butadienyl porphyrin **92** showed to be an efficient diene in Diels–Alder reactions with a wide range of dienophiles such as [60]fullerene, *N*-phenylmaleimide, 1,4-benzoquinone and 1,4-naphthoquinone and fumaronitrile affording the expected adducts and/or the dehydrogenated ones in good yields. The adduct obtained from the reaction of **92** with fumaronitrile was used as precursor to a porphyrin–phthalonitrile that gave access to a series of novel porphyrin–phthalocyanine dyads bearing a rigid arrangement of the two units in close proximity [119].

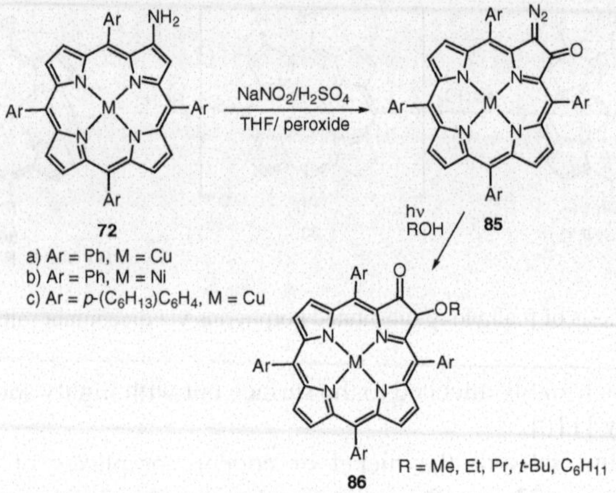

Scheme 27 Synthesis of 2-diazo-3-oxo-tetraphenylchlorins

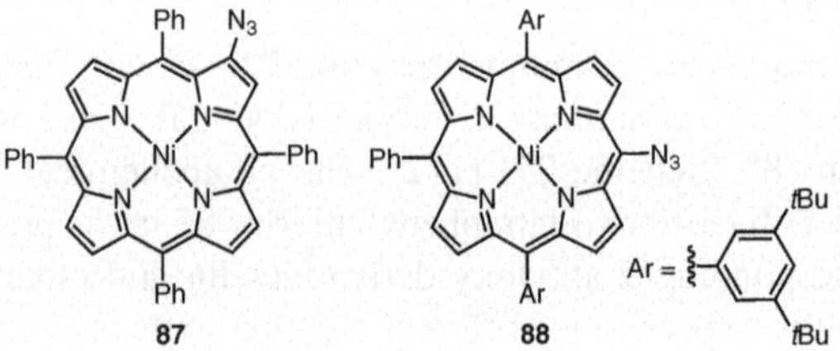

Fig. 17 Structures of β- and *meso*-azidoporphyrins

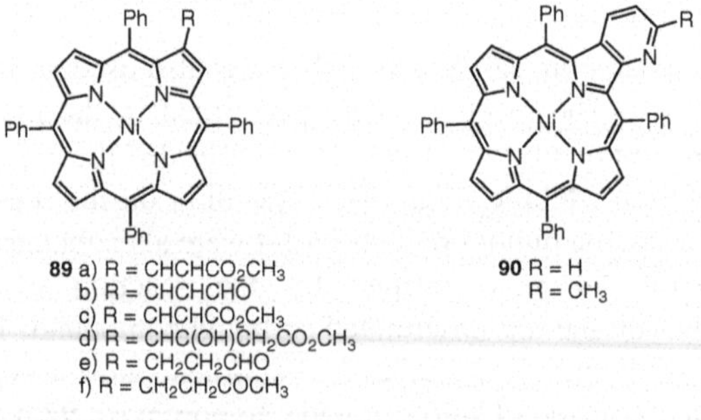

Fig. 18 Structures of porphyrins **89–90**

Using the principles of the diazotization reaction, Igarashi's group [120] developed a highly sensitive porphyrin-based spectrophotometric method for the determination of nitrite ion. This methodology uses the ability of the amino group of 5,10,15,20-tetrakis(4-aminophenyl)porphyrin to form a diazo group in the presence of nitrite ion in acidic conditions. The formation of a quinoid structure is responsible for a significant decrease in the absorbance relatively to the initial porphyrin. Latter,

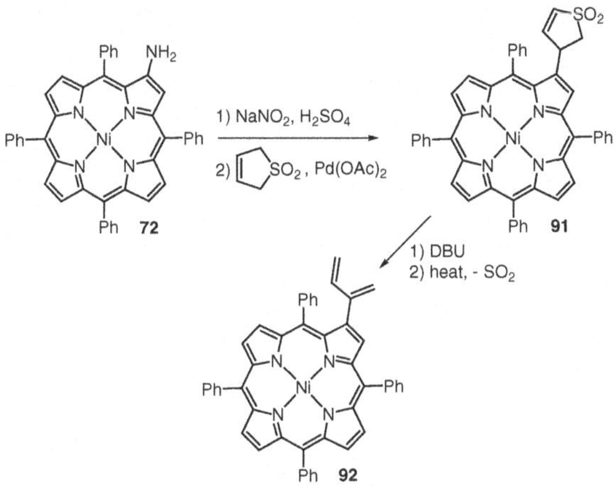

Scheme 28 Synthesis of β-butadienyl porphyrin **92**

the same group used a porphyrin with only one amino group, the [5-(4-aminophenyl)-10,15,20-tris(4-pyridyl)porphyrin, in this sensing methodology [121].

5.5 *The Dual Behavior of Aminoporphyrins*

The dual behavior of 2-aminoporphyrins to act as an aromatic amine or as enamine gave access to an interesting number of new derivatives namely heterocyclic-fused porphyrins.

In 1997 Cavaleiro and coworkers reported that the reaction of the nickel(II) complex of 2-aminoTPP **81** with propenal and methyl vinyl ketone afforded, in the presence of H_2SO_4 and $Pd(AcO)_2$, the fused pyridoporphyrins **90** [117]. The authors referred that a Michael addition, imine formation, and a dehydrogenation took place in products formation. The extension of those studies by the same group to other α,β-unsaturated carbonyl compounds such as 1,4-benzoquinone, 1,4-naphthoquinone and 2-hydroxy-1,4-naphthoquinone in the presence of catalytic amount of sulfuric acid, gave access to a plethora of new porphyrin–quinone dyads and π-extended heterocycle-fused porphyrin derivatives **93–98** (Scheme 29). The type of products obtained was dependent on the quinone used and was justified based on of dual behavior of the aminoporphyrin **81**. The aromatic character can explain the formation of products **94**, **99–101** while the others (**93**, **95–98**) can be justified by its enamine character. The adaptation of the Nenitzescu reaction allowed the authors to elucidate the formation of the π-extended heterocyclic fused porphyrin derivatives **93** and **98**.

The products obtained from the reaction of 2-amino-*meso*-tetraphenylporphyrin with acryloyl chloride, also reflects the dual behavior of 2-aminoporphyrins [122]. While the amide **102** was the result of *N*-acylation, the formation of the dihydro-2-pyridone fused porphyrin **103** was justified through the aza-annulation reaction

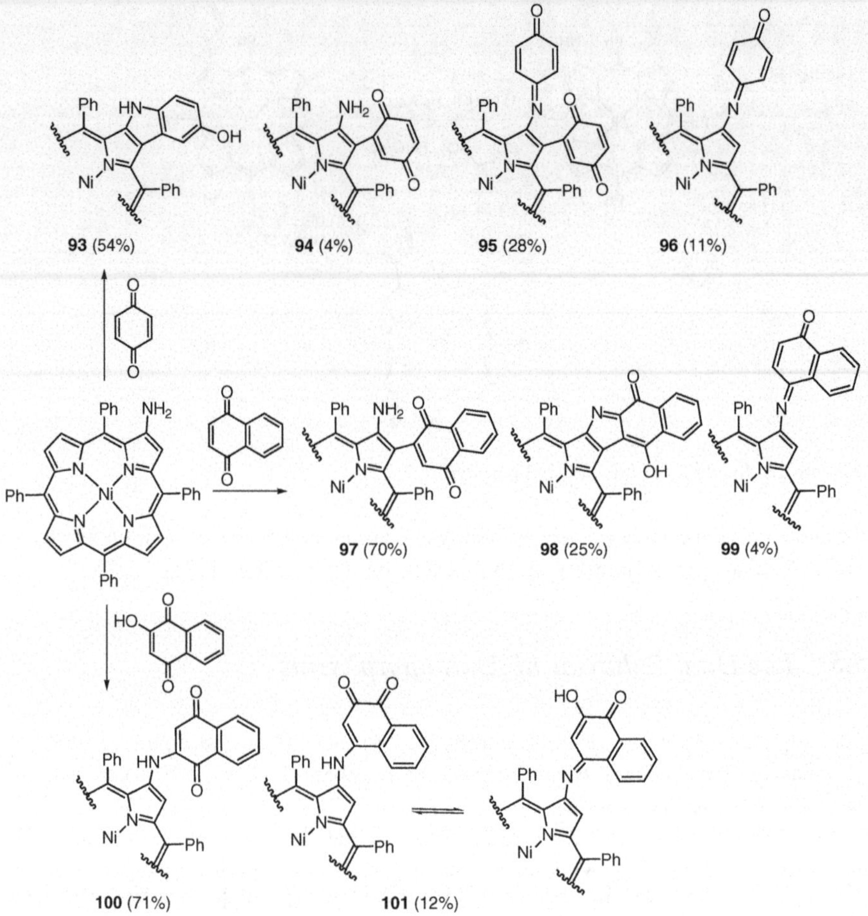

Scheme 29 Reaction of 2-aminoTPP with quinones

initiated by the Michael addition of the enamine followed by an intramolecular *N*-acylation (Scheme 30). Compound **103** is easily oxidized with DDQ to the corresponding 2-pyridone. The failure to obtain the dihydro-2-pyridone from the reaction with cinnamoyl chloride was explained by considering steric or electronic effects due to the phenyl group present on the β-carbon of the acyl chloride.

5.6 *β-Iminoporphyrins as Heterodienes*

The possibility of using the β-iminoporphyrins **104** as heterodienes in hetero-Diels–Alder reactions was reported by Cavaleiro and coworkers [123, 124]. A three component reaction involving β-aminoTPP, an aromatic aldehyde and an

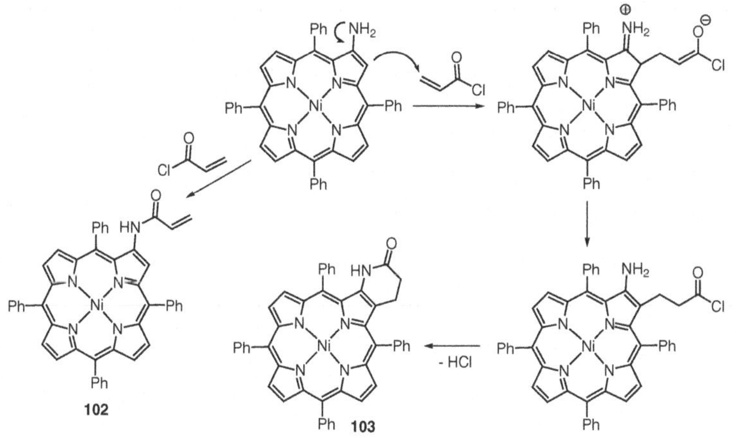

Scheme 30 Reaction of 2-aminoTPP with acryloyl chloride

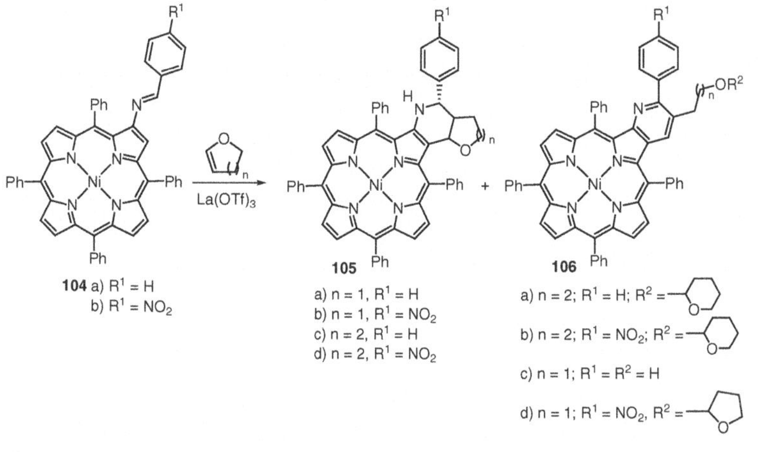

Scheme 31 Reaction of β-iminoporphyrin **104** with cyclic enol ethers

electron-rich dienophile (3,4-dihydro-2*H*-pyran or 2,3-dihydrofuran) catalyzed by lanthanum triflate leads to the expected tetrahydropyridine-fused derivatives **105** accompanied by the corresponding pyrido[2,3-*b*]porphyrins **106** (Scheme 31). A probable pathway to compounds **106** involves the aromatization of the pyridine ring, with the opening of the pyranyl ring, followed by the addition of another molecule of the enol ether. In the presence of $La(OTf)_3$ and the enol ether, compounds **105** are converted into the corresponding pyrido[2,3-*b*] porphyrins **106**.

Pyrido[2,3-*b*]porphyrins bearing two vicinal hydroxyalkyl groups (**107** and **108**) were prepared through a two component domino reaction where an enol ether was used to generate the iminic heterodiene and also to act as the dienophile

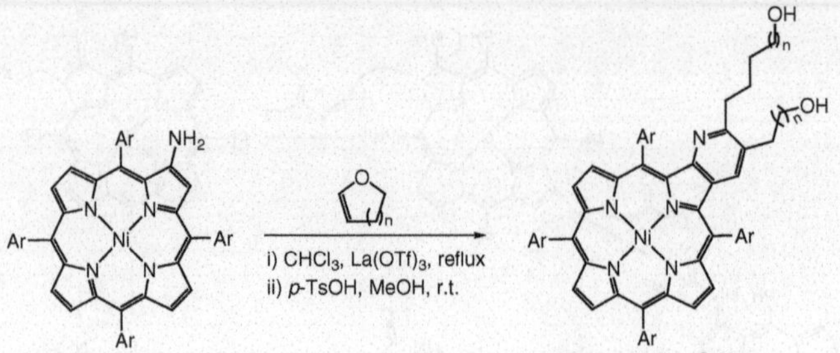

107a, n = 1; Ar = C_6H_5
107b, n = 1; Ar = *m*-$MeOC_6H_4$
107c, n = 1; Ar = *p*-BrC_6H_4
107d, n = 1; Ar = *p*-$MeO_2CC_6H_4$

108a, n = 2; Ar = C_6H_5
108b, n = 2; Ar = *m*-$MeOC_6H_4$
108c, n = 2; Ar = *p*-BrC_6H_4
108d, n = 2; Ar = *p*-$MeO_2CC_6H_4$

Scheme 32 Synthesis of pyrido[2,3-*b*]porphyrins **107–108**

(Scheme 32) [125]. Treatment of the crude reaction mixture with a methanolic solution of *p*-toluenesulfonic acid is a key step in order to improve the yield of the desired products and to facilitate the purification process. The esterification of hydroxyalkyl groups in template **107a** with succinic anhydride and dodecanoyl chloride afforded the corresponding esters in almost quantitative yields. The crystal structure of the most hydrophobic one showed that these porphyrin derivatives form one-dimensional supramolecular structure in the solid state.

5.7 *Aminoporphyrins as Carbene Acceptors*

The catalytic insertion of ethyl diazoacetate into the amino group of 5-(4-aminophenyl)-10,15,20-triphenylporphyrin in the presence of an Rh-based catalyst was recently investigated [126]. Besides the formation of one compound resulting from the insertion of two carbene units, two other unexpected amides were isolated (Scheme 33). The formation of these amides was also observed when the 2-(4-aminophenyl)porphyrin was reacted with ethyl glycolate in the presence of the same catalyst. Derivative **110** crystallizes in an unusual chiral supramolecular metalloporphyrin chain, forming a right-handed helix arrangement.

The β-iminoporphyrin **112** reacts with carbenes generated from ethyl diazoacetate in the presence of catalytic amounts of lanthanum triflate (Scheme 34) [127]. The *cis*- and *trans*-aziridine **113** were obtained as the main products and the β-amino-α,β-unsaturated esters **114** and **115** as minor products (Scheme 35).

Porphyrin-2,3-diones **116** are another example of porphyrin derivatives that can be obtained from 2-aminoporphyrins. Traditionally, the route to porphyrin-2,3-diones is the photo-oxidation of a 2-aminoporphyrin followed by hydrolysis of the resulting keto-imino chlorin [128], or the oxidation of 2-hydroxyporphyrins

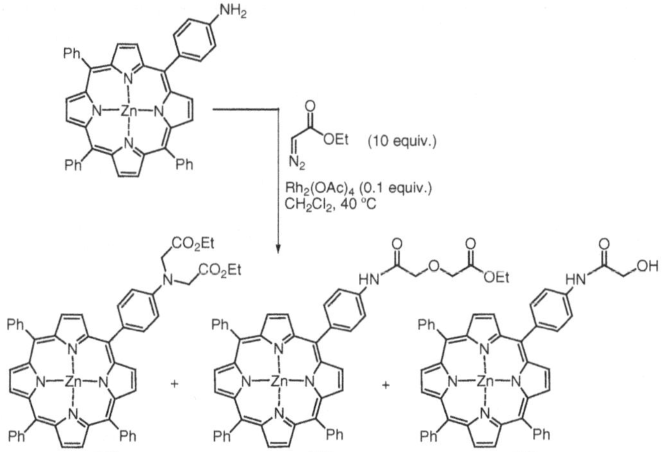

Scheme 33 Reaction of 5-(4-aminophenyl)-10,15,20-triphenylporphyrin with carbenes

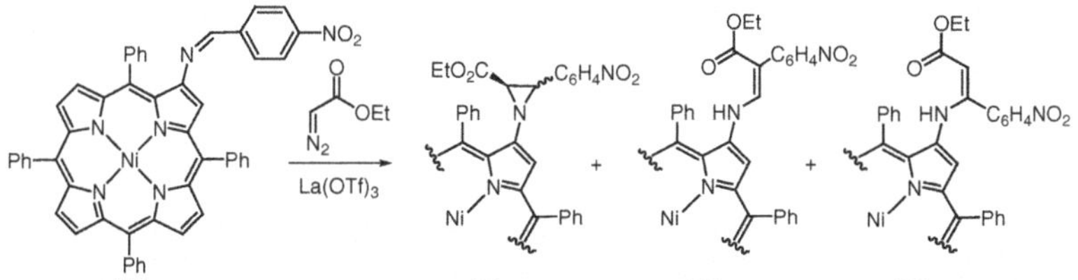

Scheme 34 Reaction of β-iminoporphyrin **112** with carbenes

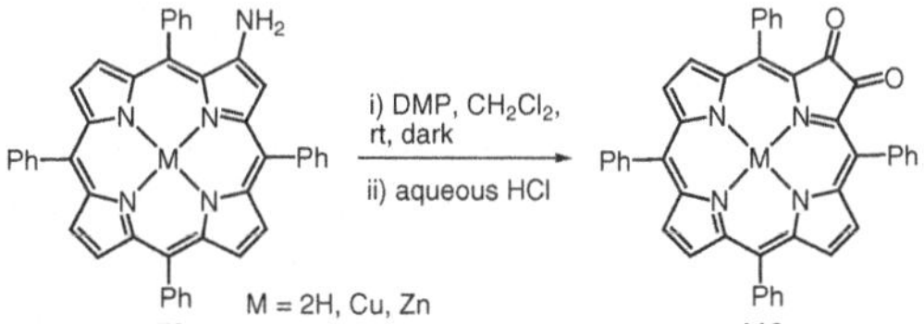

Scheme 35 Synthesis of porphyrin-2,3-diones

with the Dess–Martin periodinane (DMP) [129]. Burn reported that DMP is also efficient for the oxidation of 2-aminoporphyrins to porphyrin-2,3-diones **116** (Scheme 35) [130]. This method allows the reaction to be carried out on a large scale and it is easier than the classic photo-oxidation procedure. The porphyrin-2,3-diones are commonly used as building blocks for conjugated porphyrin arrays in the development of organic materials and molecular wires [131].

5.8 Aminoporphyrins in the Construction of New Assemblies

The design of molecular assemblies based on porphyrins self-association or aggregation is considered a simple method to afford supramolecular systems with a wide range of applications from models of enzyme active sites with relevance in catalysis to light-energy conversion and nanostructured components of electronic and optoelectronic devices.

In 1995 Gautam et al. [132] were able to conclude, from the coordination behavior of 5,10,15,20-tetrakis(3-aminophenyl)porphyrin and of the corresponding nitro precursor towards several metal(II) ions, such as Mg(II), Co(II), Zn(II) and Ag(II), that the amino derivatives are more prone to form aggregates than the nitro derivatives. These conclusions were based on the significant red shifts observed for the absorption and emission bands of the metallated amino derivatives when compared with the ones of the corresponding nitro derivatives. Similar red shifts were obtained for the nitro derivatives in the presence of dimethylaminopyridine supporting the existence of aggregated species in which metal ions are axially coordinated with the peripheral amino groups. Based on the model studies, possible structures were proposed and the authors refer that the amino group in the *meta* position is a key feature for the adequate binding of this group to the metal ion in the adjacent porphyrin.

The coordination ability of porphyrin derivatives bearing an amino group in conjugation with a keto group had also a great success in the construction of new assemblies connected by metal ions. The first studies [133, 134] involved the use of the enaminoketone porphyrins **118** obtained from the reaction of the corresponding ketone derivatives **117** with nitrogen nucleophiles bearing a potential leaving group (hydroxylamine, hydrazine, tosylhydrazine, and hydroxylamine *O*-sulfonic acid or 4-amino-4*H*-1,2,4-triazole) (Scheme 36). An alternative to the previous route used for obtaining ketone **117a** that involves the formylation of a metalloporphyrin under Vilsmeier–Haack conditions followed by acid-catalyzed cyclization of the aldehyde and demetallation of the porphyrin [135–137] was suggested for derivative **117b**. The new synthetic strategy is based on the hydrolysis of the ester group of the nickel complex of a porphyrin bearing a *o*-methoxycarbonylphenyl group and three 3,5-di-*tert*-butylphenyl groups, followed by acid chloride formation and an intramolecular Friedel–Crafts reaction. The authors refer the superiority of 4-amino-4*H*-1,2,4-triazole in the amination process when compared with the other nitrogen nucleophiles, since it can be used in the amination of free-bases and also of nickel, palladium, and copper complexes.

Porphyrin dimers **119** were assembled by several routes namely by the selective coordination of a metal ion to the external sites of two enaminoketone free-bases, followed by metallation of the internal sites or by coordination of two molecules of the metalloenaminoketone on a selected metal ion (Scheme 37). In general, this last approach showed to be more adequate due to the instability of the initial dimer obtained from the free-bases. The *trans* arrangement around the metal was confirmed by NMR experiments and the electronic spectra (intensified red-shifted Q bands) and electrochemical behavior (oxidation potentials split into two redox

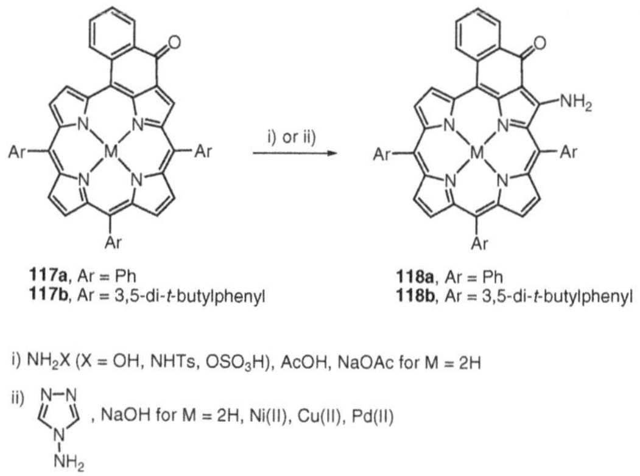

Scheme 36 Synthesis of enaminoketone porphyrins

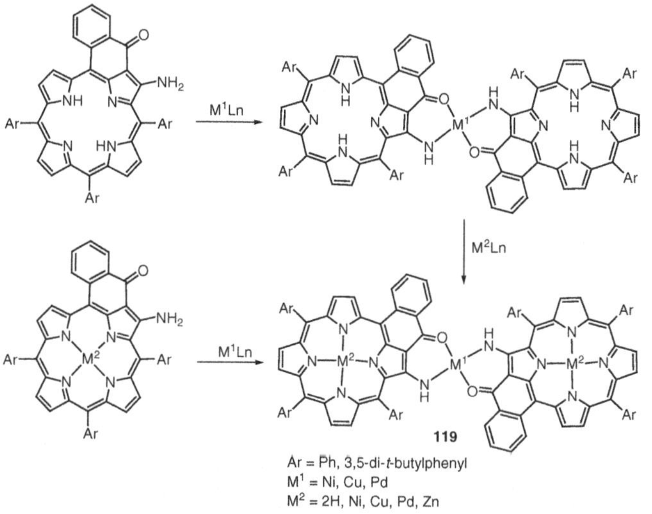

Scheme 37 Synthesis of porphyrin dimers by coordination of a metal ion to the external sites of two enaminoketones

steps substantially lowered when compared with the monomeric units) of the dimers were indicative of a strong interaction between the two units introduced through the connecting metal; the coplanar coordination to the porphyrin ring is indicated by the authors as the main reason for the large interactions displayed in the ground state.

The extension of the previous studies to bis-enaminoketones allowed a stepwise preparation of polymetallic oligomers of type **120** and **121** (Fig. 19) connected by metal centers [73, 138].

These studies were extended to the synthesis of a series of nickel bis-enaminoketone isomers and by thionation of these derivatives to bis-enaminothioketone analogues [139]. The authors refer that by controlling the amount of the Lawesson's reagent

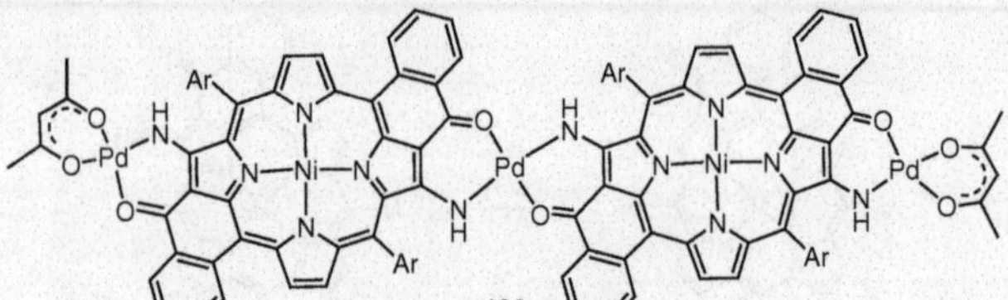

120

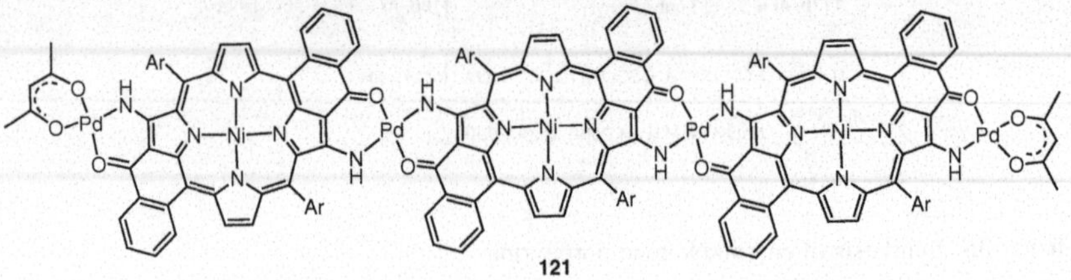

121

Fig. 19 Polymetallic porphyrin oligomers

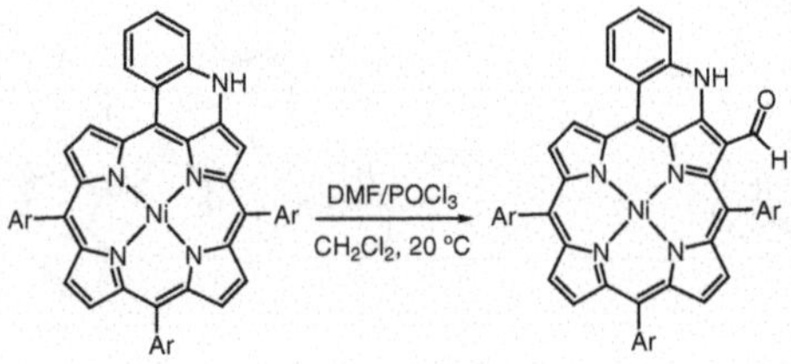

36 122

Scheme 38 Synthesis of enaminoaldehyde functionalized porphyrins

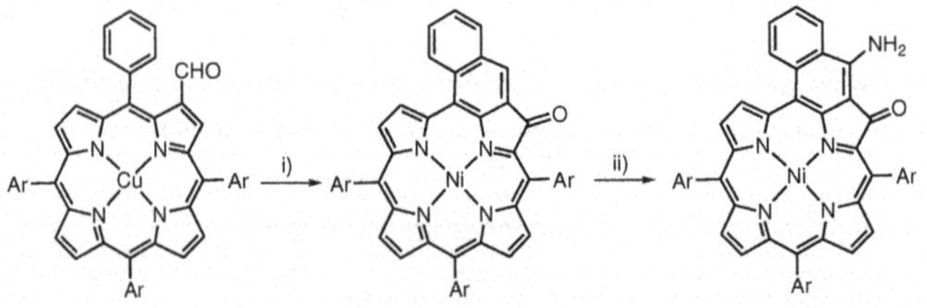

123 124 125

i) 1. CF_3CO_2H, chloranil; 2. H_2SO_4, Ni(acac)$_2$; ii) 4-amino-4*H*-1,2,4-triazole, NaOH

Scheme 39 Synthesis of enaminoketone porphyrins

used in the thionation process it was possible to isolate the monothionated analogues affording derivatives with mixed external chelating groups (NO/NS). The electrochemistry behavior of the new derivatives was also studied and it is in good agreement with the recorded electronic spectra.

Porphyrins bearing other peripheral chelating groups fully conjugated with the porphyrin core, such as the enaminoaldehydes **122** (Scheme 38) and the enaminoketones **125** (Scheme 39), were also developed by the same group and were used in the preparation of porphyrin dimers linked by metal ions [42].

The enaminoketone ligands **125** were obtained from 2-formyl-*meso*-tetraarylporphyrins **79** according to Ishkov conditions [140] followed by amination with 4-amino-4*H*-1,2,4-triazole. All ligands, namely the corresponding thioanalogues obtained by thionation with Lawesson's reagent, were metallated with palladium affording the corresponding dimers. From the structural characterization the authors were able to conclude that in the enamino aldehyde and thioaldehyde series the *cis* isomer is thermodynamically favored and strong porphyrin–porphyrin interactions were also detected in this new series of ligands.

6 Conclusions

This chapter brings an update to the synthesis and reactivity features of nitro- and aminoderivatives of *meso*-tetraarylporphyrins. Such derivatives can be obtained by well-established synthetic and derivatization methodologies. The nitro and the amino substituents play a key role in the functionalization of the corresponding porphyrin macrocycles into a variety of other ones which have already demonstrated significant applications.

Acknowledgments Authors are grateful to "Fundação para a Ciência e a Tecnologia – FCT", Lisbon, for funding QOPNA (the Research Unit in Organic Chemistry and projects). Thanks are also due to FCT for grants awarded to PhD and Post-Doc students. Thanks are also due to the University of Aveiro and to students involved in the work carried out in Aveiro.

References

1. Fisher H, Zeile K (1929) Synthese des hämatoporphyrins, protoporphyrins und hämins. Liebigs Ann Chem 468:98
2. Woodward RB, Ayer WA, Beaton JM, Bickelhaupt F, Bonnett R, Buchschacher P, Closs GL, Dutler H, Hannah J, Hauck FP, Itô S, Langemann A, Le Goff E, Leimgruber W, Lwowski W, Sauer J, Valenta Z, Voltz H (1960) The total synthesis of chlorophyll. J Am Chem Soc 82:3800
3. Pandey RK, Zheng G (2000) Porphyrins as photosensitizers in photodynamic therapy. In: Kadish KM, Smith KM, Guilard R (eds) The porphyrin handbook. Academic, San Diego
4. Aida T, Inoue S (2000) Metalloporphyrins as catalysts for precision macromolecular synthesis. In: Kadish KM, Smith KM, Guilard R (eds) The porphyrin handbook. Academic, San Diego
5. Chou J-H, Kosal ME, Nalwa HS, Rakow NA, Suslick KS (2000) Applications of porphyrins and metalloporphyrins to materials chemistry. In: Kadish KM, Smith KM, Guilard R (eds) The porphyrin handbook. Academic, San Diego
6. Ono N (2002) The nitro group in organic synthesis. Wiley-VCH, New York
7. Ricci A (2008) Amino group chemistry: from synthesis to the life sciences. Wiley-VCH, Weinheim

8. Meng GG, James BR, Skov KA (1994) Porphyrin chemistry pertaining to the design of anti-cancer drugs; part 1, the synthesis of porphyrins containing *meso*-pyridyl and *meso*-substituted phenyl functional groups. Can J Chem 72:1894
9. Thomas DW, Martell AE (1956) Tetraphenylporphine and some *para*-substituted derivatives. J Am Chem Soc 78:1335
10. Thomas DW, Martell AE (1956) Absorption spectra of *para*-substituted tetraphenylporphines. J Am Chem Soc 78:1338
11. Rothemund P (1935) Formation of porphyrins from pyrroles and aldehydes. J Am Chem Soc 57:2010
12. Lindsey JS (2000) Synthesis of meso-substituted porphyrins. In: Kadish KM, Smith KM, Guilard R (eds) The porphyrin handbook. Academic, Boston
13. Cavaleiro JAS, Tomé AC, Neves MGPMS (2010) *meso*-Tetraarylporphyrin derivatives: new synthetic methodologies. In: Kadish KM, Smith KM, Guilard R (eds) Handbook of porphyrin science. World Scientific, Singapore
14. Bettelheim A, White BA, Raybuck SA, Murray RW (1987) Electrochemical polymerization of amino-, pyrrole-, and hydroxy-substituted tetraphenylporphyrins. Inorg Chem 26:1009
15. Chauhan SMS, Sahoo BB, Srinivas KA (2001) Microwave-assisted synthesis of 5,10,15,20-tetraarylporphyrins. Synth Commun 31:33
16. Collman JP, Gagne RR, Christopher AR, Halbert TR, Lang G, Robinson WT (1975) "Picket fence porphyrins". Synthetic models for oxygen binding hemoproteins. J Am Chem Soc 97:1427
17. Haesegawa E, Nemoto J, Kangyama T, Tsuchida E (1978) Oxygenation of polymer-covalently bonded metalloporphyrins. Macromolecules 11:947
18. Haesegawa E, Nemoto J-I, Kanayama T, Tsuchida E (1978) Syntheses and properties of vinyl monomers containing a *meso*-tetraphenylporphin ring and their copolymers. Eur Polym J 14:123
19. Tsuchida E (1979) Approaches to artificial macromolecular oxygen carriers. J Macromol Sci-Chem A13:545
20. Collman JP, Brauman JI, Doxsee KM, Halbert TR, Bunnenberg E, Linder RE, LaMar GN, Del Gaudio J, Lang G, Spartalian K (1980) Synthesis and characterization of "tailed picket fence" porphyrins. J Am Chem Soc 102:4182
21. Sun Y, Martell AE, Tsutsui M (1986) The synthesis and proton nuclear magnetic resonance study of some nitro- and amino-unsymmetrically meta-substituted tetraphenylporphyrins. J Heterocycl Chem 23:561
22. Little RG (1981) Notes on the synthesis of *meso*-substituted porphyrins from pyrryl carbinols and the mechanism of the Rothemund reaction. Heterocycl Chem 18:129
23. Schiavon MA, Iwamoto LS, Ferreira AG, Iamamoto Y, Zanoni MVB, Assis MD (2000) Synthesis and characterization of a novel series of *meso*(nitrophenyl) and *meso*(carboxyphenyl) substituted porphyrins. J Braz Chem Soc 11:458
24. Kruper WJ, Chamberlin TA (1988) US Patent 4,746,735
25. Kruper WJ, Chamberlin TA, Kochanny M (1989) Regiospecific aryl nitration of *meso*-substituted tetraarylporphyrins: a simple route to bifunctional porphyrins. J Org Chem 54:2753
26. Ostrowski S, Lopuszynska B (2003) Preparation of *meso*-tetraarylporphyrins nitrated in two neighboring aromatic rings. Synth Commun 33:4101
27. Ostrowski S, Mikus A, Lopuszynska B (2004) Synthesis of highly substituted *meso*-tetraarylporphyrins. Tetrahedron 60:11951
28. Luguya R, Jaquinod L, Fronczek FR, Vicente MGH, Smith KM (2004) Synthesis and reactions of *meso*-(*p*-nitrophenyl)porphyrins. Tetrahedron 60:2757
29. Smith NW, Dzyubam SV (2010) Efficient nitration of *meso*-tetraphenylporphyrin with nitronium tetrafluoroborate. Arkivoc vii:10
30. Jaquinod L (2000) Functionalization of 5,10,15,20-tetra-substituted porphyrins. In: Kadish KM, Smith KM, Guilard R (eds) The porphyrin handbook. Academic, San Diego

31. Jonhson AW, Winter M (1975) *Meso*-addition reaction of *meso*-tetraphenylporphyrin. Chem Ind (London) 351
32. Cavaleiro JAS, Neves MGPMS, Hewlins MJE, Jackson AH (1986) Reactions of porphyrins with nitronium tetrafluoroborate in pyridine. J Chem Soc Perkin Trans 1:575
33. Catalano MM, Crossley MJ, Harding MM, King, LC (1984) Control of reactivity at the porphyrin periphery by metal ion co-ordination: a general method for specific nitration at the β-pyrrolic position of 5,10,15,20-tetraarylporphyrins. J Chem Soc Chem Commun 1535
34. Crossley MJ, Govenlock LJ, Prashar JK (1995) Synthesis of porphyrin-2,3,12,13- and −2,3,7,8-tetraones: building blocks for the synthesis of extended porphyrin arrays. J Chem Soc Chem Commun 2379
35. Dahal S, Krishnan V, Nethaji M (1998) Coordination aggregation of meso-aryl porphyrins. Proc Indian Acad Sci (Chem Sci) 110:37
36. Cavaleiro JAS, Hewlins MJE, Jackson AH, Neves MGPMS (1986) Structures of the ring-opened oxidation products from meso-tetraphenylporphyrin. J Chem Soc Chem Commun 142
37. Evans B, Smith KM, Cavaleiro JAS (1978) Bile pigment studies. Part 4. Some novel reactions of metalloporphyrins with thallium(III) and cerium(IV) salts. Ring cleavage of *meso*-tetraphenylporphyrin. J Chem Soc Perkin Trans 1 768
38. Giraudeau A, Callot HJ, Jordan J, Ezhar I, Gross M (1979) Substituent effects in the electroreduction of porphyrins and metalloporphyrins. J Am Chem Soc 101:3857
39. Hombrecher HK, Gherdan VM, Ohm S, Cavaleiro JAS, Neves MGPMS, Condesso MF (1993) Synthesis and electrochemical investigation of β-alkyloxy substituted *meso*-tetraphenylporphyrins. Tetrahedron 49:8569
40. Vicente MGH, Neves MGPMS, Cavaleiro JAS, Hombrecher HK, Koll D (1996) Electrochemical and spectroscopic properties of Cu(II) β-nitro-*meso*-tetra(pentafluorophenyl) porphyrins. Tetrahedron Lett 37:261
41. Sen A, Ray PC, Das PK, Krishnan V (1996) Metalloporphyrins for quadratic nonlinear optics. J Phys Chem 100:19611
42. Richeter S, Jeandon C, Gisselbrecht J-P, Graff R, Ruppert R, Callot H (2004) Synthesis of new porphyrins with peripheral conjugated chelates and their use for the preparation of porphyrin dimers linked by metal ions. Inorg Chem 43:251
43. Ostrowski S, Szerszen D, Ryszczuk M (2005) Electrophilic nitration of *meso*-tetraarylporphyrin complexes at the beta-pyrrolic position. Synthesis 819
44. Wyrebek P, Ostrowski S (2007) Synthesis of some β-nitro-*meso*-tetraphenylporphyrin derivatives. J Porphyr Phthalocyanines 11:822
45. Bartoli JF, Battioni P, De Poor WR, Mansuy D (1994) Synthesis and remarkable properties of iron β-polynitroporphyrins as catalysts for monooxygenation reactions. J Chem Soc Chem Commun 23
46. Ozette K, Battioni P, Leduc P, Bartoli J-F, Mansuy D (1998) A new manganese-β-heptanitroporphyrin with extreme redox potentials: spectral electrochemical and catalytic properties. Inorg Chim Acta 272:4
47. Chauhan SMS, Kumar A, Srinivas KA (2004) β-Polynitration of 5,10,15,20-tetrakis (2,6-dichlorophenyl)porphyrins with HNO_3 and $Cu(NO_3)_2$ on clay using microwave irradiation. Synth Commun 34:2673
48. Ostrowski S, Shim YK (2001) Selective derivatization of TPP on one phenyl ring by nitration and subsequent nucleophilic substitution of hydrogen. Bull Korean Chem Soc 11:1
49. Ostrowski S, Urbańska N, Mikus A (2003) Nucleophilic substitution of hydrogen in *meso*-nitroaryl-substituted porphyrins unprotected at the NH-centers in the core ring. Tetrahedron Lett 44:4373
50. Ostrowski S, Mikus A (2003) A new approach to the synthesis of porphyrin-fullerene dyads. Mol Divers 6:315

51. Ostrowski S, Grzyb S (2007) Transformation of nitro-5,10,15,20-tetraarylporphyrins into their amino/nitro-functionalized derivatives. Jordan J Chem 2:297
52. Ostrowski S, Grzyb S, Mikus A (2007) Direct amination of *meso*-tetraarylporphyrin derivatives – easy route to A3B-, A2BC-, and A2B2-type porphyrins bearing two nitrogen-containing substituents at the *meso*-positioned phenyl groups. Helvetica Chim Acta 90:2000
53. Zhi-jun S, Yang-yan O, Qin H, Dan-li Z, Zhang-ping C (2005) Novel substitution reactions of 5-(4-nitrophenyl)-10,15,20-triphenylporphyrin with nucleophilic reagents. Wuhan Univ J Nat Sci 10:919
54. Baldwin JE, Crossley MJ, DeBernardis J (1982) Efficient peripheral functionalization of porphyrins. Tetrahedron 38:685
55. Catalano MM, Crossley MJ, Harding MM, King LC (1984) Control of reactivity at the porphyrin periphery by metal ion co-ordination: a general method for specific nitration at the β-pyrrolic position of 5,10,15,20-tetraarylporphyrins. J Chem Soc Chem Commun 1537
56. Crossley MJ, King LG, Pyke SM (1987) A new and highly efficient synthesis of hydroxyporphyrins. Tetrahedron 43:4569
57. Crossley MJ, King LG (1996) Reaction of metallo-2-nitro-5,10,15,20-tetraphenylporphyrins with oxyanions. Temperature-dependent competition between nucleophilic addition and single-electron transfer processes. J Chem Soc Perkin Trans 1 1251
58. Crossley MJ, King LG (1993) A new method for regiospecific deuteration and reduction of 5,10,15,20-tetraphenylporphyrins: nucleophilic reaction of borohydride ion with 2-nitro-5,10,15,20-tetraphenylporphyrins. J Org Chem 58:4370
59. Crossley MJ, King LG, Newsom IA, Sheehan CS (1996) Investigation of a 'reverse' approach to extended porphyrin systems. Synthesis of a 2,3-diaminoporphyrin and its reactions with α-diones. J Chem Soc Perkin Trans 1 2675
60. Crossley MJ, Harding MM, Tansey CW (1994) A convenient synthesis of 2-alkyl-5,10,15,20-tetraphenylporphyrins: reaction of metallo-2-nitro-5,10,15,20-tetraphenylporphyrins with Grignard and organolithium reagents. J Org Chem 59:4433
61. Crossley MJ, Gosper JJ, King LG (1988) C-hydroxyarylation of tetraphenylporphyrin – convenient introduction of a functionality which is oriented towards the porphyrin centre. Tetrahedron Lett 29:1597
62. Crossley MJ, King LG, Simpson JL (1997) Solvent-dependent ambident nucleophilicity of phenoxide ion towards nitroporphyrins: synthesis of 2-hydroxyaryl- and 2-aryloxy-5,10,15,20-tetraphenylporphyrins by displacement of a nitro group. J Chem Soc Perkin Trans 1 3087
63. Guo Q-N, Li Z-Y, Chan W-H, Lau K-C, Crossley M (2010) Appending zinc tetraphenylporphyrin with an amine receptor at β-pyrrolic carbon for designing a selective histamine chemosensor. Supramol Chem 22:122
64. Chen Z, Hu Q, Zhai B, Jiang Z, Qin W (2007) Nucleophilic substitution on porphyrin ring: synthesis of 2-(2-hydroxynaphthyl)-5,10,15,20-tetraphenylporphyrin. Chin J Chem 55:251
65. Huang Q, Pan Z, Wang P, Chen Z, Zhang X, Xu H (2006) Zinc(II) and copper(II) complexes of β-substituted hydroxylporphyrins as tumor photosensitizers. Bioorg Med Chem Lett 16:3030
66. Pereira AMVM, Alonso CMA, Neves MGPMS, Tomé AC, Silva AMS, Paz FAA, Cavaleiro JAS (2008) A new synthetic approach to *N*-arylquinolino[2,3,4-*at*]porphyrins from β-arylaminoporphyrins. J Org Chem 73:7353
67. Jaquinod L, Gros C, Olmstead MM, Antolovich M, Smith KM (1996) First syntheses of fused pyrroloporphyrins. Chem Commun 1475
68. Jaquinod L, Gros C, Khoury RG, Smith KM (1996) A convenient synthesis of functionalized tetraphenylchlorins. Chem Commun 2581
69. Shea KM, Jaquinod L, Smith KM (1998) Dihydroporphyrin synthesis: new methodology. J Org Chem 63:7013

70. Giuntini F, Faustino MAF, Neves MGPMS, Tomé AC, Silva AMS, Cavaleiro JAS (2005) Synthesis and reactivity of 2-(porphyrin-2-yl)-1,3-dicarbonyl compounds. Tetrahedron 61: 10454
71. Lacerda PSS, Silva AMG, Tomé AC, Neves MGPMS, Silva AMS, Cavaleiro JAS, Llamas-Saiz AL (2006) [1,2,3]Triazolo[4,5-*b*]porphyrins: new building blocks for porphyrinic materials. Angew Chem Int Ed 45:5487
72. Richeter S, Hadj-Aïssa A, Taffin C, van der Lee KA, Leclercq D (2007) Synthesis and structural characterisation of the first N-heterocyclic carbene ligand fused to a porphyrin. Chem Commun 2148
73. Richeter S, Jeandon C, Ruppert R, Callot HJ (2002) A modular approach to porphyrin oligomers using metal ions as connectors. Chem Commun 266
74. Lefebvre J-F, Leclercq D, Gisselbrecht J-P, Richeter S (2010) Synthesis, characterization, and electronic properties of metalloporphyrins annulated to exocyclic imidazole and imidazolium rings. Eur J Org Chem 1912
75. Tomé AC, Lacerda PSS, Neves MGPMS, Cavaleiro JAS (1997) meso-Arylporphyrins as dienophiles in Diels–Alder reactions: a novel approach to the synthesis of chlorins, bacteriochlorins and naphthoporphyrins. Chem Commun 1199
76. Silva AMG, Tomé AC, Neves MGPMS, Silva AMS, Cavaleiro JAS (1999) *meso*-Tetraarylporphyrins as dipolarophiles in 1,3-dipolar cycloaddition reactions. Chem Commun 1767
77. Tomé AC, Neves MGPMS, Cavaleiro JAS (2009) Porphyrins and other pyrrolic macrocycles in cycloaddition reactions. J Porphyr Phthalocyanines 13:408
78. Silva AMG, Tomé AC, Neves MGPMS, Cavaleiro JAS (2002) Porphyrins in 1,3-dipolar cycloaddition reactions: synthesis of a novel pyrazoline-fused chlorin and a pyrazole-fused porphyrin. Synlett 1155
79. Ostrowski S, Wyrębek P (2006) The first example of Diels–Alder cycloaddition of ortho-xylylenes to meso-tetraarylporphyrins containing electron-deficient β, β-double bonds. Tetrahedron Lett 47:8437
80. Rose E, Kossanyi A, Quelquejeu M, Soleilhavoup M, Duwavran F, Bernard N, Lecas A (1996) Synthesis of biomimetic heme precursors: the "double picket fence" 5,10,15, 20-tetrakis(2′,6′-dinitro-4′-*tert*-butylphenyl)porphyrin. J Am Chem Soc 118:1567
81. Weimin S, Qi S, Yucheng W, Lihong L, Jingchao T (2010) An alternative approach to amino porphyrins. J Heterocycl Chem 47:1221
82. Lipinska ME, Teixeira DMD, Laia CAT, Silva AMG, Rebelo SLH, Freire C (2013) β-Functionalized zinc(II)aminoporphyrins by direct catalytic hydrogenation. Tetrahedron Lett 54:110
83. George CD, Richardson T, Hofton ME, Vale CM, Neves MGPMS, Cavaleiro JAS (1999) Chlorine gas sensing using thin films of *meso*-tetra(*p*-stearamidophenyl)porphyrin. Mater Sci Eng C 8–9:559
84. Kadish KM, Araullo-McAdams C, Han BC, Franzen MM (1990) Syntheses and spectroscopic characterization of (T(p-Me2N)F4PP)H2 and (T(p-Me2N)F4PP)M where T(p-Me2N) F4PP = the dianion of meso-tetrakis(o, o, m, m-tetrafluoro-p-(dimethylamino)phenyl) porphyrin and M = cobalt(II), copper(II), or nickel(II). Structures of (T(p-Me2N)F4PP)Co and meso-tetrakis(pentafluorophenyl)porphinatocobalt(II), (TF5PP)Co. J Am Chem Soc 112: 8364
85. Costa JIT, Tomé AC, Neves MGPMS, Cavaleiro JAS (2011) 5,10,15,20-Tetrakis (pentafluorophenyl)porphyrin: a versatile platform to novel porphyrinic materials. J Porphyr Phthalocyanines 15:1117
86. Collman JP, Gagne RR, Halbert TR, Marchon JC, Reed CA (1973) Reversible oxygen adduct formation in ferrous complexes derived from a picket fence porphyrin. Model for oxymyoglobin. J Am Chem Soc 95:7868
87. Baldwin JE, Perlmutter P (1984) Bridged, capped and fenced porphyrins. In: Vögtle F, Weber E (eds) Topics in current chemistry. Springer, Berlin

88. Tomé AC, Silva AMS, Alkorta I, Elguero J (2011) Atropoisomerism and conformational aspects of *meso*-tetraarylporphyrins and related compounds. J Porphyr Phthalocyanines 15:1
89. Sessler JL, Karnas E, Sedenberg E (2012) Porphyrins and expanded porphyrins as receptors. In: Philip A, Jonathan W (eds) Supramolecular chemistry: from molecules to nanomaterials. Wiley, Chichester
90. Collman JP, Boitrel B, Fu L, Galanter J, Straumanis A, Rapta M (1997) The chloroacetamido group as a new linker for the synthesis of hemoprotein analogues. J Org Chem 62:2308
91. Collman JP, Wang Z, Straumanis A (1998) Isocyanate as a versatile synthon for modular synthesis of functionalized porphyrins. J Org Chem 63:2424
92. Lindsey JS (1980) Increased yield of a desired isomer by equilibriums displacement on binding to silica gel, applied to *meso*-tetrakis(*o*-aminophenyl)porphyrin. J Org Chem 45:5215
93. Lee C, Lee DH, Hong J-I (2001) Colorimetric anion sensing by porphyrin-based anion receptors. Tetrahedron Lett 42:8665
94. Ladomenou K, Bonar-Law RP (2002) Urea porphyrins as simple receptors for sugars. Chem Commun 2108
95. Cormode DP, Murray SS, Cowley AR, Beer PD (2006) Sulfate selective anion recognition by a novel tetra-imidazolium zinc metalloporphyrin receptor. Dalton Trans 5135
96. Cormode DP, Drew MGB, Jagessar R, Beer PD (2008) Metalloporphyrin anion sensors: the effect of the metal centre on the anion binding properties of amide-functionalised and tetraphenyl metalloporphyrins. Dalton Trans 6732
97. Beer PD, Cormode DP, Davis JJ (2004) Zinc metalloporphyrin-functionalised nanoparticle anion sensors. Chem Commun 414
98. David P, Cormode DP, Drew MGB, Jagessar R, Beer PD (2008) Anion sensing porphyrin functionalized nanoparticles. J Inorg Organomet Polym 18:32
99. Dixon IM, Lopez F, Tejera AM, Esteve J, Blasco MA, Pratviel G, Meunier B (2007) A G-quadruplex ligand with 10000-fold selectivity over duplex DNA. J Am Chem Soc 129:1502
100. Brown A, Beer PD (2012) Porphyrin-functionalised rotaxanes for anion recognition. Dalton Trans 41:118
101. Giuntini F, Alonso CMA, Boyle RW (2011) Synthetic approaches for the conjugation of porphyrins and related macrocycles to peptides and proteins. Photochem Photobiol Sci 10:759
102. Chen B, Li A, Liang F, Zhou X, Cao X, He Z (2006) Synthesis of some multi-β-substituted cationic porphyrins and studies on their interaction with DNA. Tetrahedron 62:5487
103. Silva EMP, Serra VV, Ribeiro AO, Tomé JPC, Domingues P, Faustino MAF, Neves MGPMS, Tomé AC, Cavaleiro JAS, Ferrer-Correia AJ, Iamamoto Y, Domingues MRM (2006) Characterization of cationic glycoporphyrins by electrospray tandem mass spectrometry. Rapid Commun Mass Spectrom 20:3605
104. Khan MM, Ali H, van Lier JE (2001) Synthesis of new aminoporphyrins via palladium-catalysed cross-coupling reactions. Tetrahedron Lett 42:1615
105. Soares ARM, Martínez-Díaz MV, Bruckner A, Pereira AMVM, Tomé JPC, Alonso CMA, Faustino MAF, Neves MGPMS, Tomé AC, Silva AMS, Cavaleiro JAS, Torres T, Guldi DM (2007) Synthesis of novel N-linked porphyrin-phthalocyanine dyads. Org Lett 9:1557
106. Pereira AMVM, Neves MGPMS, Cavaleiro JAS, Jeandon C, Gisselbrecht J-P, Choua S, Ruppert R (2011) Diporphyrinylamines: synthesis and electrochemistry. Org Lett 13:4742
107. Pereira AMVM, Neves MGPMS, Cavaleiro JAS, Gisselbrecht J-P, Jeandon C, Ruppert R (2011) A new approach to N-phenylquinolino[2,3,4-at]porphyrins. Electrochemical and photochemical studies. J Porphyr Phthalocyanines 15:575
108. Fuhrhop J-H (1978) Irreversible reactions on the porphyrin periphery (excluding oxidations, reductions, and photochemical reactions). In: Dolphin D (ed) The porphyrins, vol 2. Academic, New York

109. Guo Z, Du F, Ren D, Chen Y, Zheng J, Liuc Z, Tian J (2006) Covalently porphyrin-functionalized single-walled carbon nanotubes: a novel photoactive and optical limiting donor–acceptor nanohybrid. J Mater Chem 16:3021
110. Gross AJ, Bucher C, Coche-Guerente L, Labbé P, Downard AJ, Moutet J-C (2011) Nickel (II) tetraphenylporphyrin modified surfaces via electrografting of an aryldiazonium salt. Electrochem Commun 13:1236
111. Hombrecher HK, Gerdan VM, Cavaleiro JAS, Neves MGPMS (1994) An efficient synthesis of β-alkyloxy substituted porphyrins. J Prak Chem-Chem Ztg 336:542
112. Cavaleiro JAS, Gerdan VM, Hombrecher HK, Neves MGPMS, Silva A (1997) Synthesis and characterization of new 2-diazo-3-oxo-5,10,15,20-tetraphenylchlorins. Heterocycl Commun 3:253
113. Shen D-M, Liu C, Chen Q-Y (2007) Synthesis and versatile reactions of β-azidotetraarylporphyrins. Eur J Org Chem 1419
114. Shen D-M, Liu C, Shen Q-Y (2007) Synthesis and versatile reactions of β-azidotetraarylporphyrins. Eur J Org Chem 2888
115. Séverac M, Pleux LL, Scarpaci A, Blart E, Odobel F (2007) Synthesis of new azido porphyrins and their reactivity in copper(I)-catalyzed Huisgen 1,3-dipolar cycloaddition reaction with alkynes. Tetrahedron Lett 48:6518
116. Ludke K, Alonso CMA, Neves MGPMS, Silva AMS, Cavaleiro JAS, Hombrechner HK (1997) A new approach to the synthesis of unsaturated porphyrins. Heterocycl Commun 3:503
117. Alonso CMA, Neves MGPMS, Silva AMS, Cavaleiro JAS, Hombrecher HK (1997) Reaction of β-amino-meso-tetraphenylporphyrin with α, β-unsaturated carbonyl compounds: an approach to fused pyridinoporphyrins. Tetrahedron Lett 38:2757
118. Alonso CMA, Neves MGPMS, Tomé AC, Silva AMS, Cavaleiro JAS (2000) Synthesis and Diels-Alder reactions of 2-(buta-1,3-dien-2-yl)-5,10,15,20-tetraphenylporphyrin. Tetrahedron Lett 41:5679
119. Tomé JPC, Pereira AMVM, Alonso CMA, Neves MGPMS, Tomé AC, Silva AMS, Cavaleiro JAS, Martinez-Diaz MV, Torres T, Rahman GMA, Ramey J, Guldi DM (2006) Synthesis and photophysical studies of new porphyrin–phthalocyanine dyads with hindered rotation. Eur J Org Chem 257
120. Kawakami T, Igarashi S (1996) Highly sensitive spectrophotometric determination of nitrite ion using 5,10,15,20-tetrakis(4-aminophenyl)porphine for application to natural waters. Anal Chim Acta 333:175
121. Kawakami T, Igarashi S (1997) Asymmetric porphyrin derivative having one reactive site as a highly sensitive spectrophotometric reagent for nitrite. Anal Chim Acta 354:159
122. Alonso CMA, Neves MGPMS, Tomé AC, Silva AMS, Cavaleiro JAS (2006) Reaction of 2-amino-5,10,15,20-tetraphenylporphyrinatonickel(ii) with α, β-unsaturated acyl chlorides: synthesis of 2-pyridone-fused porphyrin derivatives. J Mex Chem Soc 50:100
123. Alonso CMA, Neves MGPMS, Tomé AC, Silva AMS, Cavaleiro JAS (2001) Hetero-Diels–Alder reactions of β-imino-***meso***-tetraphenylporphyrin derivatives: a new approach to pyrido[2,3-***b***]porphyrins. Tetrahedron Lett 42:8307
124. Alonso CMA, Neves MGPMS, Tomé AC, Silva AMS, Cavaleiro JAS (2004) β-Imino-*meso*-tetraphenylporphyrin derivatives in hetero-Diels-Alder reactions. Eur J Org Chem 3233
125. Alonso CMA, Serra VV, Neves MGPMS, Tomé AC, Silva AMS, Paz FAA, Cavaleiro JAS (2007) An easy synthetic approach to pyridoporphyrins by domino reactions. Org Lett 9:2305
126. Gomes ATPC, Paz FAA, Neves MGPMS, Tomé AC, Silva AMS, Souza MCBV, Ferreira VF, Cavaleiro JAS (2011) Synthesis and characterization of new porphyrin/4-quinolone conjugates. Tetrahedron Lett 37:4741
127. Alonso CMA, Neves MGPMS, Tomé AC, Silva AMS, Cavaleiro JAS (2006) Beta-imino-*meso*-tetraphenylporphyrin derivatives: reactivity towards ethyl diazoacetate. Jordan J Chem 1:95

128. Crossley MJ, King LJ (1984) Novel heterocyclic systems from selective oxidation at the β-pyrrolic position of porphyrins. J Chem Soc Chem Commun 920
129. Beavington R, Rees PA, Burn PL (1998) A study on the oxidation of 2-hydroxyporphyrins to porphyrin-α-diones. J Chem Soc Perkin Trans 1 2847
130. Promarak V, Burn PL (2001) A new synthetic approach to porphyrin-α-diones and a-2,3,12,13-tetraone: building blocks for laterally conjugated porphyrin arrays. J Chem Soc Perkin Trans 1 14
131. Khoury T, Crossley MJ (2009) Expansion of the porphyrin π-system: stepwise annelation of porphyrin β, β′-pyrrolic faces leading to trisquinoxalinoporphyrin. New J Chem 33:1076
132. Gautam P, Krishnan V (1995) Coordination aggregation of meso-aryl substituted porphyrins. Proc Indian Acad Sci 107:477
133. Richeter S, Jeandon C, Ruppert R, Callot HJ (2001) Porphyrins acting as external and internal ligands: preparation of conjugated trimetallic dimeric porphyrins. Chem Commun 91
134. Richeter S, Jeandon C, Gisselbrecht JP, Ruppert R, Callot HJ (2002) Syntheses and optical and electrochemical properties of porphyrin dimers linked by metal ions. J Am Chem Soc 124:6168
135. Henrick K, Owston PG, Peters R, Tasker PA, Dell A (1980) Cyclisation involving a meso-phenyl substituent of a metalloporphyrin: X-ray structure of 5,10,15-triphenyl{22-oxo-benzo [2324]cyclohexa[a, b]porphinato(2-)}copper(II). Inorg Chim Acta 45:161
136. Callot HJ, Schaeffer E, Cromer R, Metz F (1990) Unexpected routes to naphtoporphyrin derivatives. Tetrahedron 46:5253
137. Ishkov YV, Zhilina ZI (1995) Porphyrins and their derivatives. 17. Intramolecular cyclization of 2-formyl-5,10,15,20-tetraphenylporphyrin complexes. Zh Org Khim 31:136
138. Richeter S, Jeandon C, Sauber C, Gisselbrecht JP, Ruppert R, Callot HJ (2002) Preparation, mass spectrometry and electrochemical studies of metal connected porphyrin oligomers. J Porphyr Phthalocyanines 6:423
139. Richeter S, Jeandon C, Gisselbrecht JP, Ruppert R, Callot HJ (2007) Synthesis, structural characterization, and electrochemical studies of nickel porphyrins bearing two peripheral conjugated chelating groups. Inorg Chem 46:10241
140. Ishkov YV (2001) Porphyrins and their derivatives: XXII. A new product of intramolecular cyclization of 5,10,15,20-tetraphenyl-2-formylporphyrin copper complex. Russ J Org Chem 37:288

Top Heterocycl Chem (2014) 33: 79–142
DOI: 10.1007/7081_2013_107

Published online: 22 May 2013

Functionalization of Corroles

Joana F.B. Barata, Carla I.M. Santos, M. Graça P.M.S. Neves, M. Amparo F. Faustino, and José A.S. Cavaleiro

Abstract Corroles have assumed an important place in tetrapyrrolic chemistry. This review highlights the reactivity features of *meso*-arylcorroles under different reaction conditions.

Keywords Aromatic substitution reactions · Corroles · Cycloaddition reactions · Functionalization · Metal-catalyzed reactions

Contents

J.F.B. Barata, C.I.M. Santos, M.G.P.M.S. Neves, M.A.F. Faustino, and J.A.S. Cavaleiro (✉)
Department of Chemistry (QOPNA and CICECO), University of Aveiro, 3810-193 Aveiro, Portugal
e-mail: jcavaleiro@ua.pt

1 Introduction

Corroles are tetrapyrrolic derivatives discovered in the 1960s when researchers were attempting to synthesize analogues of the vitamin B_{12} series containing the corrin-type macrocycle **1** [1, 2]. Contrary to what happens with other tetrapyrroles, like porphyrins, chlorins, and bacteriochlorins, which play vital functions in nature, corroles are not natural compounds.

The corrole macrocycle **2** is structurally related to the corrin **1**, with a 19-atom carbon skeleton; due to its aromaticity, photophysical properties, and coordination behavior of its metal complexes, it is also related to the porphyrin-type **3** macrocycle.

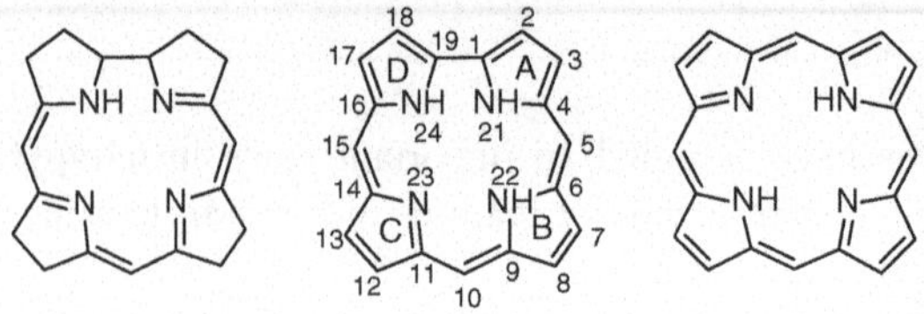

1 2 3

In fact corrole derivatives show similar electronic properties as porphyrins, such as the visible light absorption, high luminescence yields, and strong absorption features of the excited states. These macrocycles are able to stabilize high valent transition metal ions, due to the existence of three NH protons in the inner core, an interesting property amongst cyclic tetrapyrroles.

Having in mind potential applications of corroles and metallocorroles, several research groups have been looking for new methodologies leading to key corrole derivatives. As a result, mainly from 1999, efficient synthetic methodologies leading to *meso*-substituted corroles became available. The work initially performed by the groups of Gross [3] and Paolesse [4] and later by the Gryko's group [5] deserves a highlighted reference. Facing the significance of such synthetic methodologies for further work with such compounds, it will be presented here a brief description of them.

In the procedure proposed by Gross et al. [3], the condensation of pyrrole and aldehydes took place in the absence of solvent and proved to be particularly suitable for the synthesis of A_3-type corroles from aldehydes with strong electron-withdrawing substituents. For instance, the 5,10,15-tris(pentafluorophenyl)corrole **4** was obtained in 11% yield, from the direct condensation, under heterogeneous conditions, of pentafluorobenzaldehyde with pyrrole (Scheme 1).

The route developed by Paolesse et al. [4] was based on the acidic conditions developed by Adler for the synthesis of *meso*-tetraarylporphyrins [6]. The condensation of pyrrole with an aldehyde, in a 3:1 ratio, instead of 1:1 usually used for porphyrins, took place in acetic acid. For instance, under these conditions, the condensation of pyrrole with benzaldehyde afforded the *meso*-triphenylcorrole **5** in 6% yield, accompanied by the corresponding *meso*-tetraphenylporphyrin, TPP (Scheme 2).

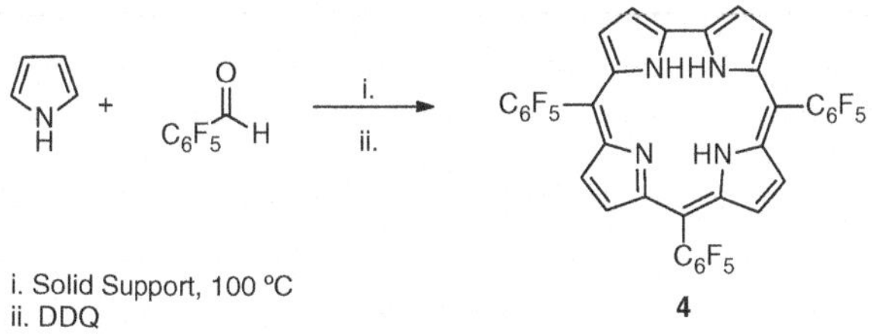

Scheme 1 Synthesis of 5,10,15-tris(pentafluorophenyl)corrole under heterogeneous conditions

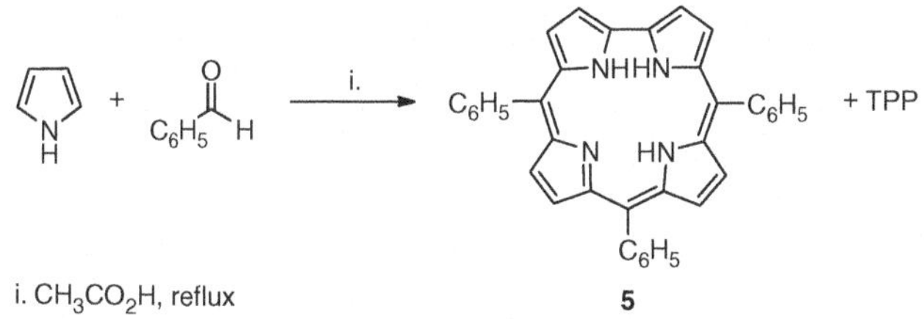

Scheme 2 Synthesis of 5,10,15-triphenylcorrole by Adler modified method

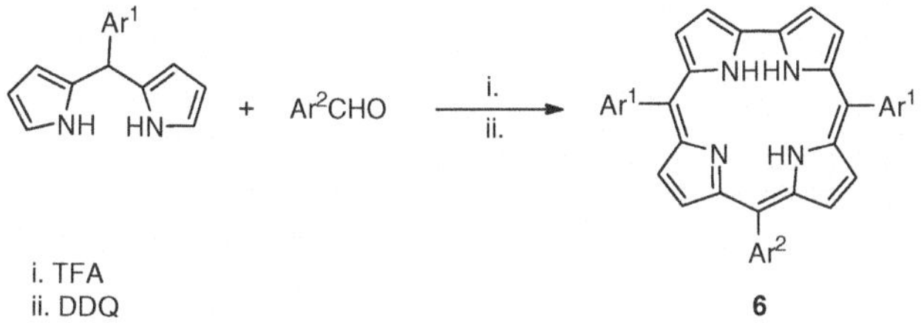

Scheme 3 Synthesis of *trans*-A_2B-triarylcorroles in acidic conditions

These two landmark approaches were followed by other publications namely by the Gryko's group, considering new improvements in the experimental conditions. This group was able to refine the methods giving access to *meso*-substituted A_3- and *trans*-A_2B-corroles [5]. After a careful examination of various reaction parameters (reactivity of the aldehyde, catalyst, solvent, concentration, time, etc.), authors elaborated three different sets of conditions depending on the aromatic aldehyde reactivity type and steric hindrance to afford *meso*-A_3-corroles. It was observed that small amounts of TFA promoted the bilane formation, the direct precursor of corrole, and it was made clear that the amount of pyrrole to be used depends on the reactivity of the aldehyde.

The access to a series of *meso*-substituted *trans*-A_2B-corroles **6** with different functionalities was also possible by the TFA-catalyzed condensation of a dipyrromethane with an aldehyde, followed by oxidation with 2,3-dichloro-5,6-dicyano-1,4-benzoquinone (DDQ) (Scheme 3).

In 2006, the same group considered the possibility of performing the condensation of the aldehyde with pyrrole, in a water–methanol mixture in the presence of HCl. The optimization of the various reaction parameters, such as co-solvent, reagent, and acid concentration, allowed to obtain high yields of bilanes that were isolated and subsequently reacted and oxidized. Under these conditions triphenylcorrole was obtained in 32% yield [7]. The authors extended also these reaction conditions to other dipyrromethanes affording *trans*-A_2B-corroles in yields ranging from 13% to 56% [7].

Certain variations were also considered by other authors. The use of microwave irradiation under Gross conditions was also considered in the synthesis of tri-aryl- and tri-pyrimidyl-corroles (type A_3) by Collman and Decréau [8]. Compared with the conventional heating methodology, the microwave irradiation afforded an increase in corrole yields of ca. 30% and led to noticeably cleaner reaction mixtures. Chauhan and Kumari [9], in 2008, reported that corroles can also be obtained by the condensation of aryl aldehydes with pyrrole, in the presence of Amberlyst 15 catalyst and under solvent-free conditions. With this method it was possible to isolate the 5,10,15-triphenylcorrole in 15% yield and the 5,10,15-tris (pentafluorophenyl)corrole in 30% yield.

The use of the ionic liquid [Bmim][BF_4] was also reported as a suitable reaction medium in the preparation of *meso*-substituted *trans*-A_2B-corroles. The main advantage of this protocol is the reduction of the amount of organic solvents in response to the demand of green chemistry methodologies [10].

Recently, Nocera and co-workers [11] carried out the synthesis of 5,10,15-tris (pentafluorophenyl)corrole in large scale. This was performed by condensation of pyrrole, pentafluorobenzaldehyde, and paraformaldehyde under Lindsey conditions. The role of paraformaldehyde in the corrole forming reaction is likely to adjust the acid concentration for the formation of the bilane precursor that delivers corrole rather than the corresponding porphyrin.

All these studies allow the research community to have access not only to corroles of type A_3 but also to corroles of types ABC and *trans*-A_2B. This topic has been nicely reported in several publications [7, 12–22]. Considering such synthetic approaches developed for corroles, the chemical modification in the periphery and in the inner core of the macrocycles can afford new compounds, which can then be available for further studies and applications, namely as catalysts, sensors, dyes for solar energy conversion, and in medicine [23–25].

This review highlights synthetic methodologies available for the functionalization of *meso*-triarylcorroles, mainly those published after 1999.

1.1 Corrole Inner Core Reactions

A corrole macrocycle is susceptible to react at the inner nitrogen atoms, at the β-pyrrolic positions (2,3,7,8,12,13,17,18), at the *meso*-positions (5,10,15), and at its aryl substituents.

Considering the inner core, corroles can be protonated and deprotonated; their relative high acidity is particularly relevant, when compared with other tetrapyrrolic macrocycles. The steric relief obtained after the removal of one proton from the corrole-free base leading to the corresponding mono-anionic species is an obvious explanation for this particular feature. The coordination of these ligands with different metal ions is also distinctive from porphyrins by providing a trianionic coordination sphere that is able to stabilize metal ions in high oxidation state. Comprehensive reviews dedicated to the coordination chemistry of these ligands were recently published [23, 26–28].

Thinking on further functionalization on the corrole periphery the protection of the inner core by metallation is also an important issue, to prevent the occurrence of reactions in the inner nitrogens. As an example of a side reaction is the formation of bridged derivative **7** under Vilsmeier–Haack conditions (vide infra Scheme 14) [29].

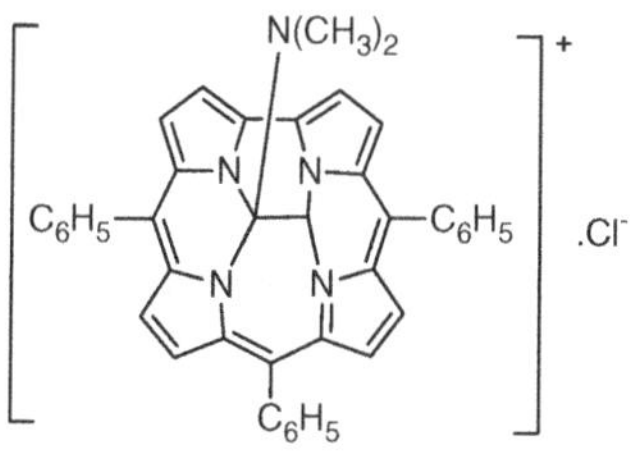

7

Also, Gross et al. have reported the unusual reactivity of the inner nitrogen atoms during attempts to insert a carboxylic acid moiety at the periphery of 5,10,15-tris(pentafluorophenyl)corrole **4** using phosgene. Authors found that depending on the base used, $COCl_2$ reacts at the *N*-21 atom of the inner core of the macrocycle affording, the *N*-acylated product **8** or the bridged keto chiral adduct **9** (vide infra Scheme 16) [30].

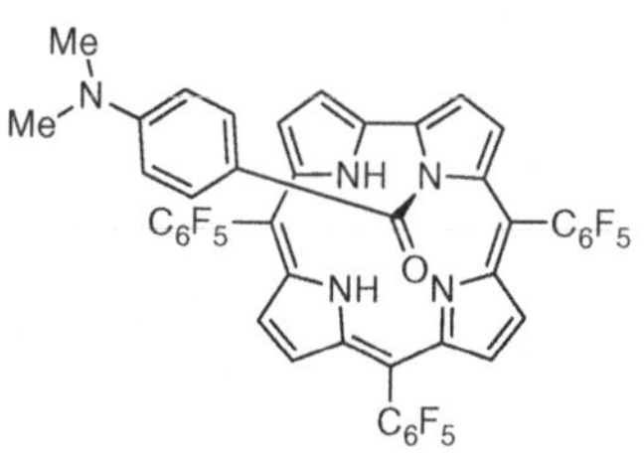

8

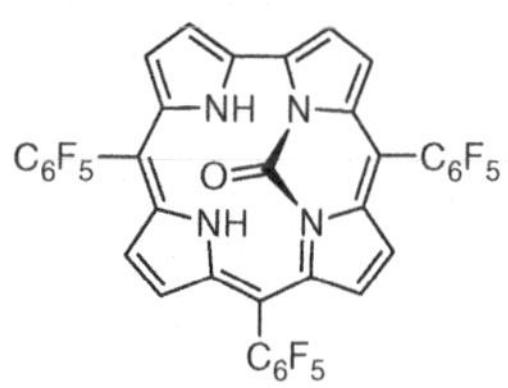

9

The high reactivity of the nitrogens in the inner core was indeed demonstrated in 1965, when *N*-alkylated corroles were obtained from β-substituted corroles and alkylating reagents [2]. Considering *meso*-arylcorroles, Gross et al. reported the access to chiral *N*-substituted corroles by reaction of 5,10,15-tris(pentafluorophenyl)

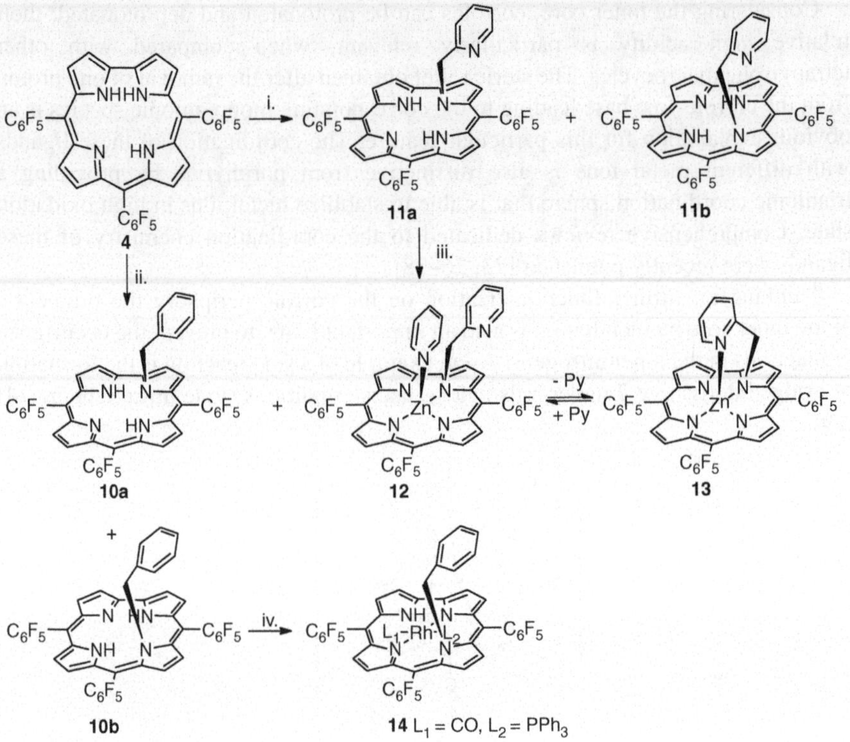

i) 2-PyCH$_2$Cl.HCl, dry toluene, dry K$_2$CO$_3$, 4 h
ii) C$_6$H$_5$CH$_2$Br, dry toluene, dry K$_2$CO$_3$, 4 h
iii) Zn(OAc)$_2$, dry pyridine, reflux, 10 min
iv) PPh$_3$, dry K$_2$CO$_3$, dry NaOAc, dry benzene, Rh$_2$(CO)$_4$Cl$_2$, reflux, 5 h

Scheme 4 Synthesis of chiral *N*-substituted corroles

corrole **4** with either 2-(chloromethyl)pyridine or benzyl halide (chloride or bromide) [31]. This reaction afforded the corresponding *N*-21 and *N*-22 isomers **10** and **11** in high overall yields (Scheme 4).

Reaction of *N*-21-picolylcorrole **11a** with zinc acetate, in pyridine, gave rise to the zinc(II) complex **12**. This complex **12** has an axial pyridine ligand coordinated to the metal center, whereas after purification by column chromatography such ligand is lost with formation of the new zinc(II) complex **13**. The reaction is reversible by adding or removing the pyridine (Scheme 4) [31].

The extension of these studies to other isomers and to monovalent rhodium(I) salts afforded the rhodium(I) corrole complex **14** from the reaction of *N*-22-benzyl derivative **10b** with $Rh_2(CO)_4Cl_2$ in the presence of triphenylphosphine. The structure of **14** was confirmed by X-ray crystallography; it is shown that only two of the core nitrogen atoms coordinate to the rhodium(I) (Scheme 4) [32, 33]. These results have demonstrated that the *N*-21 substitution creates a less crowded environment, which is reflected in the lower deformation of the corrole ring.

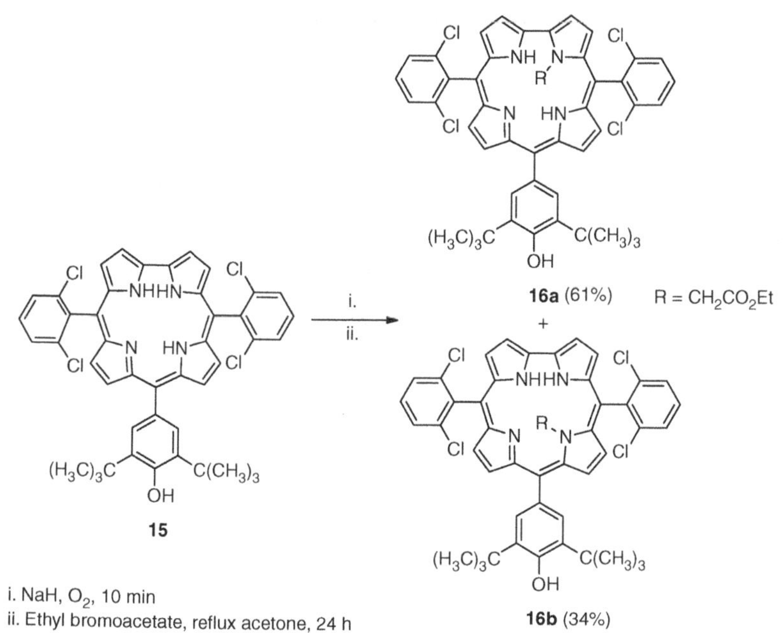

Scheme 5 *N*-alkylation of *trans*-A_2B-triarylcorrole

Dehaen and co-workers reported that the *N*-alkylation of the *trans*-A_2B-corrole **15** with ethyl bromoacetate under oxidative conditions afforded the two isomeric *N*-alkylated non-oxidized corroles **16a** and **16b** (Scheme 5) [20]. Contrary to what happens with the porphyrin macrocycle, only one substituent is introduced in the corrole macrocycle; this is presumably due to the smaller size of the corrole cavity. Moreover the distribution of the substituent groups over the core nitrogen atoms (i.e., the ratio of both isomers) is in accordance with previous *N*-alkylation corrole studies [31, 33]. Once more, the authors have demonstrated that substitution on the *N*-22 center of triarylcorroles results in a more crowded environment (as compared to the *N*-21 substituted isomer), this being reflected in the higher deformation of the corrole ring planarity and of the *meso*-aryl groups deviation from perpendicular orientation, rendering the isomer **16b** (*N*-22 substituted isomer) less stable.

Corrole complexes are usually used to perform further work on macrocyclic functionalization. This is due to the lability of the free bases toward O_2 and light, affording in general open-chain biliverdin-type species [34–37]. Even the 5,10,15-tris(pentafluorophenyl)corrole **4**, which is considered to be one of the most stable corroles, suffers polymerization in solution, when it is left under daylight, affording the dimer **17** as the major product and the trimer **18** as the minor one [38]. This type of oligomers was also obtained via oxidation of **4** with *p*-chloranil and *p*-fluoranil, by Osuka and co-workers [39].

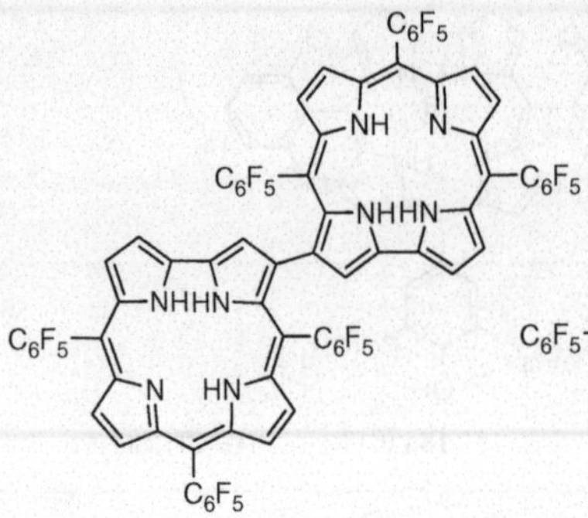

17

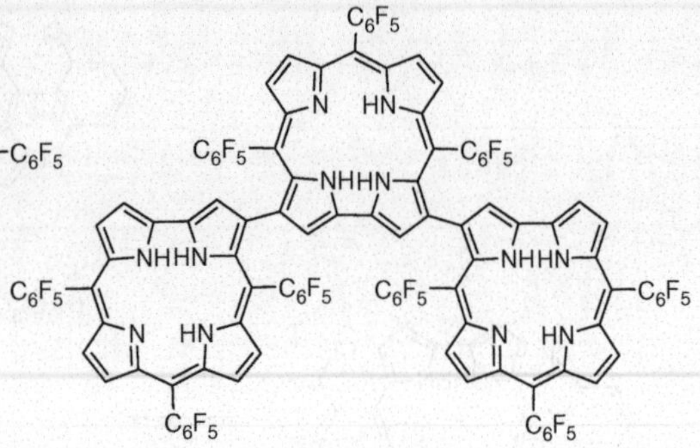

18

Most of the functionalizations are performed on metallocorrole complexes; it is then necessary at a later stage to have access to the free bases. Demetallation protocols for that are now available. For instance, the reductive demetallation of Ag-, Cu-, and Mn-corroles [40–42] allows that such metallation/demetallation processes can be used as a strategy to obtain more structurally elaborated free-base corroles. Some of those protocols will be mentioned in the following sections.

2 Key Substituents for Further Functionalization of Corroles

Since 1999 there has been a considerable research directed to the development of synthetic strategies to functionalize readily available *meso*-triarylcorroles. Some of this work is related to the introduction of key groups into the periphery of the macrocycle. The most recent procedures to perform target functionalizations will then be highlighted.

2.1 *Halo Group*

Several halogenation procedures are now available for the insertion of halogen atoms (Br, I, Cl) into the β-pyrrolic positions of the corrole macrocycle.

In 2001, Paolesse and co-workers [16, 43] reported that the bromination of 5,10,15-triphenylcorrole **5** and 5,10,15-tris(4-nitrophenyl)corrole **19**, using *N*-bromosuccinimide (NBS), afforded, as the main products, the isocorroles **20** and **21**, respectively. Authors pointed out that treatment of **20** and **21** with cobalt(II) acetate and triphenylphosphine afforded the correspondent Co(III) corrole complexes **Co22** and **Co23** (Scheme 6). The formation of isocorroles **20** and **21** was explained by the attempt of the corrole ring to reduce the sterical strain induced by the NH protons. In fact, X-ray crystallographic characterization of

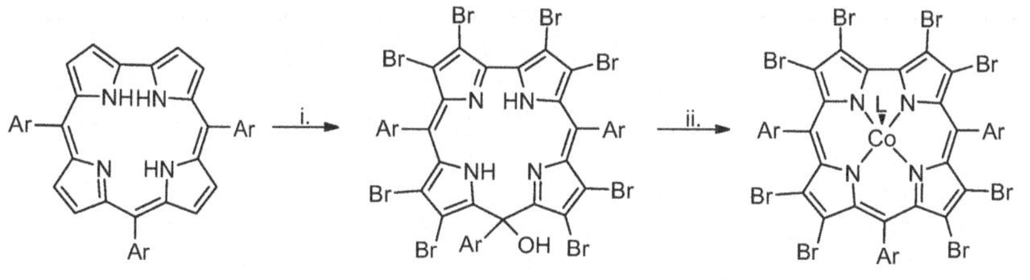

L=PPh$_3$

5, Ar = C_6H_5
19, Ar = 4-NO_2-C_6H_5

20, Ar = C_6H_5
21, Ar = 4-NO_2-C_6H_5

Co22, Ar = C_6H_5
Co23, Ar = 4-NO_2-C_6H_5

i. NBS, $CHCl_3$, rt, 12 h
ii. $Co(AcO)_2$, PPh_3

Scheme 6 Bromination of *meso*-triarylcorroles

the cobalt complex has shown only slight deviations from the planarity of the ring, and this is in agreement with the steric relief due to the metal complex formation.

Chen and co-workers [44] have also performed similar studies with 5,10,15-tris(pentafluorophenyl)corrole **4**; they have shown that the amount of used NBS rules the type of product being formed. For instance, when 1.1 equiv. of NBS dissolved in MeCN was added dropwise over 8 h into a CH_2Cl_2 solution of 5,10,15-tris(pentafluorophenyl)corrole **4**, a mixture of two monobrominated corrole isomers (55%) assigned as the 3-bromo-5,10,15-tris(pentafluorophenyl)corrole **24** and as the 2-bromo-5,10,15-tris(pentafluorophenyl)corrole **25** was obtained. Attempts to separate the two isomers, either as free bases or as cobalt complexes, were not successful. Increasing the amount of NBS to 2.1, 3.1, and 4.4 equiv., the dibrominated-**26**, tribrominated-**27**, and tetrabrominated-**28** corroles were isolated in moderate to good yields.

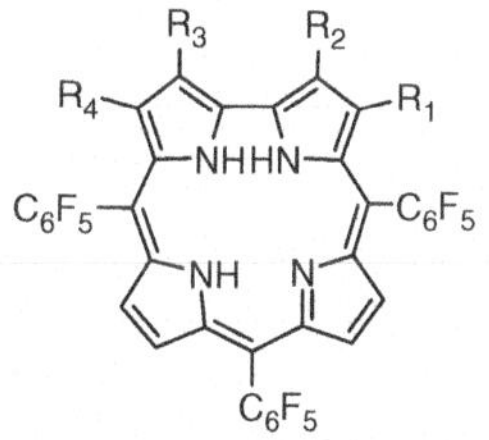

24, R_1= Br, R_2=R_3=R_4=H
25, R_2= Br, R_1=R_3=R_4=H
26, R_1= R_2= Br, R_3=R_4=H
27, R_1= R_2= R_4=Br, R_3=H
28, R_1= R_2= R_3=R_4=Br

The access to the mono-brominated corroles **32** and **33** was described by Paolesse and co-workers [45] during studies concerning the synthesis of stable isocorrole macrocycles, via oxidation of 5,10,15-tritolylcorrole **29** with DDQ followed by addition of EtMgBr (Scheme 7). Besides the expected 5- and 10-ethyltriarylisocorroles **30** and **31**, two monobrominated regioisomers **32** and **33** were isolated.

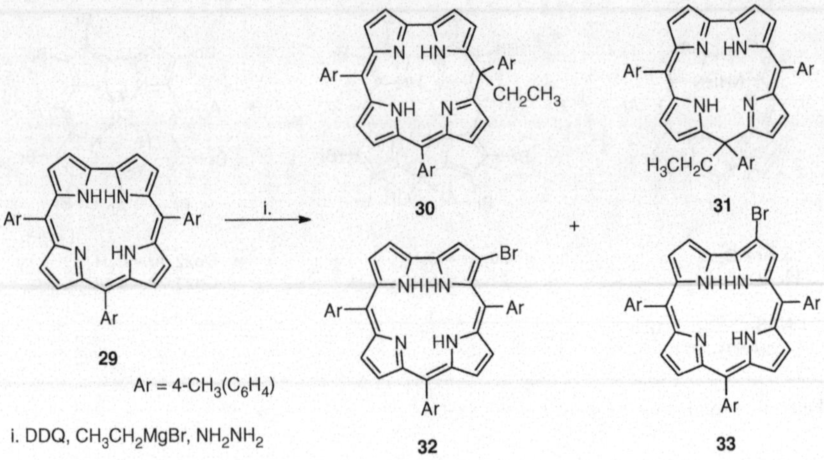

Scheme 7 Mono-bromination conditions of 5,10,15-tritolylcorrole

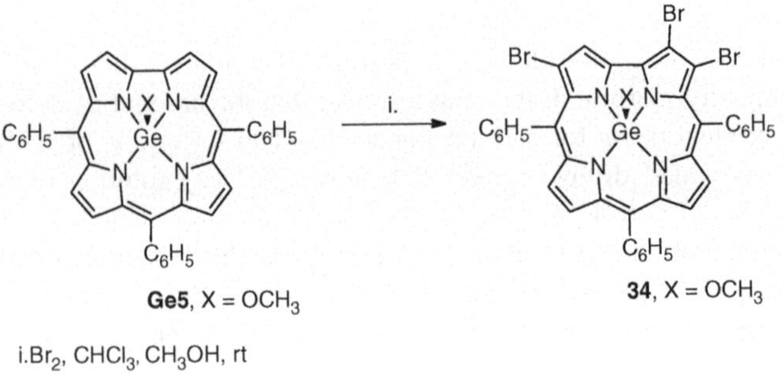

Scheme 8 Tri-bromination of germanium(IV) corrole complex

The partial bromination of germanium complex **Ge5** was also studied [46]. Authors reported that, at room temperature, in the presence of a slight excess of bromine in $CHCl_3$, with methanol as co-solvent, the tribrominated corrole **34** was isolated (Scheme 8). The access to pentabrominated and hexabrominated corroles was also possible, when higher proportions of Br_2 toward the germanium complex were used. The photophysical characterization of the germanium corrolates namely of **Ge5**, **34** and of a μ-oxo dimer, obtained in the synthesis of germanium complex **Ge5**, was described. Authors highlighted the high phosphorescence emissions of these complexes.

Gross and co-workers also reported the synthesis of a series of fully brominated metallocorroles **35** by using a high excess of Br_2 [47–54]. The effect of the bromine substitution in, for instance, the redox potential and photophysical properties of such products is reported in those publications.

The same authors also reported the octabromination of 5,10,15-tris-(pentafluorophenyl)corrolate-oxochromium(V), by using NBS [55].

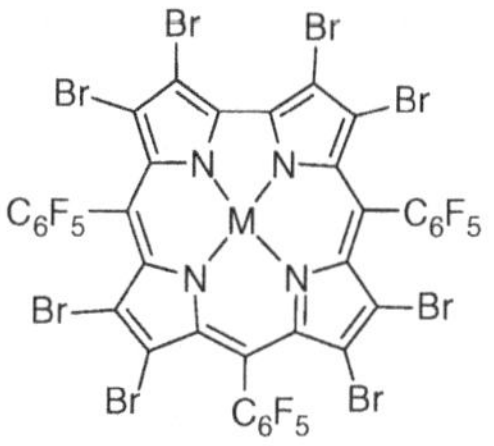

35

M = Ga(py), Al(py)$_2$,Ir((NCH$_3$)$_3$)$_2$, Fe,Mn,Co, Cu,Cr(O)

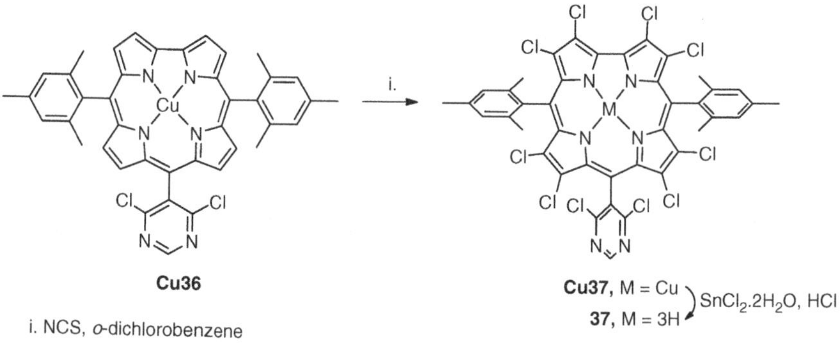

Scheme 9 Preparation of octa-chlorinated corrole derivatives

The formation of the mono-, di-, tri-, and tetra-brominated corroles confirmed the higher reactivity of the β-pyrrolic positions of the directly linked pyrrole units A and D toward B and C.

The synthesis of a fully chlorinated corrole was reported by Maes and co-workers [42, 56] during their studies concerning the functionalization of pyrimidinyl porphyrins (Scheme 9). Treatment of the copper complex **Cu36** with *N*-chlorosuccinimide (NCS), at high temperature, afforded corrole **Cu37** in 46% yield. The access to the corresponding free-base **37** (20%) was also reported under reductive demetallation conditions developed by that group.

The access to the fully chlorinated cobalt(III) corrole **38** was described by Gross and co-workers [57]. Authors have selected for these studies the cobalt(III) complex of 5,10,15-tris(pentafluorophenyl)corrole **4** (Scheme 10); they also refer that the addition of an excess of Cl_2 in a benzene solution induced a color change from green to yellow, which was accompanied by a strong intensity decrease of the Soret band and complete disappearance of the Q-bands. The addition of pyridine and $NaBH_4$ to the reaction mixture caused an immediate color change from yellow to green and the appearance of the characteristic electronic spectrum of a hexa-coordinated cobalt(III) corrole; complex **38** was obtained in about 90% yield after a simple work-up.

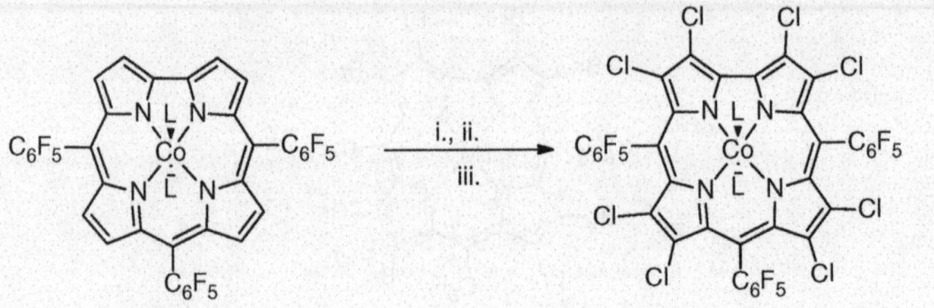

Co4 38

i. Cl_2, benzene
ii. Pyridine
iii. $NaBH_4$

Scheme 10 Chlorination of cobalt(III) complex of 5,10,15-tris(pentafluorophenyl)corrole

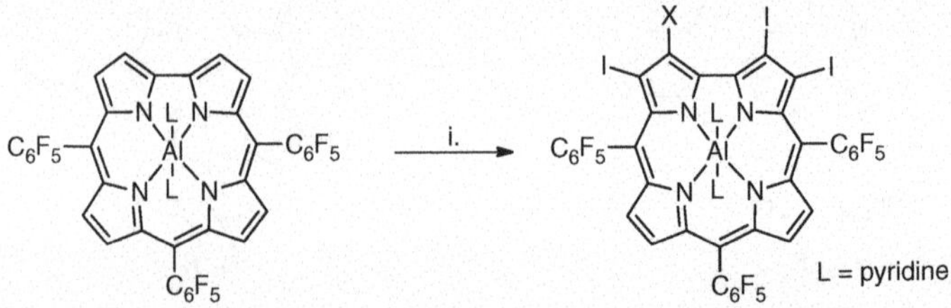

Al4

39, X=H
40, X=I

i. NIS, CH_3CN, rt, 1 h

Scheme 11 Iodination of aluminium(III) complex of 5,10,15-tris(pentafluorophenyl)corrole

Authors referred that the addition of TFA to complex **38** induces dissociation of the axial ligands, which is responsible for the changes also observed in the electrochemical and spectroscopic properties.

Gross and co-workers [58] have also reported the synthesis of corroles **39** and **40**, starting with the Al(III) complex of corrole **4**, using *N*-iodosuccinimide (NIS) or 1,3-diiodo-5,5-dimethylhydantoin (DIH) as the iodinating agent, Scheme 11. The best results were obtained with NIS and it was reported that the tetra-iodinated corrole possesses both short- and long-lived excited states.

2.2 Chlorosulfonic Group

Chlorosulfonation of corroles proved to be highly selective either when free bases or the corresponding complexes are used.

In the first report from Gross and co-workers [59], the reaction was carried out by dissolving 5,10,15-tris(pentafluorophenyl)corrole **4** in chlorosulfonic acid (Scheme 12). The 2,17-substituted derivative **41** was obtained as the main product and it was accompanied by a minor amount of the 3,17-isomer [60].

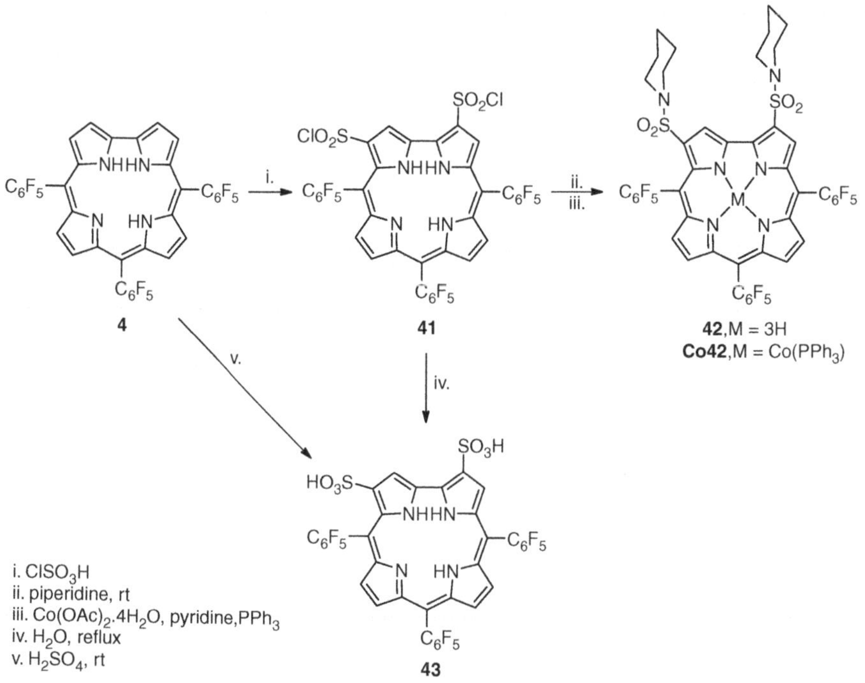

Scheme 12 Chlorosulfonation of 5,10,15-tris(pentafluorophenyl)corrole

Authors mentioned that treatment of the crude reaction mixture with piperidine afforded quantitatively the sulfonamide **42**. The subsequent metallation of **42** with cobalt(II) acetate/triphenylphosphine afforded the (triphenylphosphine)cobalt(III) complex **Co42** (Scheme 12). In the same publication it was reported that the hydrolysis of **41** provided the bis-sulfonic acid derivative **43**, which can also be obtained by direct sulfonation, with concentrated sulfuric acid, at room temperature [61, 62].

The 2,17-disubstituted corrole **41** has been extensively used in medicinal studies. Gross and co-workers have put a huge effort concerning the corroles medical applications. The amphiphilicity of corrole **41** enables a unique approach to bioconjugate formation, whereby the carrier and drug form a stable complex by noncovalent assembly [63]. This brightly fluorescent complex confers promising cytotoxic and antitumor activities, among other properties [63–70].

The chlorosulfonated corrole derivatives have being used as highly versatile precursors for the preparation of corroles with functional groups that may be used for advanced applications and materials. The syntheses of water-soluble carbohydrate corroles **44**, pegylated corroles **45**, donor/acceptor derivatives **46**, amino acid corroles **47**, and corrole–biotin conjugates **48** are examples of such approaches (Scheme 13) [60].

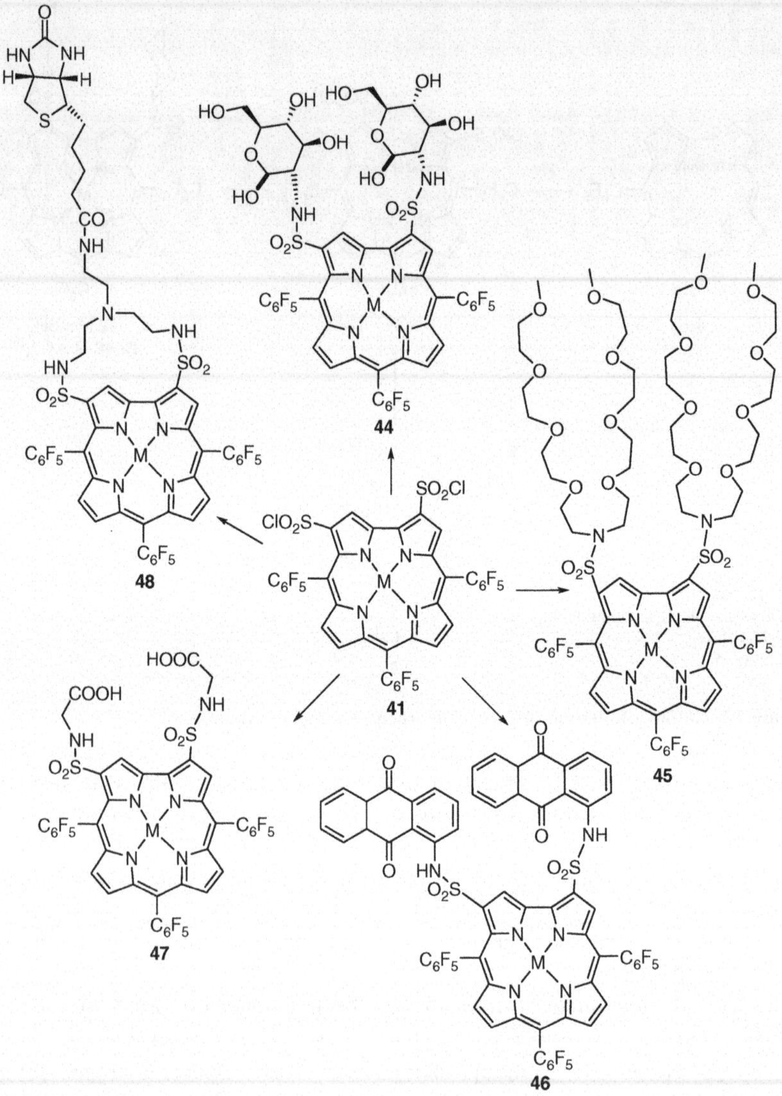

Scheme 13 Functional corroles from chlorosulfonated corrole **41**

2.3 Formyl Group

The work concerning the introduction of a formyl group in a corrole core is based on the use of Vilsmeier–Haack conditions. In fact, the efficiency of the Vilsmeier–Haack reaction is well established giving access to β-formylcorroles in excellent yields.

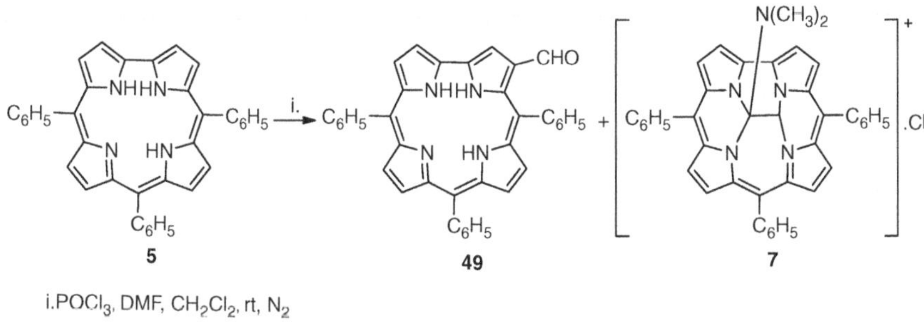

Scheme 14 Vilsmeier formylation of 5,10,15-triphenylcorrole

The formylation of a free base corrole was first reported by Paolesse et al. [29]. The reaction was carried out by adding the Vilsmeier reagent $POCl_3$/DMF (1:110 molar ratio) to a solution of 5,10,15-triphenylcorrole **5** in dry dichloromethane, at room temperature under an inert atmosphere. The expected 3-formyl-5,10,15-triphenylcorrole **49** was obtained as the major product (58%) (Scheme 14). In this reaction, it was also isolated a more polar green compound in 15% yield, whose structure was assigned as being **7**, by X-ray crystallographic characterization. Compound **7** became the major reaction product (60% yield) if a higher amount of DMF was used; under these conditions the formation of **49** was almost suppressed (3% yield). This peculiar reactivity was justified by considering the high acidity of the inner NH, promoting the formation of the corrole anion, which drives the attack of the Vilsmeier reagent to the macrocyclic core, as mentioned before.

This study revealed that free base corroles can be directly formylated, while in the case of a porphyrin it is necessary to carry out the reaction on appropriate metal complexes to avoid the deactivation due to the formation of the macrocycle dication. Giribabu and co-workers extended this reaction to other free base *meso*-arylcorroles [71].

Gross and co-workers [62] described the Vilsmeier formylation of gallium(III) complex of 5,10,15-tris(pentafluorophenyl)corrole **Ga4**. Depending on the corrole: Vilsmeier reagent molar ratio, the 3-formyl derivative **50** and the 2,17-diformyl derivative **52** were obtained as main products, Scheme 15. However, following the same formylation procedure, Cavaleiro and co-workers [72] were able to isolate, together with 3-formyl and 2,17-diformyl derivatives, the 2-formyl derivative **51**. This compound has a slightly higher R_f in silica than **50**. In the same work, X-ray studies confirmed the structure of compound **50** unambiguously.

In this reaction the different selectivity of the β-pyrrolic positions of the corrole macrocycle was once again confirmed. Position 3 of the corrole ring is the most reactive toward electrophilic substitution reactions, even more than the less hindered position 2, followed by position C-18 for the second substitution.

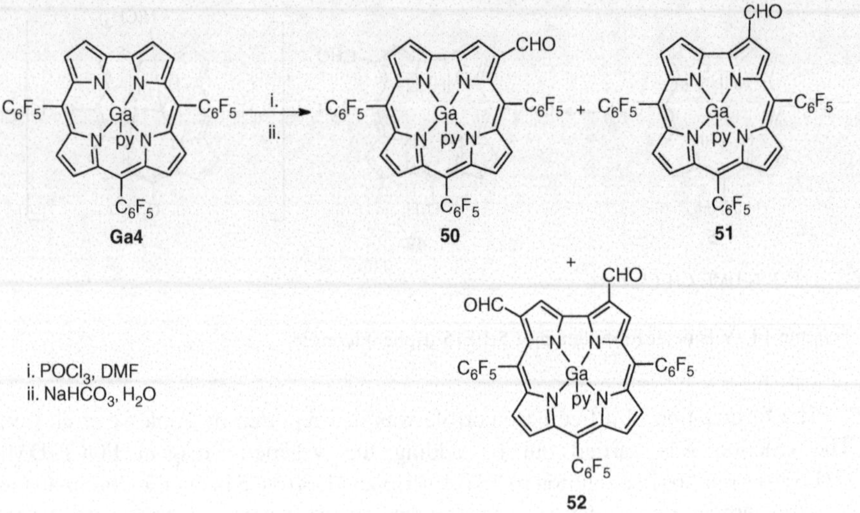

Scheme 15 Vilsmeier formylation of gallium(III) complex of 5,10,15-tris(pentafluorophenyl) corrole

2.4 *Carboxylic Group*

In the first attempt to introduce the carboxylic group in the corrole periphery, Gross and co-workers used phosgene and corrole-free base **4** (Scheme 16) [30]. Under these conditions, as it was already mentioned, the preferred reaction site was the corrole inner core, affording the *N*-substituted corrole **8** or the bridged carbamide **9**, depending on the solvent used (Scheme 16); the *N*-acyl chloride intermediate **53** was considered to be the potential intermediate of these derivatives.

This fact prompted the authors to perform the reaction with the corresponding gallium complex **Ga4**, affording the monocarboxylated derivative **54**, selectively substituted in C-3, in 58% yield, after its separation from bis-substituted corrole derivatives (Scheme 17) [30].

More recently, Giribabu and co-workers [71] described an efficient one-pot conversion of 3-formyl-5,10,15-triarylcorroles **49**, **55–57** and their Cu complexes to the corresponding carboxylic acid corroles **58–61**, as free bases or Cu complexes, using hydroxylamine hydrochloride and phthalic anhydride (Scheme 18).

The products were obtained in yields ranging from 70% to 80%; it was mentioned that these derivatives have potential applications in dye-sensitized solar cells.

Gross and co-workers developed another approach, aiming to introduce polar head groups on the corrole periphery. In this case, 3-formyl corrole derivatives **Ga50** and **Al50** reacted with $CH_2(CN)(CO_2H)$ affording the amphiphilic complexes **62** (Scheme 19) [73]. The authors described the red-shifted absorptions and emissions spectra of such products, which demonstrated to have an excited-state stabilization due to the electron-withdrawing $CH{=}C(CN)(CO_2H)$ group.

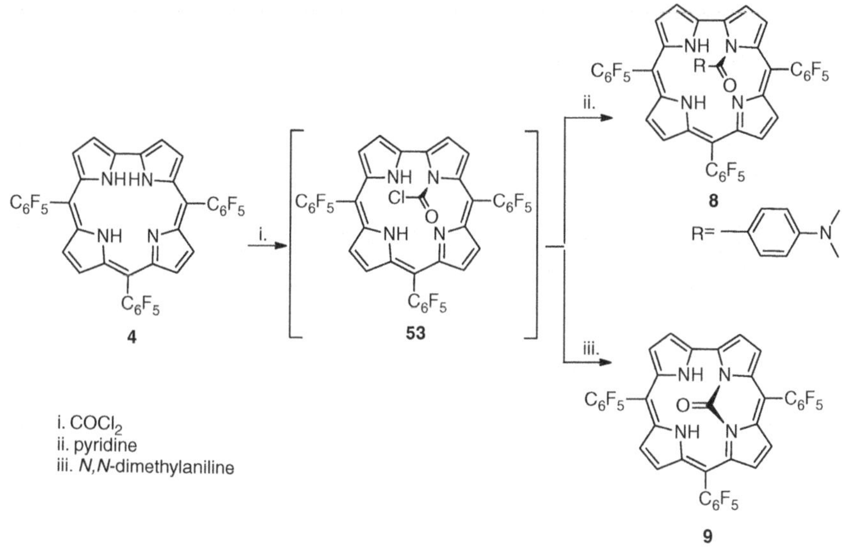

Scheme 16 Attempt of carboxylation reaction with phosgene

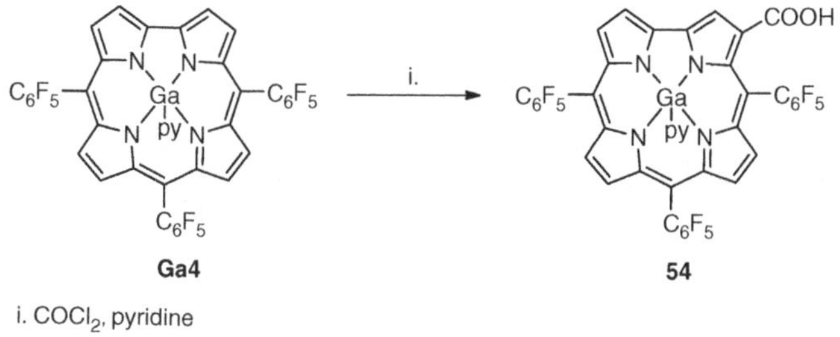

Scheme 17 Synthesis of β-carboxylated gallium(III) complex of 5,10,15-tris(pentafluorophenyl)-corrole

2.5 Vinyl Group

The introduction of a vinyl group in the corrole macrocycle was considered by Santos et al. [74]. The approach was based on a Wittig reaction using the 3-formyl derivative **Ga50** as the precursor. The reaction was performed in dry THF in the presence of methyltriphenylphosphonium bromide and sodium hydride, affording the gallium(III) complex of 3-vinyl-5,10,15-tris(pentafluorophenyl) corrole **63** (Scheme 20).

The structure of the novel corrole **63**, obtained in 74% yield, was unambiguously established by spectroscopic data, namely NMR, UV–visible, elemental analysis; its reactivity as a diene in Diels–Alder reactions will be described in Sect. 5.

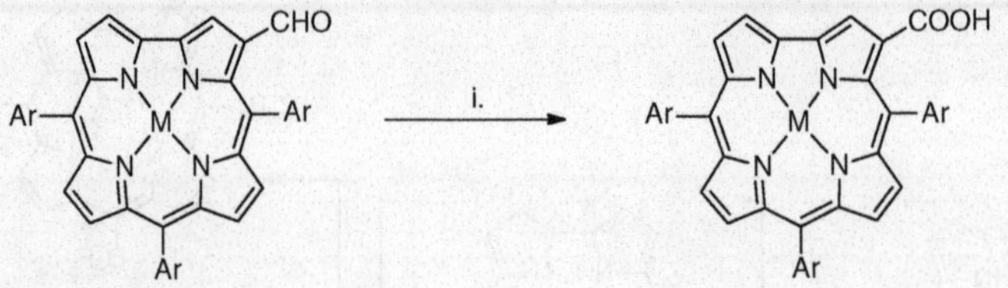

49, M = 3H, Ar = C_6H_5
Cu49, M = Cu(III), Ar = C_6H_5
55, Ar = 4-$CH_3(C_6H_4)$
Cu55, M = Cu(III), Ar = 4-$CH_3(C_6H_4)$
56, Ar = 4-$OCH_3(C_6H_4)$
Cu56, M = Cu(III), Ar = 4-$OCH_3(C_6H_4)$
57, Ar = 3,5-$(CH_3)_2C_6H_3$
Cu57, M =Cu(III), Ar = 3,5-$(CH_3)_2C_6H_3$

58, M = 3H, Ar = C_6H_5
Cu58, M = Cu(III), Ar = C_6H_5
59, Ar = 4-$CH_3(C_6H_4)$
Cu59, M = Cu(III), Ar = 4-$CH_3(C_6H_4)$
60, Ar = 4-$OCH_3(C_6H_4)$
Cu60, M = Cu(III), Ar = 4-$OCH_3(C_6H_4)$
61, Ar = 3,5-$(CH_3)_2C_6H_3$
Cu61, M =Cu(III), Ar = 3,5-$(CH_3)_2C_6H_3$

i. $NH_2OH.HCl$; $N(C_2H_5)_3$, phthalic anhydride

Scheme 18 One-pot conversion of 3-formyl corrole derivatives into the corresponding carboxylic acid corroles

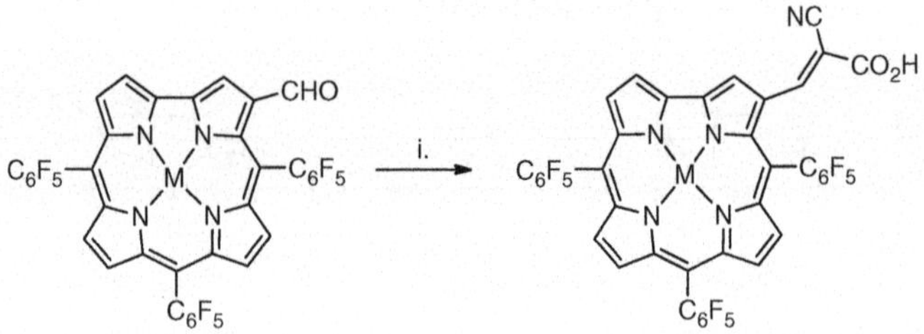

Ga50, M = Ga(py)
Al50, M = $Al(py)_2$

Ga62, M = Ga(py)
Al62, M = $Al(py)_2$

i. $CH_2(CN)(CO_2H)$, DMF, reflux

Scheme 19 Synthesis of amphiphilic derivatives **62**

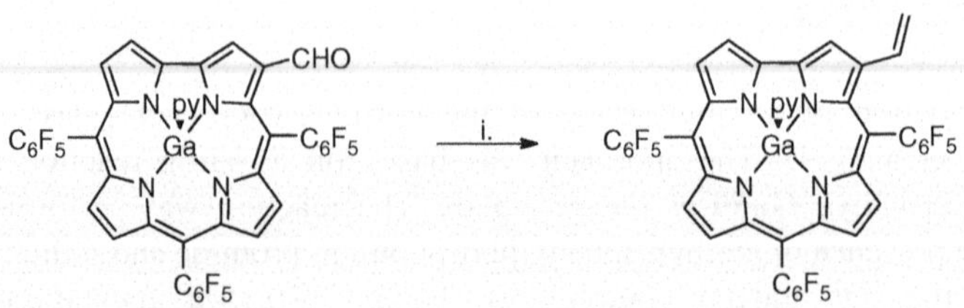

Ga50 **63**

i. CH_3PPh_3Br, NaH, THF, rt, N_2

Scheme 20 Wittig reaction from the gallium(III) complex of 3-formyl corrole derivative

2.6 Nitro Group

Among the possible β-corrole functionalizations, nitration is considered an appealing one, because the nitro group is an useful starting function for further developments.

In 2002 Gross and co-workers [62] reported that the reaction of the Ga(III) complex of 5,10,15-tris(pentafluorophenyl)corrole **Ga4** with $NaNO_2$ in acetonitrile, followed by the addition of a hexachloroantimonate salt, as a one-electron oxidant, afforded the 3-nitro **64**, the 3,17-dinitro **65**, and the 2,3,17-trinitro **66** derivatives (Scheme 21), depending on the amount of the nitrating agent.

Authors also highlighted that, even with only 75 mol% of the hexachloroantimonate salt, the starting material was fully consumed, and the mono and dinitro corroles **64** and **65** were isolated in 84% and 9% yield, respectively, which is clearly indicative of a chain reaction. Accordingly, the authors suggested that under these reaction conditions, where a very large excess of $NaNO_2$ relatively to corrole is used, the hexachloroantimonate oxidizes NO_2^- to NO_2 rather than the Gallium(III) complex to the corresponding π-cation radical.

In the same publication it was also mentioned that nitrating systems successfully used in the case of tetraarylporphyrins, such as HNO_3/H_2SO_4, N_2O_4, or $AgNO_2/I_2$, in the corrole case led to significant decomposition of the corrole ring and to mixtures of polynitrated derivatives.

The nitration of metal complexes, namely of germanium(IV) complex of 5,10,15-triphenylcorrole **5**, was also achieved by Paolesse and co-workers using two different methods. In one approach, authors reported that the use of the mild nitrating system, $LiNO_3/Ac_2O/AcOH$ in CH_2Cl_2, led to the formation of the monosubstituted germanium μ-oxo dimer **67** and of the corresponding Ge(IV) monosubstituted monomer **68a** bearing a hydroxyl group at the axial position, Scheme 22. Treatment of the μ-oxo dimer **67** with dilute HCl also led to a mixture of monomeric corroles **68**, with Cl^- and OH^- as axial ligands. The crystallization of this mixture from CH_2Cl_2/CH_3OH gave the corresponding methoxy corrole **68c**, in 45% yield. The other approach involved the use of $NaNO_3$ in $Ac_2O/AcOH$, a more severe nitrating mixture, aiming to obtain polysubstituted products. Under these conditions, the two main products identified were the μ-oxo dimers **67** and **69**, Scheme 22. An additional fraction also isolated, contained the 3,17 disubstituted monomers **70a**, **b** (X = OH or Cl), which were directly converted into the corresponding methoxy derivative **70c**, by crystallization from CH_2Cl_2/CH_3OH.

The substitution is highly regioselective in each case, giving only the 3-nitro or 3,17-dinitro derivatives among the different possible isomers [75]. Studies on the electrochemistry and spectroelectrochemistry of the Ge(IV) corrolate complexes were also carried out by the same authors.

Paolesse and co-workers selected the combination of silver nitrite/sodium nitrite to nitrate 5,10,15-tris(4-*tert*-buthylphenyl)corrole **71** (Scheme 23). The first step involves the reaction with $Cu(OAc)_2$ in refluxing pyridine to obtain the corresponding metal complex, followed by the addition of a mixture of $AgNO_2$

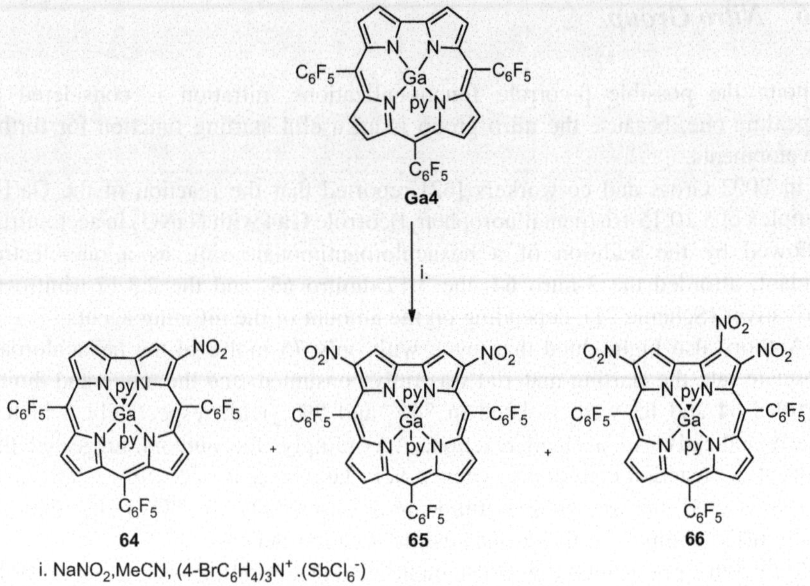

Scheme 21 Nitration of gallium(III) complex of 5,10,15-tris(pentafluorophenyl)corrole

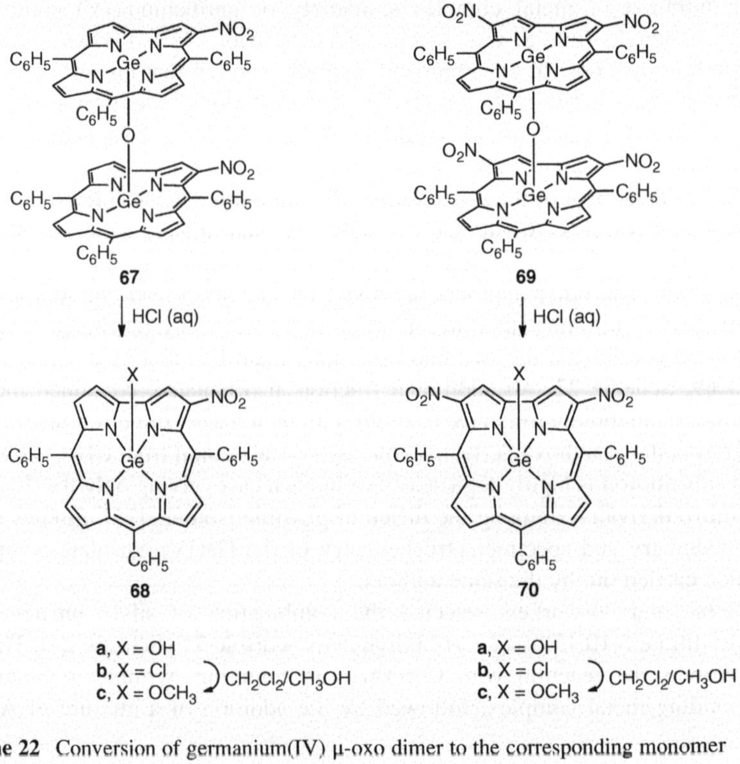

Scheme 22 Conversion of germanium(IV) μ-oxo dimer to the corresponding monomer

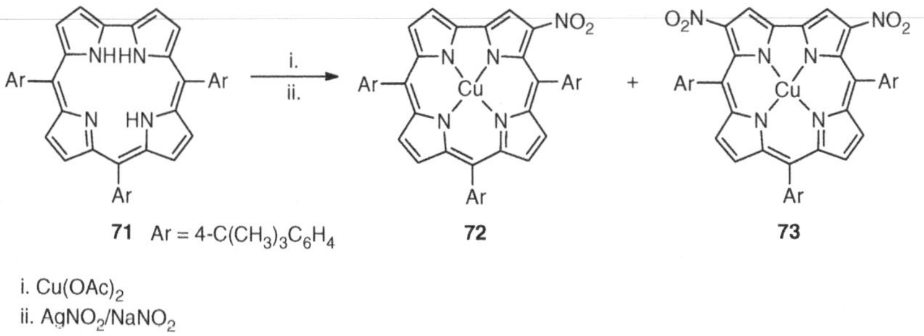

Scheme 23 Nitration of 5,10,15-tris(4-*tert*-buthylphenyl)corrole using $AgNO_2/NaNO_2$

and $NaNO_2$. Authors found that the best reaction condition to obtain corrole **72** was the use of Cu complex/silver nitrite/sodium nitrite in a molar ratio of 1:1:9. Under these conditions the 3-nitrocorrole **72** was obtained as the main product, together with the corresponding 3,17-dinitroderivative **73**, in 75% and 15% yields, respectively [76].

The same authors verified that the use of $TFA/NaNO_2$ or $HCl/NaNO_2$ systems afforded, initially, β-nitroisocorroles that can be converted into the corresponding aromatic corrole complexes by cobalt insertion [77]. Corrole **5** was dissolved in TFA and 10 or 100 equiv. of $NaNO_2$ were added. At this stage, the purification of the reaction products, identified as nitro-substituted isocorroles, was quite difficult. In this way and because the metal coordination to the macrocycle drives the re-aromatization of isocorrole to corrole [78], the authors performed the metallation of the mixture with $Co(AcO)_2$ and PPh_3, after preliminary work-up. Compounds **74–76** were isolated in low yields, Scheme 24.

However, the large decomposition seen in the reactions represented a serious drawback of this method, which led the authors to exploit a milder nitrating system. In this way, corrole-free base was reacted with 32 equiv. of $NaNO_2$ in the presence of HCl, for 5 min (Scheme 25), affording 3-NO_2-5-OH-isocorrole **77** and 3-nitrosubstituted Co(III) corrole complex **Co68**, besides several decomposition products.

In 2007, Paolesse and co-workers [79] studied the nitration of a series of corroles in the presence of $AgNO_2/I_2$ (Scheme 26). It was verified that in the presence of a large excess of $AgNO_2$, the nitration and metallation of the macrocycle occurred simultaneously, affording the corresponding Ag(III) 3-nitrocorrole **Ag68**, **Ag72**, and **Ag79** as the main reaction products. The authors put clear that the nitration of the macrocycle was clearly favored by the presence of electron-releasing groups on the three phenyl rings. In the case of **5**, **29**, and **71**, good yields of the corresponding mononitro derivatives **Ag68**, **Ag72**, and **Ag79** were obtained accompanied by traces of the non-substituted silver complexes **Ag5**, **Ag29**, and **Ag71** (Scheme 26). However, with corroles **4** and **80** bearing electron-withdrawing groups, the principal occurrence was the decomposition of the corrole ring, affording open chain derivatives.

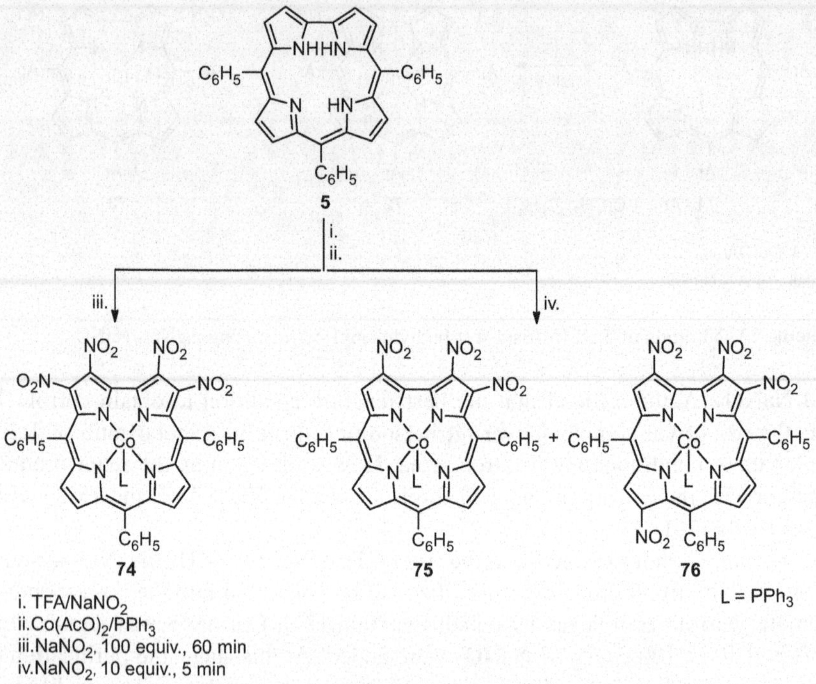

Scheme 24 Nitration of 5,10,15-triphenylcorrole using TFA/$NaNO_2$

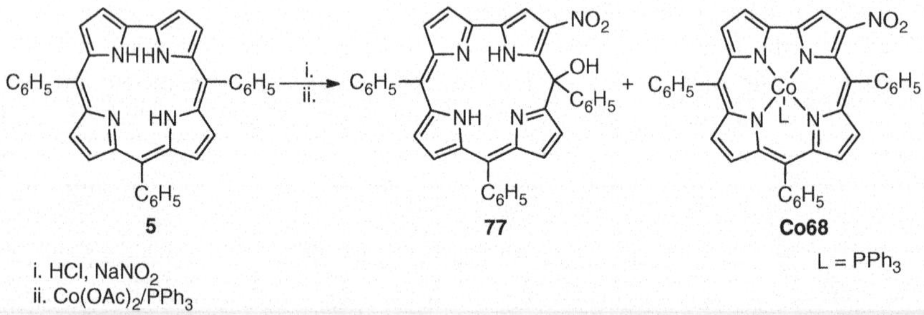

Scheme 25 Nitration of 5,10,15-triphenylcorrole using HCl/$NaNO_2$

During this study the influence of certain reagents was also investigated, and the authors proved that the presence of the silver ion is crucial for the success of the reaction.

The above results support the proposed reaction pathway shown in Scheme 27 where the nitrating agent is the NO_2^- ion which attacks the Ag(III) π-cation radical **Ag5⁺**, formed by oxidation with excess of the Ag^+ ion. Then, a second one-electron oxidation takes place and the loss of a proton restores the corrole aromaticity, affording **Ag68**. The role of the silver ion in this macrocycle oxidation

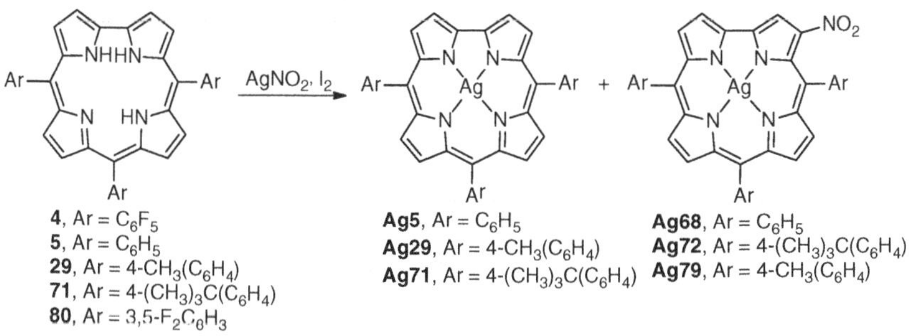

Scheme 26 Nitration of *meso*-triarylcorroles with $AgNO_2/I_2$

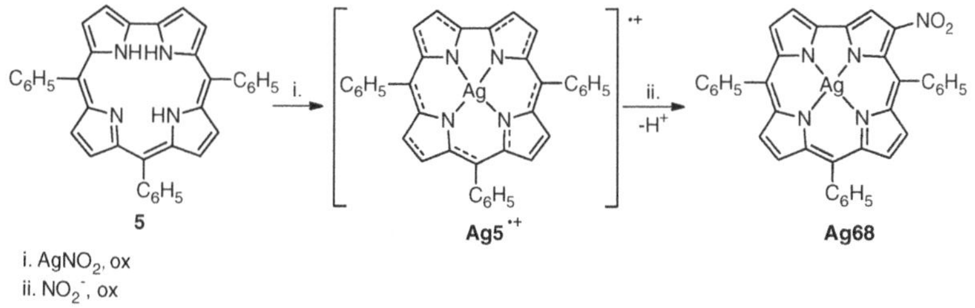

Scheme 27 Proposed nitration pathway

has earlier been reported for the nitration of β-octaalkylporphyrins at the *meso* positions using $AgNO_2$ [80].

The preparation of nitrocorrole-free bases by reductive demetallation procedures has been envisaged by Paolesse and co-workers. For instance, 3-(NO_2) corrolates were obtained via reductive demetallation of silver corrole complexes under basic conditions (DBU/THF) [40].

In another publication, the same group reported that the reaction of 5,10,15-tritolylcorrole **29** with a mixture of sodium and silver nitrites in a molar ratio of 1:1:9 (corrole/$AgNO_2$/$NaNO_2$) in refluxing DMF gave access to the 3-nitro-substituted free base corrole **79** as the main product (52% yield), and to 2-NO_2-corrole **81** as a minor one (Scheme 28) [81]. Changing the molar ratio to 1:2:8 (corrole/$AgNO_2$/$NaNO_2$) the dinitro 3,17-disubstituted corrole **82** was the main product obtained, which was accompanied by traces of 2,3- and 3,12-corrole derivatives **83** and **84**.

Authors highlighted that the use of the $AgNO_2$/$NaNO_2$ system allowed the β-functionalization of the macrocycle to take place, without inducing the concomitant metallation, which occurs when the reaction is carried out with an excess of silver nitrite. In the former case, $AgNO_2$ acts only as an oxidant, and if an adequate amount is chosen, there will be no metallation of the macrocycle.

Authors studied the influence of the β-nitro substituents on the corrole properties; this was performed by UV–visible spectroscopy and electrochemical,

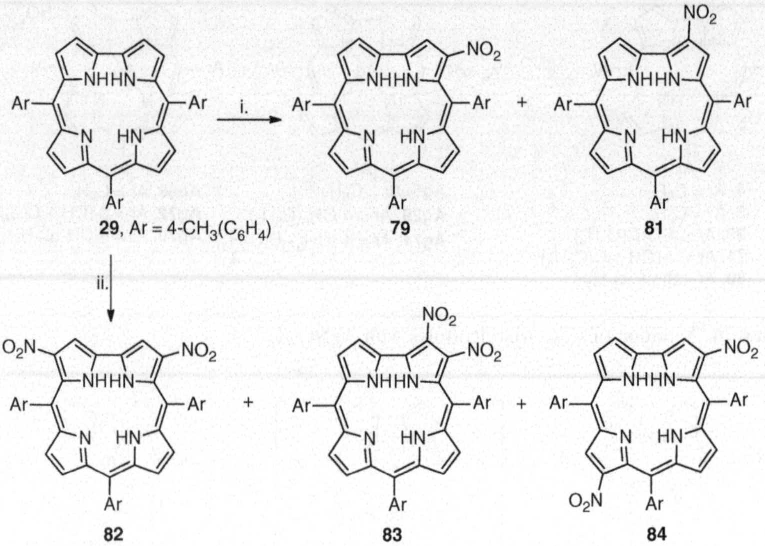

Scheme 28 Preparation of β-nitro corrole derivatives

spectroelectrochemical studies of these functionalized corroles. Paolesse and co-workers concluded that the introduction of nitro substituents at the β-pyrrole positions of the corrole ring strongly influences the chemical and spectroscopic behavior of the macrocycle. The strong electron-withdrawing character of the nitro group leads to a positive shift of the $E_{1/2}$ of the redox processes of corrole and to an increase in the macrocycle acidity. The optical absorption spectra of β-nitrocorroles are strongly influenced by the peripheral nitro groups, which increase the number of bands, and give rise to significant red shifts.

From the theoretical results from density functional theory (DFT) and time-dependent DFT (TDDFT) calculations of the ground and excited states, the authors concluded that the β-NO_2 substituents conjugate with the π-aromatic system of the macrocycle, resulting in significant changes in both the spectroscopic and redox properties of such functionalized corroles. This effect is more pronounced when the nitro group is introduced at the 2-position, because in this case the conjugation is, for steric reasons, more efficient than that in the 3-nitro isomer.

The regioselective nitration of free base corroles **4**, **5**, **19**, **29**, **85**, and **86** leading to the corresponding β-substituted nitrocorrole iron complexes **Fe65**, **Fe 68**, **Fe70**, **Fe79**, **Fe82**, and **Fe(87–91)** have also been reported (Scheme 29) [82].

Authors mentioned that the two-step procedure involving that synthesis of the iron complexes did afford three types of Fe(III) nitrosyl products, the unsubstituted iron complexes **Fe4**, **Fe5**, **Fe19**, **Fe29**, **Fe85**, and **Fe86**, the 3-nitrocorroles **Fe68**, **Fe79**, **Fe87**, and **Fe88**, and the 3,17-dinitrocorroles **Fe65**, **Fe70**, **Fe82**, and **Fe(89–91)**. In contrast, the one-pot synthetic approach drives the reaction almost exclusively to the formation of the iron nitrosyl 3,17-dinitrocorrole **Fe65**, **Fe70**,

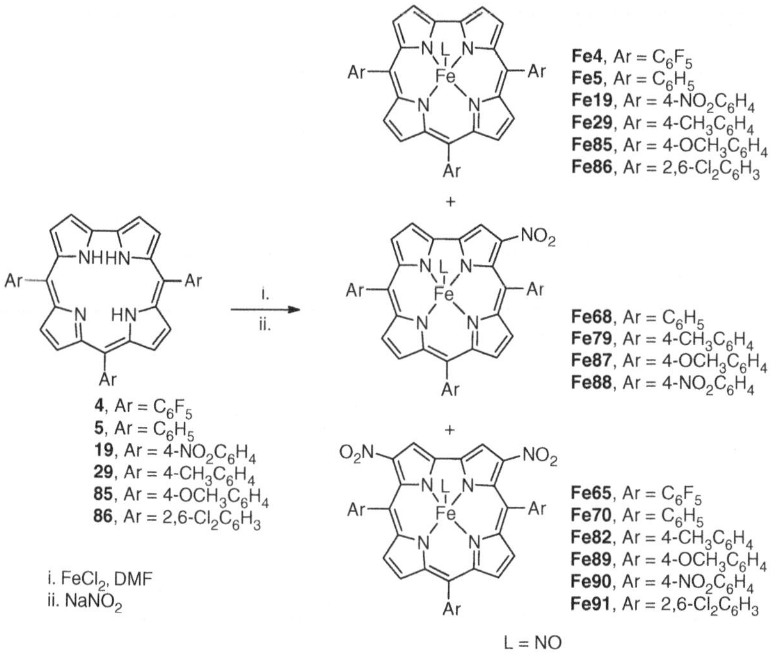

Scheme 29 Preparation of iron(III) nitrosyl products

Fe82, and **Fe(89–91)**. Electron-releasing substituents on the *meso*-aryl groups of the triarylcorroles induce higher yields and longer reaction times than for the synthesis of similar triarylcorroles with electron-withdrawing functionalities; these results can be attributed to the facile formation and stabilization of an intermediate iron corrole π-cation radical. Electron-withdrawing substituents on the *meso*-aryl groups of triarylcorrole also seem to labilize the axial nitrosyl group which, in the case of the pentafluorophenylcorrole derivative **4**, results in the direct formation of a disubstituted iron μ-oxo dimer complex. The influence of *meso*-aryl substituents on the progress and products of the nitration reaction was also investigated. In addition, to elucidate the most important factors which influence the redox reactivity of these different iron nitrosyl complexes, a detailed study on electrochemistry and spectroelectrochemistry of β-nitro substituted iron corroles was performed, thus elucidating the site of electron transfer and the influence of the peripheral nitro groups.

According to the literature data, the results obtained in the substitution reactions described above reveal that the first replacement always occurs at C-3 and the second substitution occurs at C-17 (nitration) or C-18 (formylation and chlorosulfonation). According to crystallographic data, the discrepancy between C-2 and C-3 is related to electronic effects, making position C-3 richer in electrons in relation with the other one [62].

2.7 Amino Group

The possibility of using corroles bearing amino groups as starting materials for further functionalization has also been considered by some research groups. In analogy with the work developed in the porphyrin field (see [83]), certain procedures leading to amino-substituted compounds are based on the reduction of nitro groups. For instance, Collman and Decréau [84] have selected the 5,10,15-tris (2-aminophenyl)corrole **93** (Scheme 30) for the synthesis of a series of free base hemoprotein corrole derivatives. Corrole **93** was obtained as a mixture of atropisomers (αβα, ααβ, ααα) in excellent yield by reduction of the corresponding 2-nitrophenylcorrole **92** with $SnCl_2.2H_2O$ in HCl (Scheme 30). The reaction of the ααα atropisomer obtained after enrichment of the statistical mixture with the adequate acyl chlorides (pivaloyl chloride and *N*-methyl imidazole acyl chloride) afforded the picket fence corroles **94** and **95**. Using a similar strategy the synthesis of capped and strapped corroles was also reported.

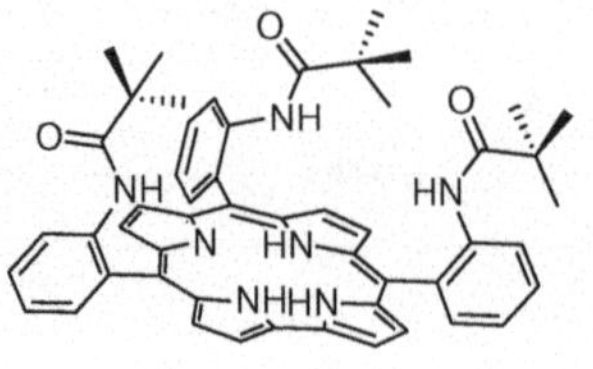

94

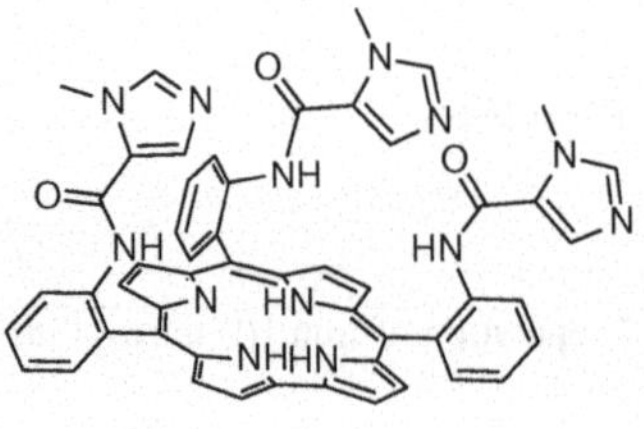

95

The same group explored the methodology to prepare the 5,15-bis(pentafluorophenyl) corroles **96** bearing an imidazole ligand covalently attached to the corrole macrocycle. The efficiency of these metallocorroles in O_2 electroreduction using the rotating ring disk electrode method was reported [85].

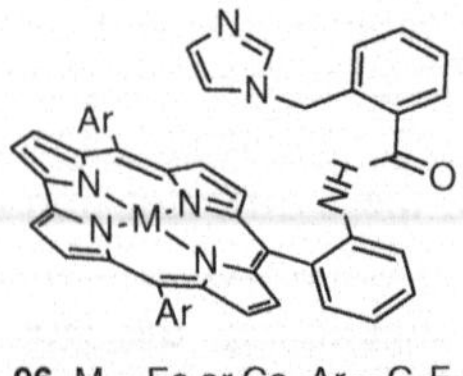

96, M = Fe or Co, Ar = C_6F_5

Corroles bearing amino groups in *para* positions of the phenyl substituents were also considered in the design of organic–inorganic hybrid materials. For instance, Barbe et al. [86] selected the mono-functionalized amino-corrole **97** to introduce the triethoxysilyl function in order to obtain hybrid materials through sol–gel processes (Scheme 31). Corrole **97** was obtained by reduction of the nitro group in the corresponding 5,15-dimesityl-10-(4-nitrophenyl)corrole with $H_2/Pd/C$; its reaction

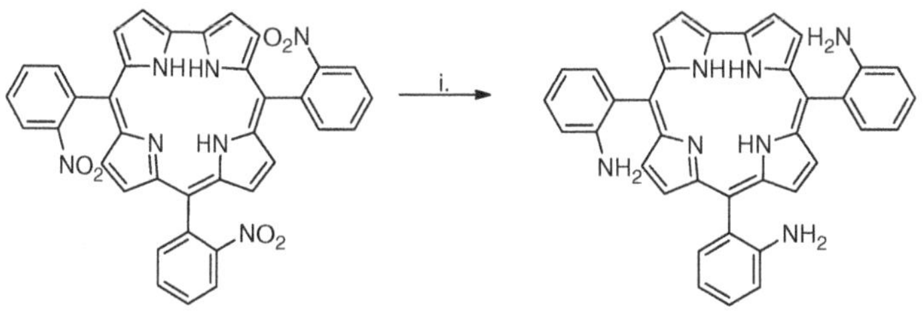

92

93
as a mixture of
αβα (25%), ααβ (60%), ααα (5%)

i. $SnCl_2.2H_2O$, HCl 37%, 70 °C

Scheme 30 Preparation of amino-substituted corroles

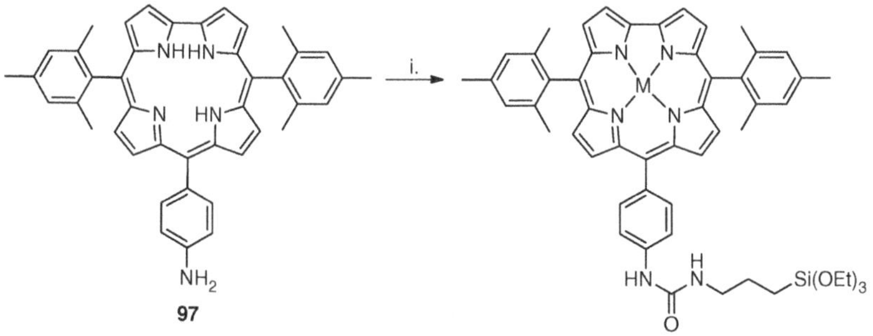

97

98, M = 3H
Co98, M = Co

i. $Si(OEt)_3(CH_2)_3NCO, CH_3CN, NEt_3$, 4 days
ii. $Co(acac)_2, CH_2Cl_2$, MeOH

Scheme 31 Functionalization of aminocorrole with (3-isocyanotopropyl)triethoxysilane

with (3-isocyanatopropyl)triethoxysilane afforded, after 4 days, compound **98** in 68% yield. The free base **98** and the corresponding cobalt complex **Co98** obtained after metallation with cobalt(II) acetylacetonate were reacted with tetraethoxysilane (TEOS) and methyltriethoxysilane (MTEOS) by following well-established sol–gel process conditions. The hybrid materials, with the Co(III) corrole incorporated, have shown very high affinity for carbon monoxide when used as gas sensors.

The same group [87] extended the studies to the synthesis of corroles **101** bearing the more reactive benzylamino group at the *meso*-position (Scheme 32). The synthetic strategy for the synthesis of corroles **100** involved the coupling of the aldehyde bearing the azido group with the adequate dipyrromethanes **99** in the presence of TFA, followed by oxidation with DDQ. The reduction of the azido group with triphenylphosphine afforded, after hydrolysis, the desired corroles **101a, b.**

It was highlighted that the condensation of corroles **101** with (3-isocyanatopropyl) triethoxysilane to afford the desired precursors **102** of organic–inorganic hybrid materials was much faster (12 h of reaction) than the one observed with corrole **98.**

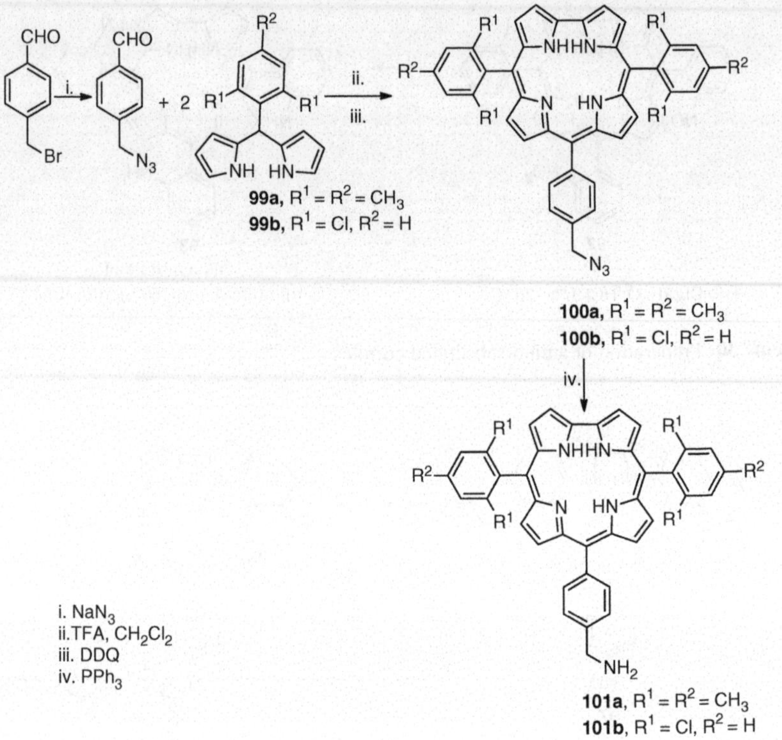

Scheme 32 Synthesis of corrole derivatives bearing benzylamino group

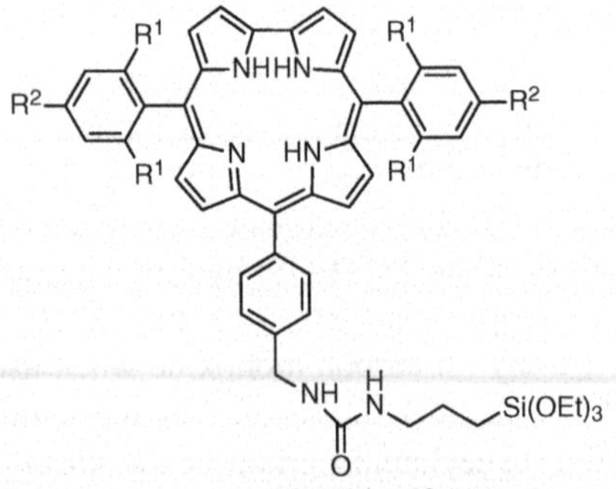

102a, $R^1 = R^2 = CH_3$
102b, $R^1 = Cl, R^2 = H$

In the same publication it was also reported the synthesis of the bis-functionalized corrole **103** as a precursor of other organic–inorganic hybrid materials. These hybrids were obtained by a sol–gel process or by grafting the cobalt(III) complexes of these corroles onto mesostructured silicas and also of others in which the triethoxysilyl chain was linked to a hydroxy group. These materials have also shown a high selectivity for CO [88].

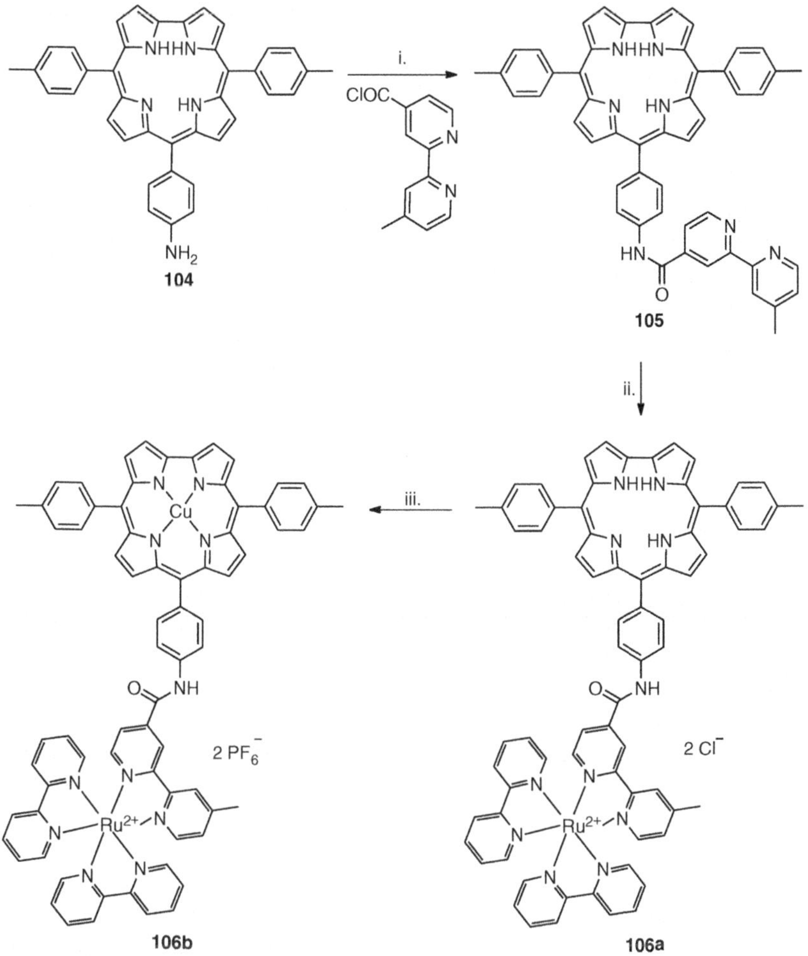

i. Et_3N, CH_2Cl_2, 0°C
ii. $Ru(bpy)_2Cl_2$,AcOH, reflux
iii. $Cu(OAc)_2.H_2O$, pyridine, NH_4PF_6

Scheme 33 Synthesis of photoactive dye bearing a ruthenium(II)tris(bipyridine)

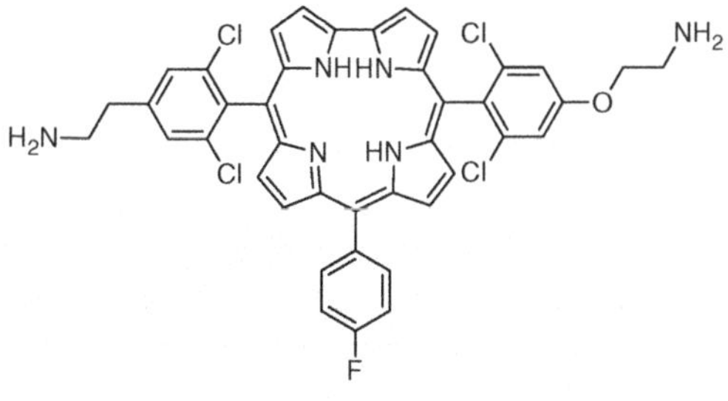

103

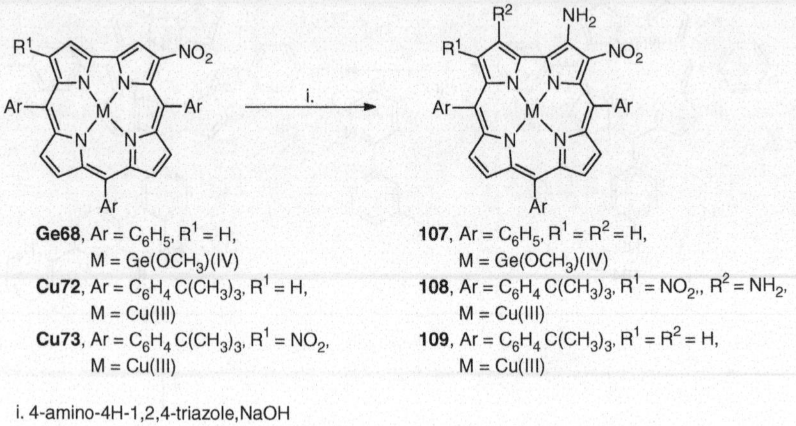

Scheme 34 Amination on 3-nitrocorrole with 4-amino-4*H*-1,2,4-triazole

Corrole **104** bearing the amino group in the *para* position of the phenyl substituent was a key synthon in the synthesis of the photoactive-dyad **106a**, **b** bearing a ruthenium(II)tris(bipyridine) moiety [89]. The electrochemical and photophysical properties of **106b** suggest that an electron transfer from the Cu(III)-corrole unit to a photogenerated ruthenium species is thermodynamically possible (Scheme 33).

The introduction of the amino functionality directly in electrophilic centers of the corrole core by following Callot procedures for porphyrins [90] was reported by Paolesse and co-workers [76]. The authors performed the amination on 3-nitrocorrole **Cu72** with 4-amino-4*H*-1,2,4-triazole in the presence of NaOH using a molar ratio of 1:12:5, according to the protocol employed for the preparation of 2-amino-3-nitroporphyrins [91]. In that reaction, after 45 min, the **Cu72** was completely consumed, and the crude reaction mixture was purified by column chromatography affording a single brownish fraction that was identified as being the desired 2-amino-3-nitrocorrole **109**. The extension of such experimental conditions to 3,17-dinitrocorrole **Cu73** afforded the tetra-substituted 2,18-diamino-3,17-dinitrocorrole **108** (Scheme 34).

Authors also reported that in the amination of germanium(IV) nitrocorrole complex **Ge68** the derivative **107** was obtained in 50%. The structure of this derivative was unambiguously confirmed by X-ray crystallography.

3 Nucleophilic Substitutions

The nucleophilic aromatic substitution methodology of the *para*-F substituents widely explored to functionalize 5,10,15,20-tetrakis(pentafluorophenyl)porphyrin [92] has also been considered for 5,10,15-tris(pentafluorophenyl)corrole **4** to introduce several functionalities on that corrole derivative.

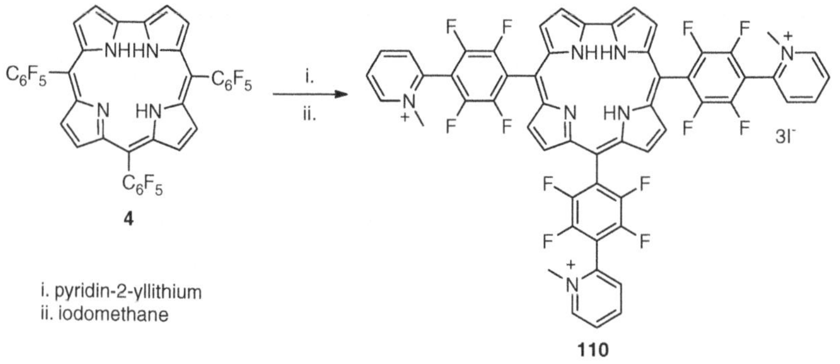

Scheme 35 Synthesis of a cationic trisubstituted corrole

For instance, Gross et al. [3] showed that the reaction of **4** with 2-pyridyl-lithium gives, after methylation with iodomethane, the cationic trisubstituted product **110** (Scheme 35). This cationic corrole revealed to be very efficient in terms of inhibiting endothelial cell proliferation and tumor progression [93].

Recently Osuka and co-workers [94] examined the nucleophilic substitution reactions of 5,10,15-tris(pentafluorophenyl)corrole **4** as a post-modification route to obtain new functionalized corroles. The group reported that corrole **4** reacts with an excess of primary or secondary amines affording the corresponding 5,10,15-tris(4-amino-2,3,5,6-tetrafluorophenyl)-substituted corroles **111** in good yields (Scheme 36). However, diisopropylamine and dibenzylamine did not afford the expected tris(amino)derivative, probably due to their steric hindrance. Through this work it was also proved that the 10- and 15-pentafluorophenyl groups are less reactive than the one at the 5-position. This selectivity was explained by DFT calculations.

Cavaleiro's group explored the approach to introduce galactose residues in the *para* positions of the pentafluorophenyl rings of corrole **4** (Scheme 37). The synthetic strategy involved the reaction of **4** with the commercially available 1,2:3,4-di-*O*-isopropylidene-α-D-galactopyranose, in dry toluene and in the presence of a base (NaH or K_2CO_3). The new corrole–galactose conjugates **112** were isolated in moderate yields. The photodynamic potentialities of these compounds were tested in Jurkat cells, and it was found that the presence of the sugar moieties increase the uptake by the cells [95].

Using this same approach the gallium(III) complex of 5,10,15-tris(pentafluorophenyl) corrole **Ga4** was covalently linked to silica particles, previously functionalized with APTES. This was one of the first reported examples on corrole particles [96].

Maes and co-workers considered the nucleophilic aromatic substitution of chloro atoms in 4,6-dichloropyrimidin-5-yl units of *meso*-pyrimidinyl-substituted corroles as a post-functionalization of this type of platform. The first studies were performed with the copper complex **Cu36** of type *trans*-A_2B (B = 4,6-dichloropyrimidin-5-yl) due to the inherent low stability of the corresponding free base [56, 97].

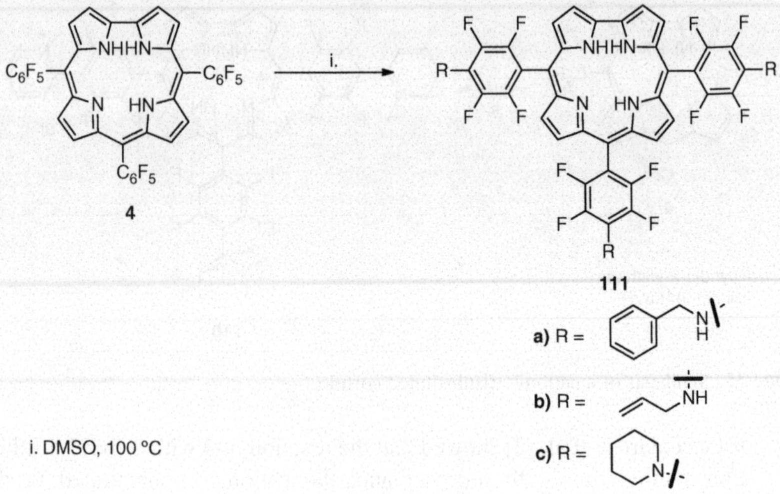

Scheme 36 Synthesis of 5,10,15-tris(4-amino-2,3,5,6-tetrafluorophenyl)-substituted corroles

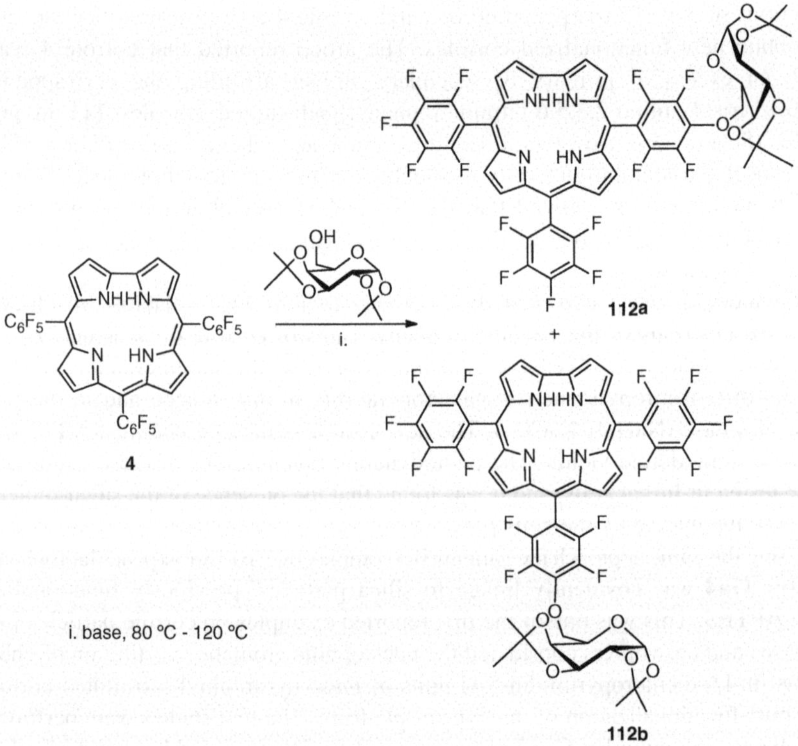

Scheme 37 Synthesis of corrole-galactose conjugates

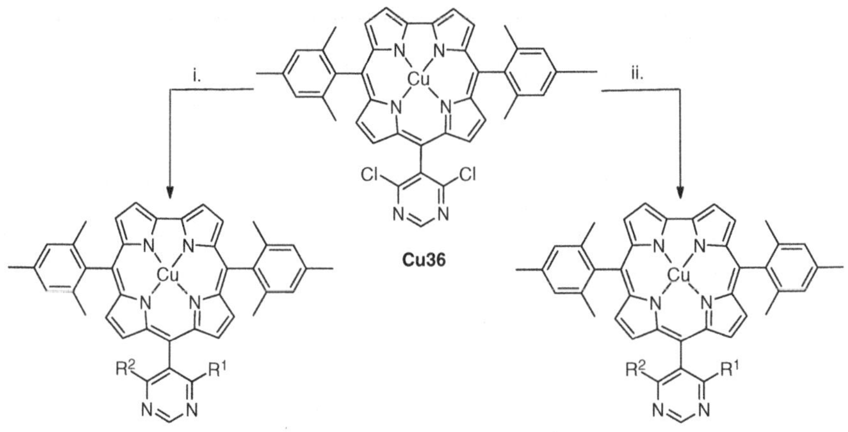

113, R^1 = Cl; R^2 = 4-*t*Bu-C_6H_4O **114,** R^1 = R^2 = 4-*t*Bu-C_6H_4O

i. 4-*tert*-butylphenol, K_2CO_3, DMF
ii. 4-*tert*-butylphenol, K_2CO_3, DMSO, MW

Scheme 38 Nucleophilic aromatic substitution of 4,6-dichloropyrimidin-5-yl units of *meso*-pyrimidinyl-substituted corroles

Authors verified that the reaction of corrole **Cu36** with 4-*tert*-butylphenol (4 equiv.) at 90°C in DMF with K_2CO_3, in a 48 h reaction, afforded the monosubstituted *trans*-A_2B-corrole **113** as the only product in 87% yield. Only under harsh substitution conditions – DMSO, K_2CO_3, 6 equiv. of 4-*tert*-butylphenol, 175°C (microwave irradiation) – the desired substituted A_2B-corrole **114** was obtained in 85% yield (Scheme 38).

The work was extended to other nucleophiles like 3-cyanophenol and 4-chlorothiophenol; the corresponding substituted derivatives **115–119** were obtained. The possibility of introducing two different phenolic units was also considered by reacting the mono-substituted corrole **115** with 4-bromophenol. The low yield (28%) obtained for the asymmetrically substituted corrole **120** was justified by the reversible character of the S_NAr process since mono- and bis-(4-bromophenoxy)-substituted corroles were also isolated. This problem can be minimized if the introduction of one of the substituents takes place by Suzuki cross-coupling (vide infra) and the other one by S_NAr. The access to representative functionalized free base corroles was also considered by reductive demetallation with tin(II) chloride in acidic medium. The absorption spectra and the photophysical properties of the studied free bases suggested that in some cases a partial charge-transfer character from the corrole to the pyrimidinyl group can contribute to the emissive properties.

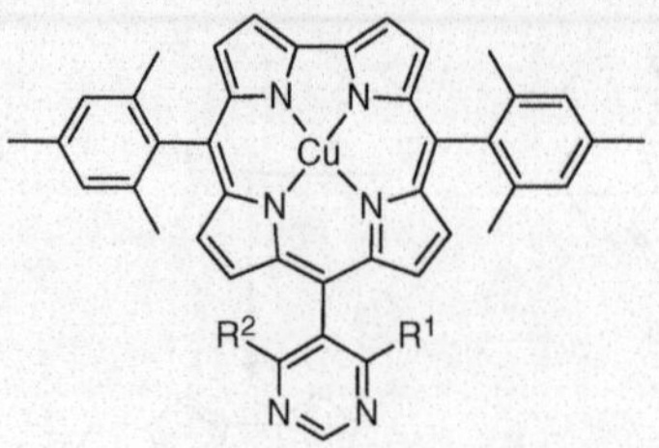

115, R^1 = Cl; R^2 = 3-CN-C_6H_4O
116, R^1 = R^2 = 3-CN-C_6H_4O
117, R^1 = Cl; R^2 = 4-Br-C_6H_4O
118, R^1 = R^2 = 4-Br-C_6H_4O
119, R^1 = Cl; R^2 = 4-Cl-C_6H_4S

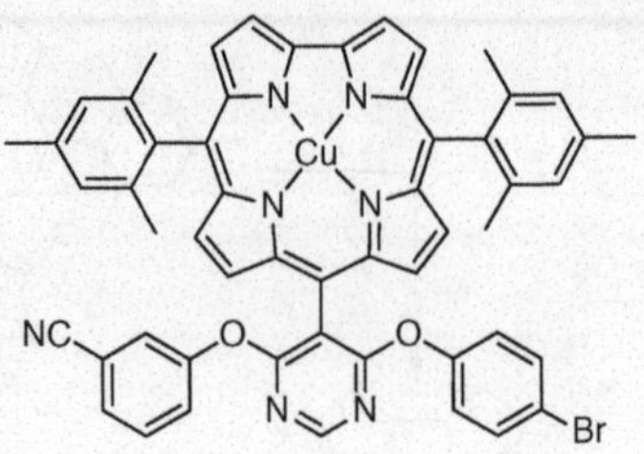

120

The same group of researchers extended the studies to *meso*-pyrimidinyl-substituted corroles of types A_2B and A_3 (A = 4,6-dichloropyrimidin-5-yl). It was shown that due to the higher stability of these corroles the products **121** and **122** were obtained in excellent yields directly from the free bases by reaction with 4-*tert*-butylphenol in DMF under microwave irradiation [98].

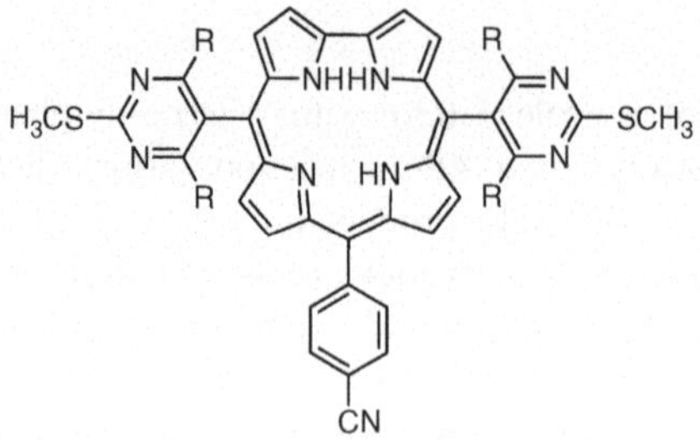

121, R = 4-*t*Bu-C_6H_4O

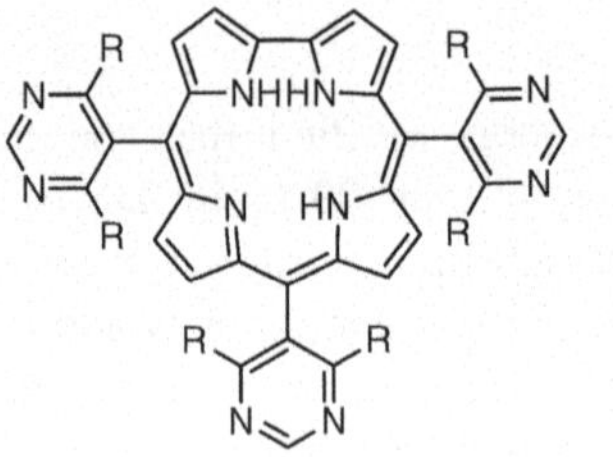

122, R = 4-*t*Bu-C_6H_4O

4 Cycloaddition Reactions

The functionalization of the corrole core via cycloaddition reactions is another important approach that can give access to a wide variety of novel corrole derivatives. The great success of such cycloaddition methodologies in the porphyrin field [99] led some researchers to study the possibility of using corroles as dienophiles, dienes, 1,3-dipoles and as dipolarophiles.

In 2004, as an extension of the pioneering studies carried out by Cavaleiro and co-workers in the porphyrin field [99–102], it was reported that 5,10,15-tris (pentafluorophenyl)corrole **4**, in the presence of pentacene, can act either as a 2π or as a 4π component (Scheme 39). The reaction was performed in 1,2,4-trichlorobenzene at 200°C affording, after 6 h, the corresponding dehydrogenated

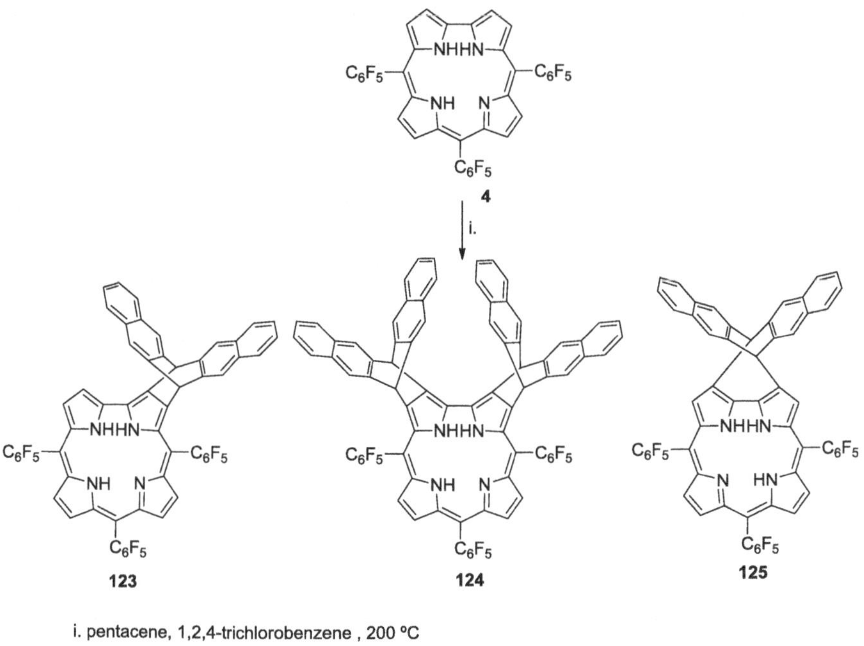

Scheme 39 Cycloaddition reactions of 5,10,15-tris(pentafluorophenyl)corrole with pentacene

Diels–Alder [4+2] cycloadducts **123** and **124** and the symmetry-forbidden [4+4] cycloadduct **125**, in moderate yields [103].

These studies were extended to other polycyclic aromatic dienes, such as anthracene, tetracene, 9,10-dimethylanthracene, and naphtho[2,3-*a*]pirene, affording the corresponding dehydrogenated Diels–Alder [4+2] cycloadducts (Fig. 1) and also minor compounds suggested to be dimers by mass spectrometry [104].

Barata et al. [105] also have shown that in the absence of diene, the formation of the dimers could be optimized by carrying out the dimerization of corrole in a very small volume of solvent (0.01 mL/mg of corrole). In such studies the dimers **130** and **17** linked respectively by the 2,3′ and 3,3′ carbons were isolated and also the dimer **131** with a double linkage by the 2,2′,18,18′ carbons (Fig. 2). It was verified that the reaction can be scaled up and the formation of radicals was considered to be a plausible pathway for the formation of the dyads. Another important feature of these results is the access, in one step, to the eight-membered ring dimer **132**, which was also described by Osuka and co-workers [106].

The possibility of using free-base corroles as regioselective 4π dienes in Diels–Alder reactions was reported by Kräutler and co-workers [107]. The authors were able to synthesize the tetra-β,β′-sulfoleno-corrole **135** by condensation of dipyromethane **134** (obtained from β,β′-sulfolenopyrrole **133** and di-*tert*-butylbenzaldehyde **132**) with tolylaldehyde (Scheme 40). The SO_2 extrusion carried out at 140°C in the presence of a large excess of C_{60} (30 equiv.) in

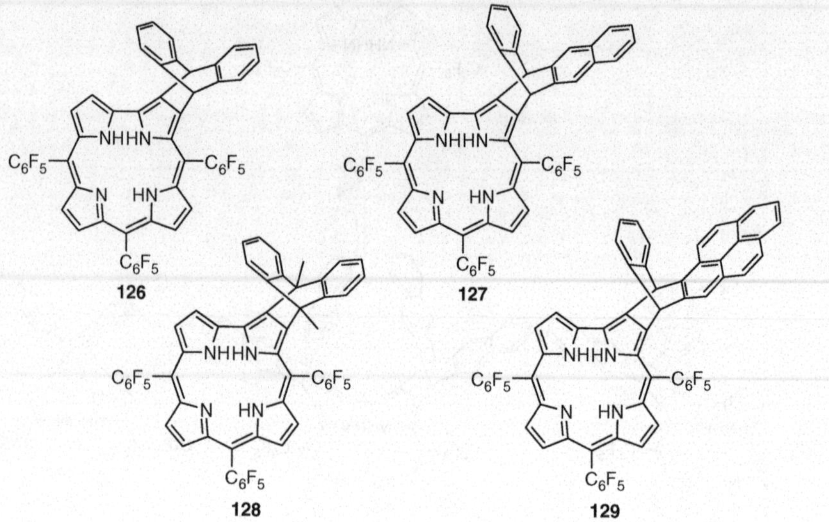

Fig. 1 Dehydrogenated cycloadducts obtained in the cycloaddition of 5,10,15-tris(pentafluorophenyl) corrole **4** with anthracene (**126**), tetracene (**127**), 9,10-dimethylanthracene (**128**), and naphtho [2,3-*a*]pirene (**129**)

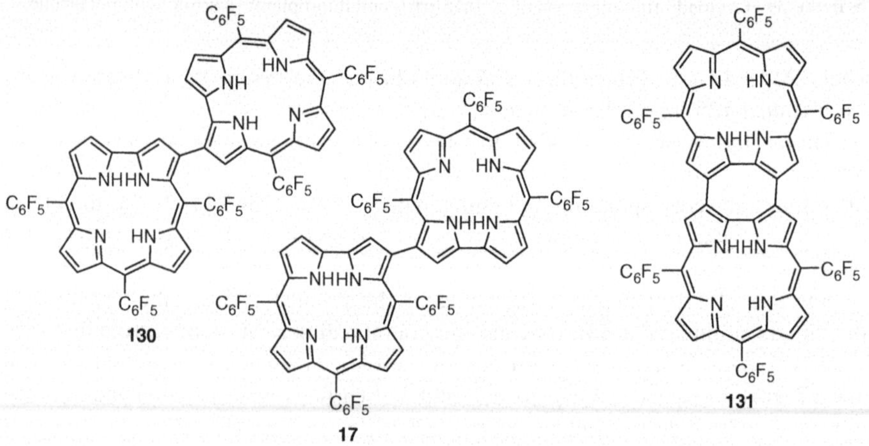

Fig. 2 Dimer derivatives obtained directly under cycloaddition conditions

deoxygenated 1,2-dichlorobenzene for 9 min leads to the difullereno-corrole **136** in 83% yield. The structure of adduct **136** indicates that the directly linked pyrrole units (rings A and D) in corrole **135** undergoes SO_2 extrusion with a higher relative rate than the opposite counterparts. This fact prompted the authors to consider this type of corroles as a heteroaromatic system "programmed" for regioselective cycloaddition chemistry at the periphery [107].

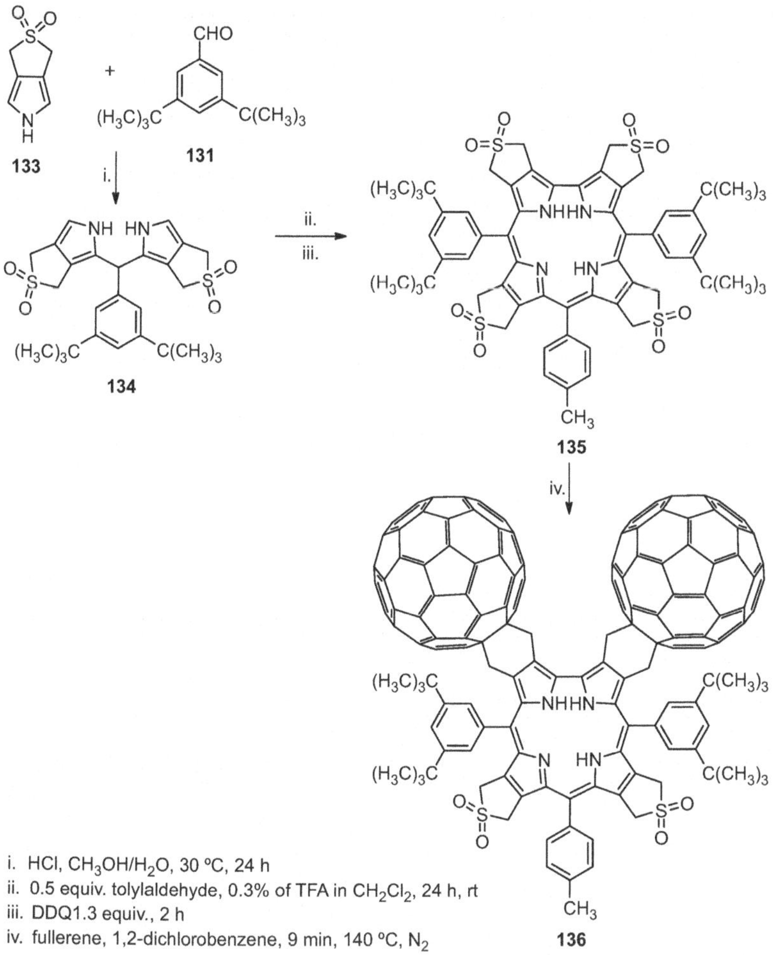

Scheme 40 Preparation of difullereno-corrole derivative

The possibility of using β-vinylcorroles as dienes in Diels–Alder reaction was also investigated [74]. The reactivity of the 3-vinyl-corrole **63** (Scheme 20) as the 4π component was studied in the presence of 1,4-benzoquinone and 1,4-naphthoquinone in refluxing toluene (Scheme 41).

The main products were isolated in yields of 64% and 76%, respectively, which were characterized as being the corresponding dehydrogenated cycloadducts **137** and **138**.

It was also demonstrated the sensing ability of the dehydrogenated cycloadducts **137** and **138** as well as their precursors **4, 50**, and **63** toward several anions and amines. Compound **4** (free base), **50**, and **138** showed to be particularly sensitive to fluoride ion. In the interaction with amines, **Ga4** showed significant changes in the emission spectrum, in the presence 4,4′-bypiridine. Compound **138** was really

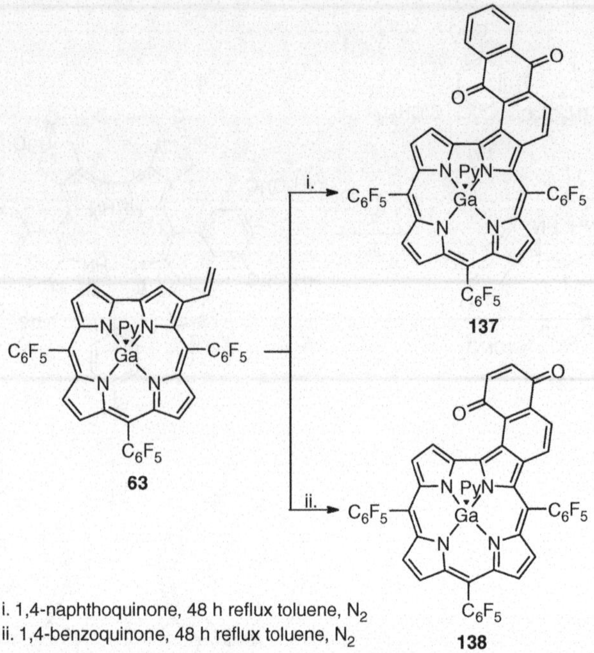

Scheme 41 3-Vinyl-corrole as diene in Diels-Alder reactions

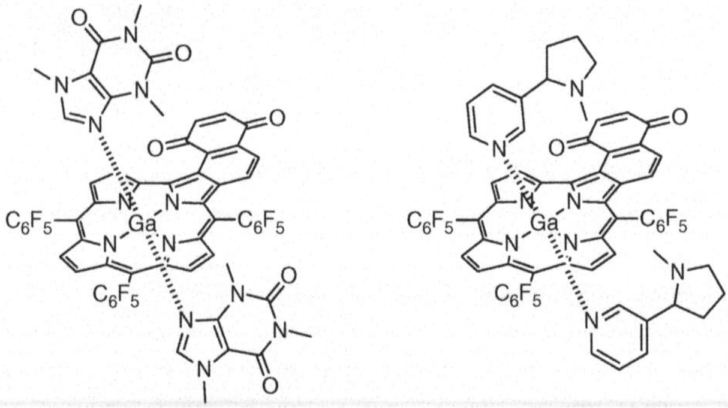

Fig. 3 Interaction of corrole **138** with caffeine and nicotine

effective to detect nicotine and caffeine (Fig. 3). Polyacrylamide gel films doped with compound **4** or with **138** were able to show a very strong emission in the presence of water samples containing, respectively fluoride ion and caffeine or nicotine [74].

The 5,10,15-tris(pentafluorophenyl)corrole-3-carbaldehyde **50** was also used as a precursor of the azomethine ylide **139** (Scheme 42). This 1,3-dipole, obtained from the reaction of corrole **50** with *N*-methylglycine in refluxing toluene, was trapped by several dipolarophiles like quinones (1,4-benzoquinone,

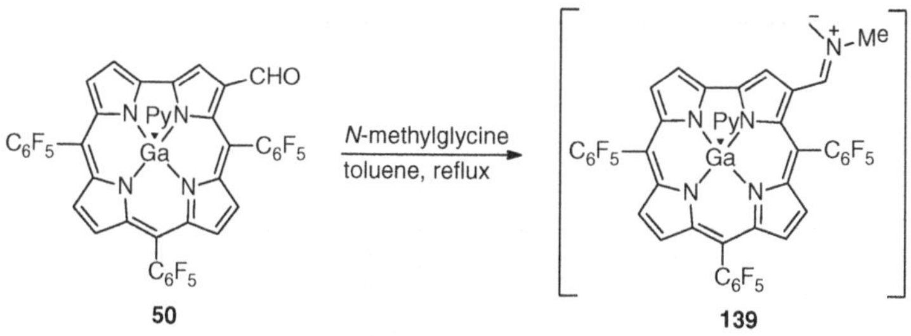

Scheme 42 Preparation of corrole azomethine ylide

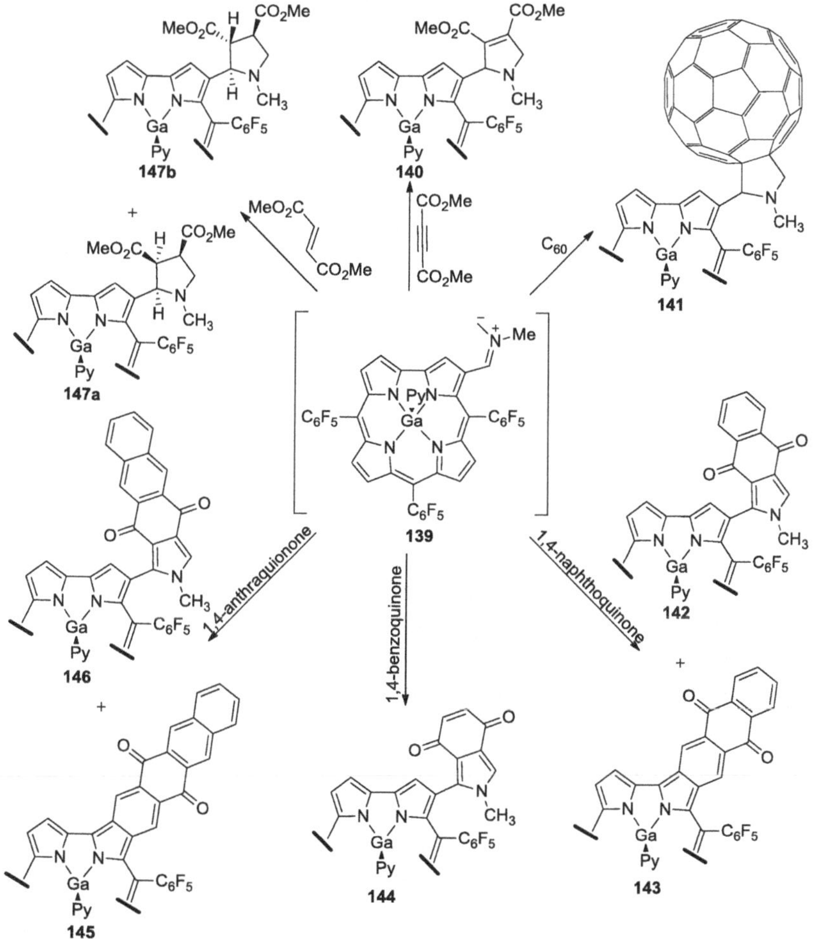

Scheme 43 1,3-Dipolar cycloaddition reaction of corrole azomethine ylide with several dipolarophiles

1,4-naphthoquinone and 1,4-anthraquinone) [108] and by C_{60}, dimethyl fumarate, and dimethyl acetylenedicarboxylate (Scheme 43) [72]. New derivatives **140–147** have been obtained.

In the particular case of 1,4-naphthoquinone and 1,4-anthraquinone, besides the expected dehydrogenated cycloadducts **142** and **146**, the quinone-fused corroles **143** and **145** have been isolated, in moderate yields (18% and 46%, respectively) [108].

Such studies have shown that the best dipolarophiles to trap the azomethine ylide (**139**) were 1,4-benzoquinone, C_{60}, and dimethyl fumarate with adducts reaching ca. 90% yield; dimethyl acetylenedicarboxylate showed to be the less effective dipolarophile affording the expected adduct **140** in 33% yield [72].

Cavaleiro and co-workers also found that 5,10,15-tris(pentafluorophenyl)correlate gallium(III) complex **Ga4** in the presence of the azomethine ylide generated *in situ* from the reaction of *p*-formaldehyde and *N*-methylglycine afforded the dimethylaminomethylcorroles **148** instead of the expected 1,3-cycloadducts [109].

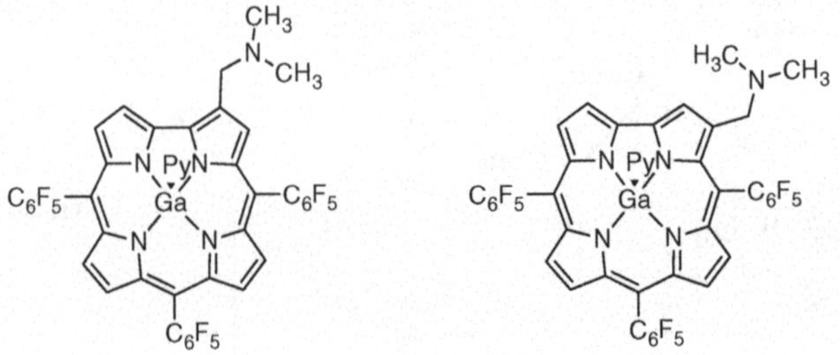

148a **148b**

The same group also extended the studies to the azomethine ylide generated from the gallium(III)(pyridine) complex of 5,10,15-tris(pentafluorophenyl)corrole-2-carbaldehyde **51**. It was found that in the presence of 1,4-naththoquinone, the expected dehydrogenated cycloadduct **150** was once more accompanied by the π-extended chromophore **143** (Scheme 44) [72].

A similar approach was followed by Gryko and co-workers to prepare free base corrole-C_{60} dyads covalently linked in a *meso*-position through rigid and semi-rigid spacers [110, 111]. Authors referred that the corrole-fullerene dyads **156** and **157** were designed, in order to study the effect of the distance between the donor and acceptor units on the rates of charge separation and charge recombination. The synthetic methodology used to prepare *trans*-A_2B-corroles **154** and **155** bearing the phenyl *p*-formyl group involved the condensation of the suitable di-aldehydes **151** and **152** with the 5-pentafluorophenyldipyrromethane **153** in H_2O/MeOH/HCl followed by oxidation with DDQ (Scheme 45) [7]. The subsequent reaction with *N*-methylglycine followed by addition of C_{60} gave rise to the desired dyads **156** and **157** in good yields. The photophysical and theoretical studies performed in polar and nonpolar solvents suggested the possibility of electron transfer from excited singlet state of corrole to the fullerene unit. An efficient fluorescence quenching of the corrole entity took place with the dyads.

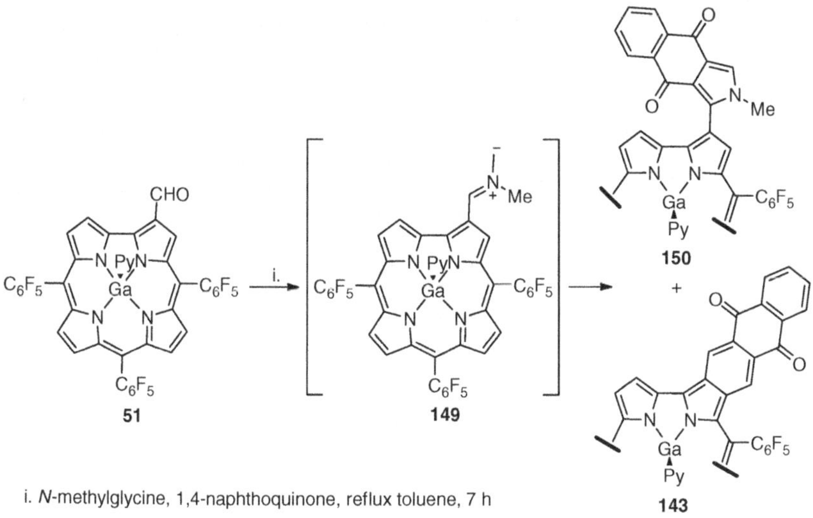

Scheme 44 Synthesis of π-extended chromophore

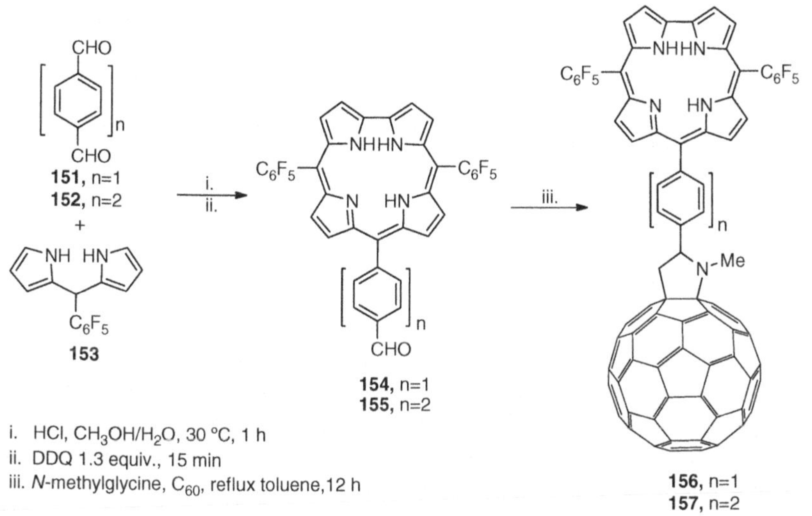

Scheme 45 Synthesis of corrole-fullerene dyads

Authors have extended this methodology to the preparation of dyad **161** bearing a semi-rigid spacer linking both chromophores (Scheme 46) [111]. The selected synthetic methodology involved the synthesis of aldehyde **158** from pentafluorobenzaldehyde and 4-hydroxybenzaldehyde *via* nucleophilic aromatic substitution in the presence of cesium fluoride. The macrocyclization in the presence of 5-pentafluorophenyldipyrromethane **153** was carried out, affording the two

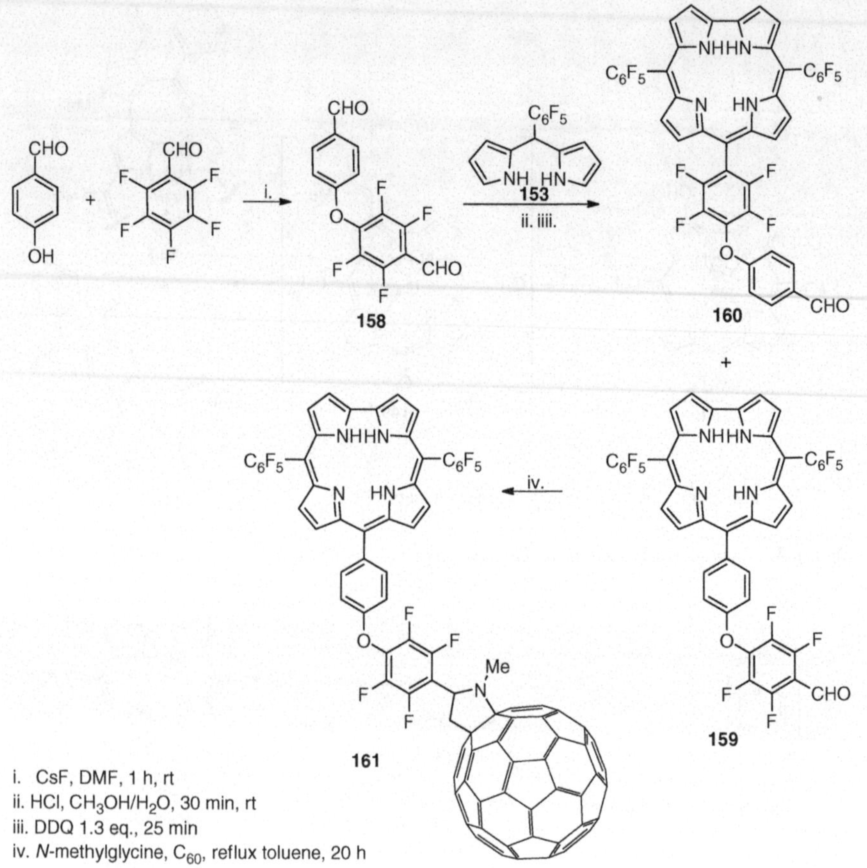

Scheme 46 Synthesis of corrole-fullerene dyad with a semi-rigid spacer

expected regioisomeric corroles **159** (obtained in higher yield) and **160**, both bearing the free formyl group. The subsequent reaction of corrole **159** with *N*-methylglycine and the dipolarophile C_{60} afforded the desired dyad **161** in 15% yield [111]. Authors have used the dyad **161** and its precursors **159** to propose the assignment of all IR and Raman spectroscopic bands.

5 Functionalization of Corroles Mediated by Transition Metal Complexes

Modern approaches involving reagents or catalysts based on transition metal compounds are also being considered as new approaches for modifications of the corrole macrocycle. Most of them are centered on transformations catalyzed by

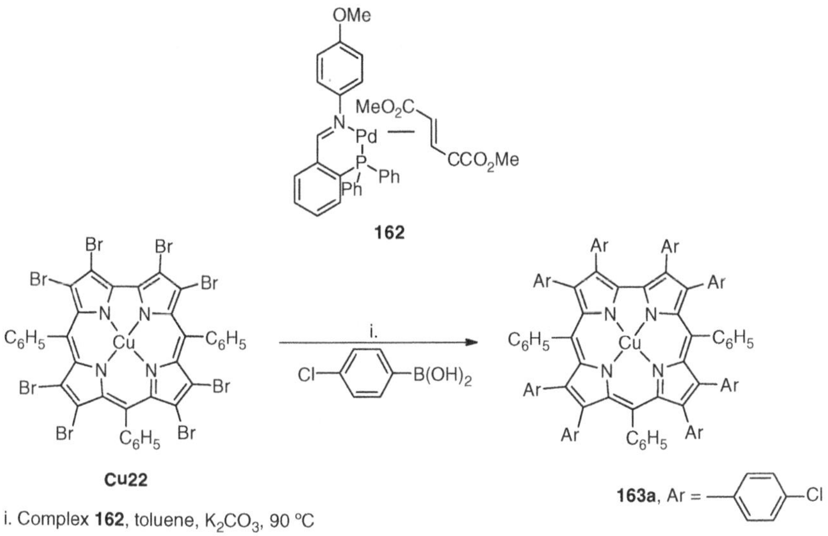

Scheme 47 Preparation of undecaaryl substituted corrole

palladium(0) using *meso*-4,6-dichloropyrimidin-5-yl-substituted corroles and *meso*-arylcorroles bearing bromine substituents in beta pyrrolic positions or in the phenyl groups.

In 2004, Scrivanti et al. [112] during their studies concerning the activity of the iminophosphine–palladium(0) complex **162** for Suzuki coupling reported that this complex, in the presence of *p*-chlorophenylboronic acid, was able to catalyze the complete substitution of the eight peripheral bromine atoms in the Cu(III) complex **Cu22** affording the undecaaryl substituted corrole **163a** in 55% yield (Scheme 47). The reaction was performed in toluene at 90°C using a corrole/catalyst ratio of 40,000:1. The low activity of the catalyst toward aryl chlorides was exploited to couple the selected chloro-substituted arylboronic acid.

It has also been reported an efficient access to different *para*-substituted undecaphenylcorroles **163b–d** via Suzuki–Miyaura coupling of corrole **Cu22** with the appropriate arylboronic acid in the presence of potassium carbonate and $Pd_2(dba)_3{\cdot}CH_2Cl_2$ [113]. From the single-crystal X-ray structure of **163b** the authors concluded that the degree of saddling of these corroles is higher than the one observed for the β-unsubstituted **Cu5** derivatives and slightly lower for corrole **Cu22** bearing eight peripheral bromine atoms. However, electronic absorption measurements show that Soret maxima of these undecaphenylcorroles are comparable to those due to corrole **Cu22**. The electrochemical studies allowed to conclude that the *para* substituents on the β-phenyl groups tune the redox potentials of copper corroles more effectively than those on *meso*-phenyl substituents.

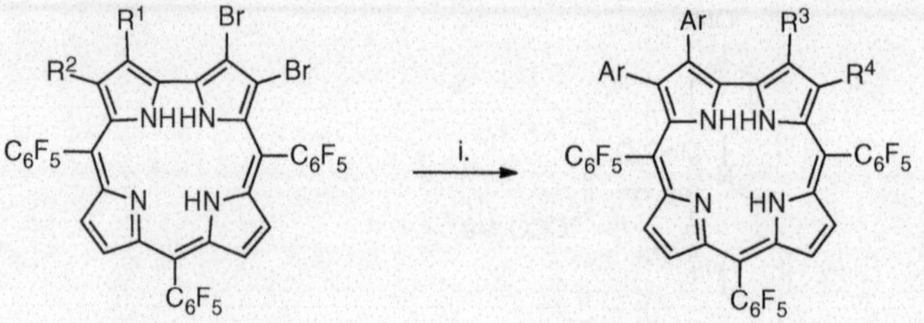

26, R^1 = H, R^2 = H
27, R^1 = H, R^2 = Br
28, R^1 = Br, R^2 = Br

164a, Ar = C_6H_5, R^3 = R^4 = H; 80%
164b, Ar = 4-$CH_3OC_6H_4$, R^3 = R^4 = H; 62%
165a, Ar = C_6H_5, R^3 = H, R^4 = C_6H_5; 89%
165b, Ar = 4-$CH_3OC_6H_4$, R^3 = H, R^4 = 4-$MeOC_6H_4$; 65%
166a, Ar = R^3 = R^4 = C_6H_5; 81%
166b, Ar = R^3 = R^4 = 4-$CH_3OC_6H_5$; 84%

i. $ArB(OH)_2$ with Ar = C_6H_5 or 4-$CH_3OC_6H_4$, $Pd_2(dba)_3$, toluene reflux, K_2CO_3

Scheme 48 Brominated corroles in Suzuki cross-coupling reactions

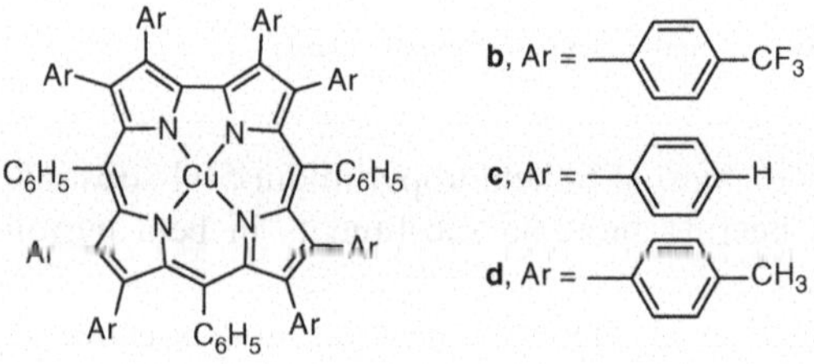

163

Chen and co-workers [44] used the Suzuki coupling methodology to functionalize the brominated derivatives **26–28** of 5,10,15-tris(pentafluorophenyl) corrole (Scheme 48). These derivatives were obtained by controlling the amount of diluted NBS added to 5,10,15-tris(pentafluorophenyl)corrole **4** as it was mentioned before. The Suzuki cross-coupling reactions involving those corroles and phenyl and *p*-methoxyphenylboronic acids led to the corresponding β-substituted corroles **164–166** in good yields.

The same publication highlights that the starting free base corrole **4** can be used to give access to the fluoroalkylated corroles **167a**, **b** and **168a** via the adequate reagents ClC_4F_8I and $I(CF_2)_nI$ ($n = 3, 4$) in the presence of $Na_2S_2O_4$ (Scheme 49).

Studies on the functionalization of *trans*-A_2B-corroles by using the Suzuki–Miyaura cross-coupling conditions have been reported by Schoefberger and co-workers [114]. Such studies involved the Cu(III) complex of 10-(4-bromophenyl)-5,15-bis-(pentafluorophenyl)corrole **Cu169** and the boron derivatives **170** (Fig. 4 and Scheme 50). In this wide variety of boronic reagents the authors included sterically hindered, inactivated and heteroaromatic boronic acids and esters, alkenylboronic acids and benzofused five-membered heterocyclic boronic acids.

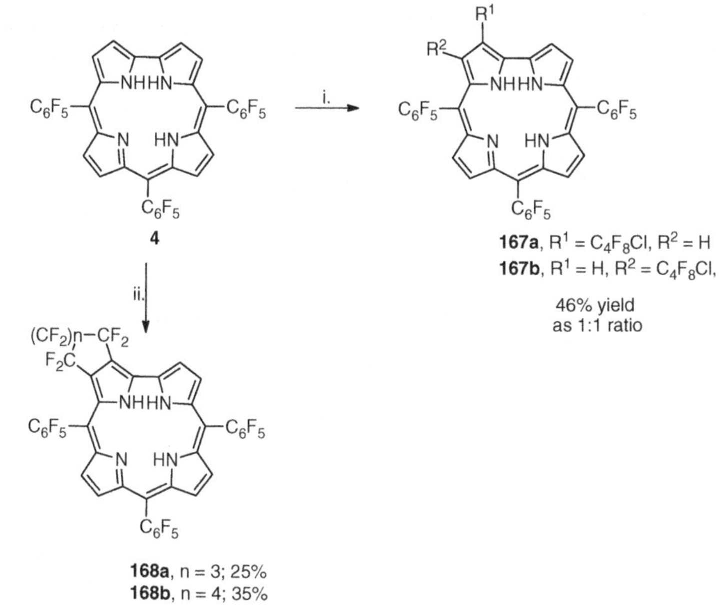

i. ClC_4F_8I, $Na_2S_2O_4$, DMSO, rt
ii. $I(CF_2)_nI$ with n = 3 and 4, $Na_2S_2O_4$, DMSO, rt

Scheme 49 Synthesis of fluoroalkylated corroles

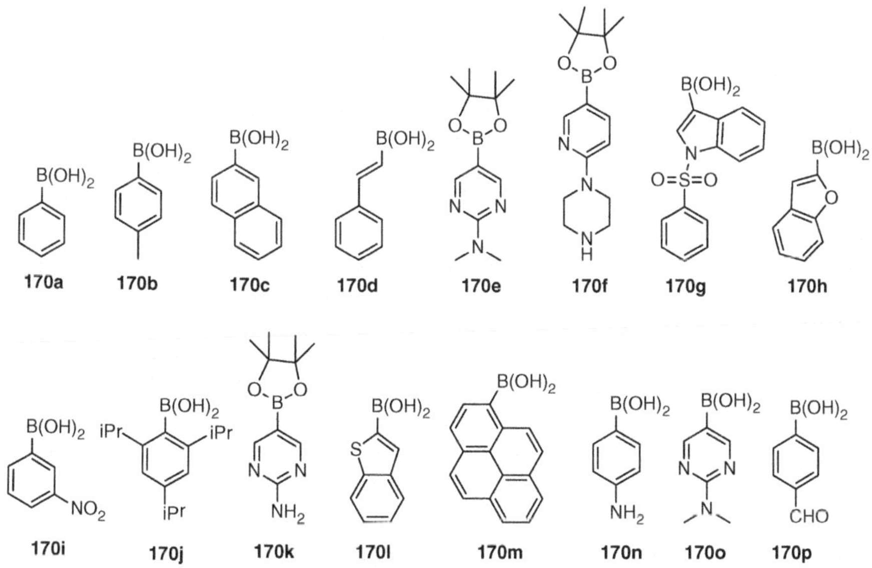

Fig. 4 Structures of boronic acids and esters **170** used in coupling reactions with the copper complex **Cu169** and bismuth complex **Bi169**

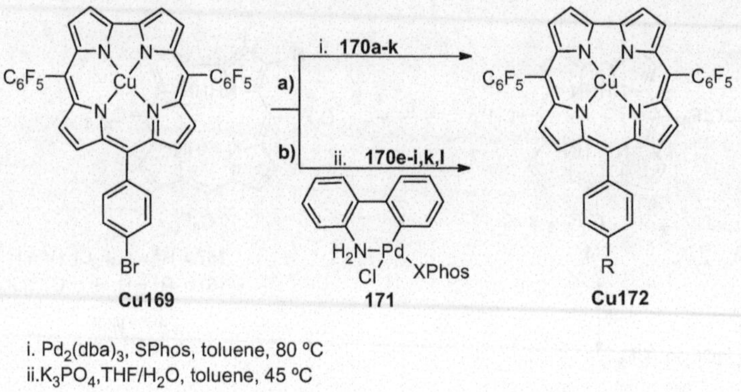

Scheme 50 Copper(III) bromocorrole complex in Suzuki-Miyaura cross coupling reactions

It became clear that for the success of the coupling a careful selection of the catalytic system is required. At 80°C, using SPhos (2-dicyclohexyl-phosphino-2′,6′-dimethoxybiphenyl) as ligand, $Pd_2(0)dba_3$ as catalyst, and K_3PO_4 as base, authors were able to obtain in high yields (~80%) the coupling products **Cu172a–d** from the reaction between corrole **Cu169** and the adequate boronic acids **170a–d**. This system showed to be less efficient for coupling the corrole complex with **170e** (35%) and **170k** (14%) and no other product was detected with the remaining boronic reagents. Authors found that the use of the precatalytic system **171** (Scheme 50b) recently reported by Buchwald and co-workers [115] brought an improvement in the coupling of the less stable boron derivatives **170e–i** and **170l, k** when compared with the previous conditions. The reactions were performed in THF/H_2O, at 45°C, and compounds **Cu172e–i** were obtained in yields higher than 68%, **Cu172l** in 45%, and **Cu172k** in 23%. The authors also referred that although most of the coupled products form stable Cu(III) corrole complexes in solution and in the solid state, some of them exhibit significant Cu(II) character. This behavior was studied by UV/vis and NMR spectroscopic techniques and by time-dependent density functional theoretical calculations.

Some time afterwards, the same group studied the possibility of using the Bi(III) complex of 10-(4-bromophenyl)-5,15-bis-(pentafluorophenyl)corrole **169** in Suzuki cross-coupling reactions in order to get functionalized free-base corroles [116]. The insertion of bismuth into the free base **169** was performed with $Bi\{N(SiMe_3)_2\}_3$ [117]. The Suzuki cross-coupling reactions were carried out in the presence of several boronic acids (**170a–d**, **170g–i**, and **170m–p**) (Scheme 51). Studies were performed in a THF/H_2O mixture using K_3PO_4 as base and a combination of SPhos (5 mol%) and $Pd(OAc)_2$ (2.5 mol%). The expected derivatives **Bi172** have been obtained in yields ranging from 69% to 87%; such products, after treatment with HCl/THF, gave rise to the corresponding demetallated corroles **172**.

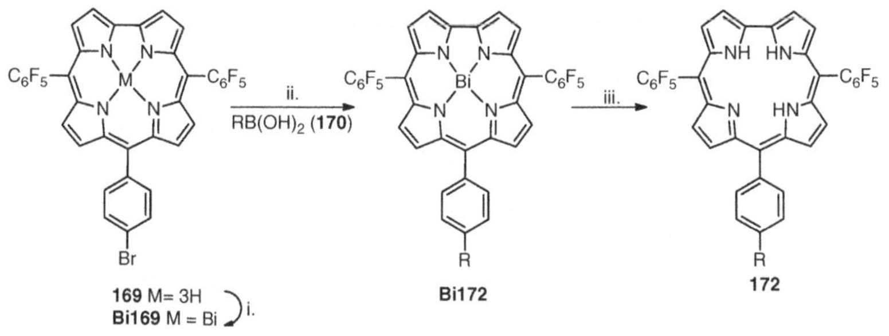

i. $Bi\{N(SiMe_3)_2\}_3$, THF
ii. 5% SPhos, 2.5% $Pd(AcO)_2$, K_3PO_4, THF/H_2O, 25-40 °C
iii. HCl/THF

Scheme 51 Bismuth(III) bromocorrole complex in Suzuki-Miyaura cross coupling reactions

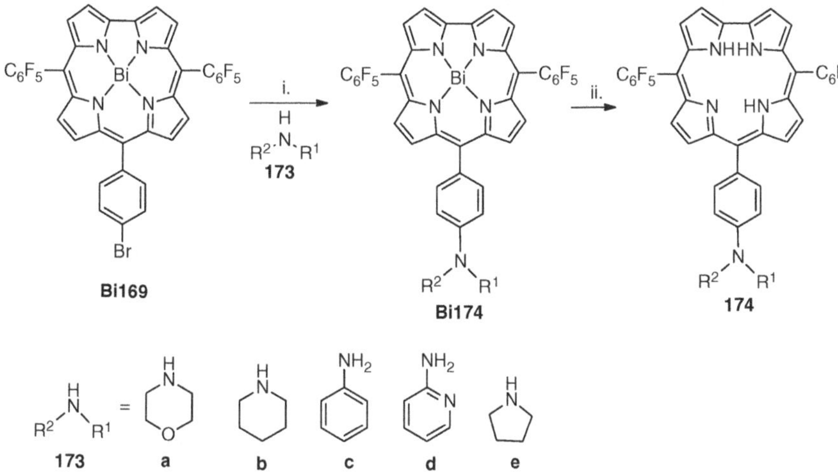

i. 5% XPhos, 2.5% $Pd_2(dba)_3$, Cs_2CO_3, *t*-BuOH, 80 °C
ii.HCl/THF

Scheme 52 Bismuth(III) bromocorrole complex in Buchwald-Hartwig cross-coupling reactions

Authors also studied the possibility of using the same Bi(III) complex in cross-coupling reactions with amines **173a–e** (Scheme 52). The Buchwald–Hartwig coupling reactions were performed in the presence of XPhos and $Pd_2(dba)_3$ and $CsCO_3$ or K_3PO_4 as bases. The expected amines **Bi174** were obtained in moderate yields, ranging from 27% to 54%, with recovery of the unreacted reagents. The demetallation of the product derivatives was performed using diluted HCl in THF. The expected free bases **174** were obtained in excellent yields (between 67% and 94%). However **Bi174e** did not support such conditions and its free base was only detected in trace amounts.

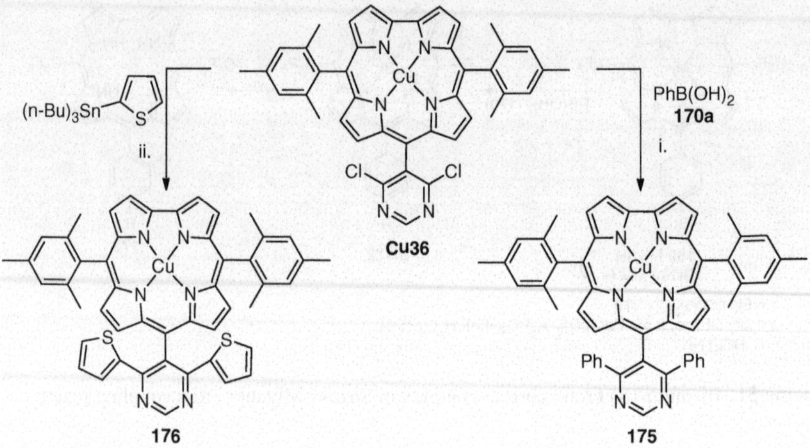

i. $Pd(PPh_3)_4$, Na_2CO_3, toluene, reflux
ii. $Pd(PPh_3)_4$, CsF, Na_2CO_3, toluene, reflux

Scheme 53 Pd-catalyzed cross-coupling reactions of *meso*-pyrimidinyl-substituted corroles

Based on ^{19}F NMR studies as well as quantum chemical calculations, the authors were able to propose a dome-shaped for the Bi(III)corrole derivatives in which the Bi ion shows a significant out-of-plane displacement; this might be the explanation for the demetallation of the Bi-corrole complexes to be much easier than the other metallocorroles.

Maes and co-workers considered the Pd-catalyzed cross-coupling reactions to functionalize a variety of *meso*-pyrimidinyl-substituted corroles of type A_2B (B = 4,6-dichloropyrimidin-5-yl) that were obtained efficiently through the condensation of 5-mesityldipyrromethane **99a** with 4,6-dichloropyrimidine-5-carbaldehydes [56, 97]. The reaction of the copper complex **Cu36** with phenylboronic acid under Suzuki conditions afforded the desired disubstituted A_2B-corrole **175** in 75% yield (Scheme 53).

A similar yield was also referred for corrole **176** obtained from the coupling of **Cu36** with 2-(tributylstannyl)thiophene under Stille conditions.

The introduction of functional moieties at the corrole stage by Pd-catalyzed Liebeskind–Srogl cross-coupling reactions on thiomethyl functionalized corrole **177** was also considered (Scheme 54) [56, 97]; under such conditions compound **178** was obtained and no products from Suzuki-type reaction were detected.

The derivatization of the more accessible *meso*-pyrimidinyl-substituted corroles of types A_2B and A_3 under Suzuki conditions did afford the sterically encumbered triarylcorroles **179** and **180** in excellent yields [98].

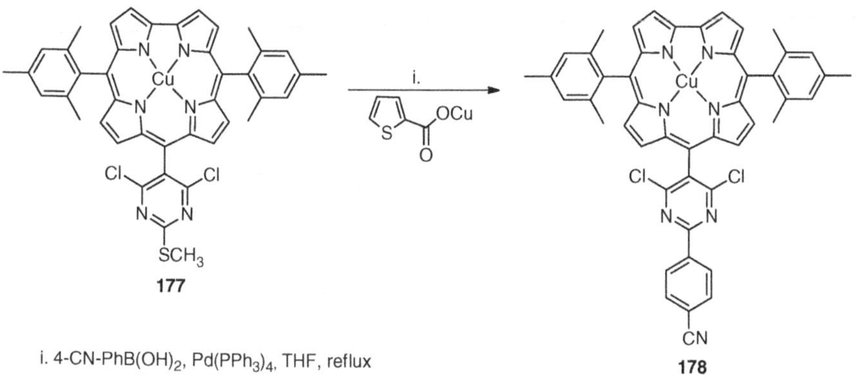

177

178

Scheme 54 Pd-catalyzed Liebeskind–Srogl cross-coupling reaction on thiomethyl functionalized corrole

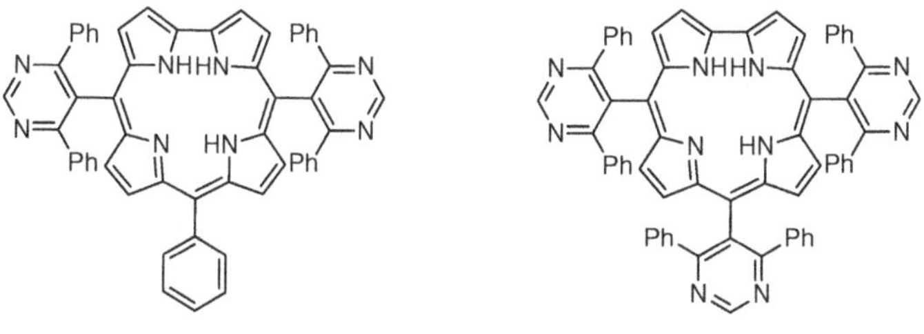

179

180

Sterically demanding A_2B-corroles were obtained through palladium-catalyzed amidation of readily accessible 5,15-bis(2,6-dibromophenyl)corroles **181a–e** with pivalamide **182** (Scheme 55) [18]. The starting corroles **181** were obtained using adequate benzaldehydes and 2,6-dibromophenyldipyrromethane. Under the best conditions established for the C–N coupling, the formation of the desired tetra-substituted derivatives **184a–e** were accompanied by the corresponding di- and trisubstituted corroles **183a–e** and **185a–e**.

The cyclopropanation of styrene with ethyl diazoacetate in the presence of nitroso iron complex of corrole **Fe184a** occurs in a diastereoselective fashion. This complex was obtained from the insertion of iron ion into **184a** [R^1 = C $(CH_3)_3$] followed by in situ treatment with sodium nitrite.

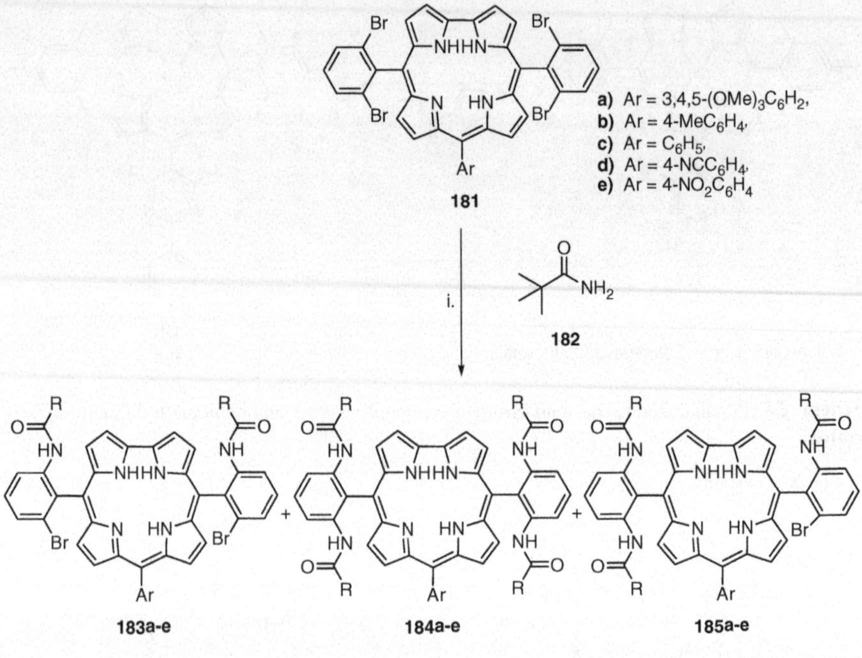

i. Pd(OAc)$_2$, Xantphos, Cs$_2$CO$_3$ R = C(CH$_3$)$_3$

Scheme 55 Pd-catalyzed amidation of 5,15-bis(2,6-dibromophenyl)corroles

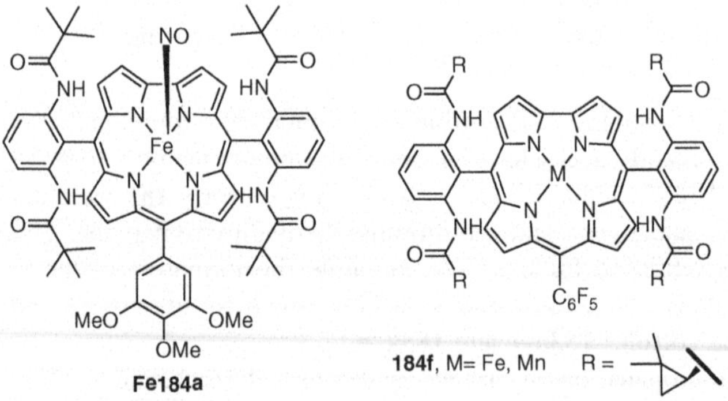

It has also been reported that attempts to introduce the pivaloylamido moieties in the *ortho*-positions of the *meso*-aryl substituents of the dipyrromethanes, before the oxidative macrocyclization, failed to afford the desired picket fence corroles [118]. A possible explanation for this fact considers sterical reasons and that was based on the stretched conformation adopted by the precursors. So, the key to a successful and versatile synthesis of such chiral and achiral sterically encumbered corrole ligands requires that the palladium-catalyzed amidations of *o*-bromophenyl groups should be performed as the last step.

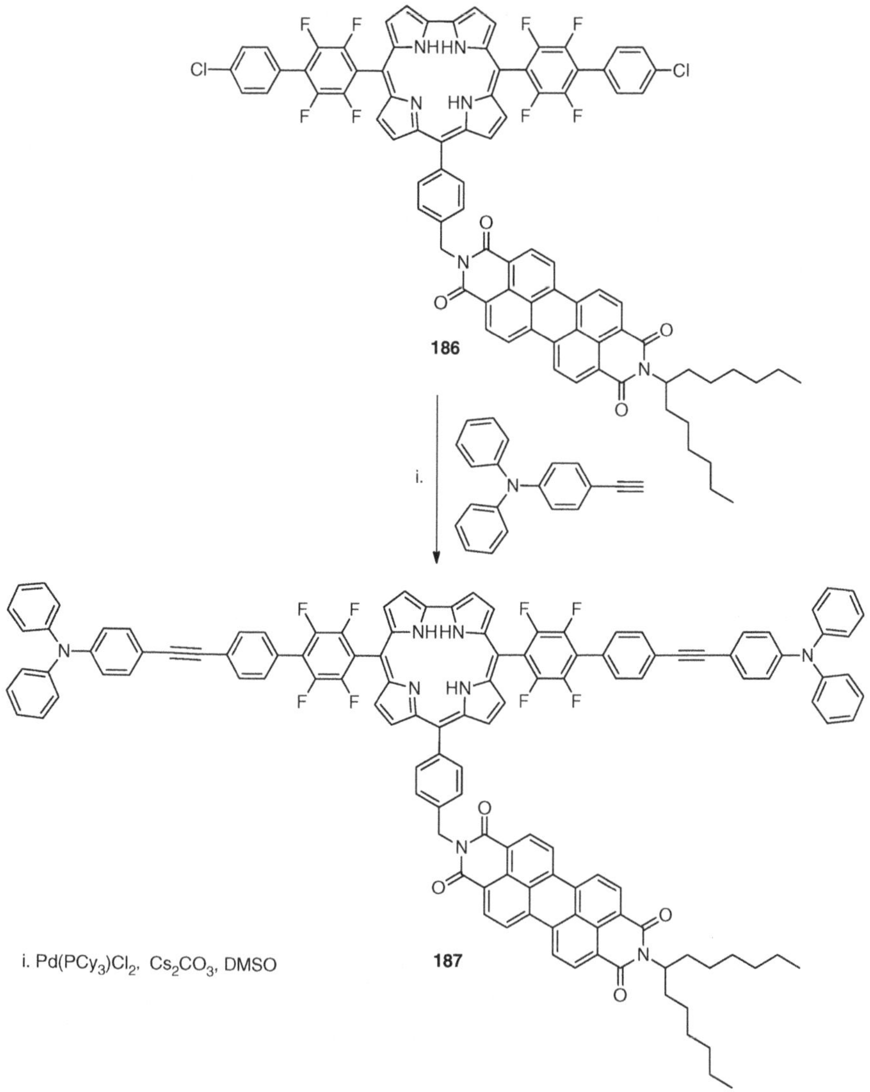

Scheme 56 Preparation of multichromophoric system by Sonogashira methodology

Gross and co-workers also considered the synthesis of the palladium-catalyzed amidation of C–Br bonds in the preparation of chiral manganese(III) and iron(III) corrole complexes **Fe184f** and **Mn184f**. Such catalysts have a low stability under the oxidative conditions, and as a result low conversions have been obtained [119].

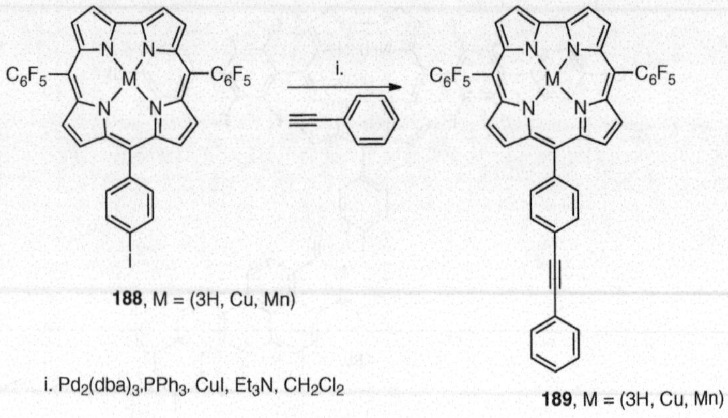

Scheme 57 Preparation of alkynylcorrole derivatives

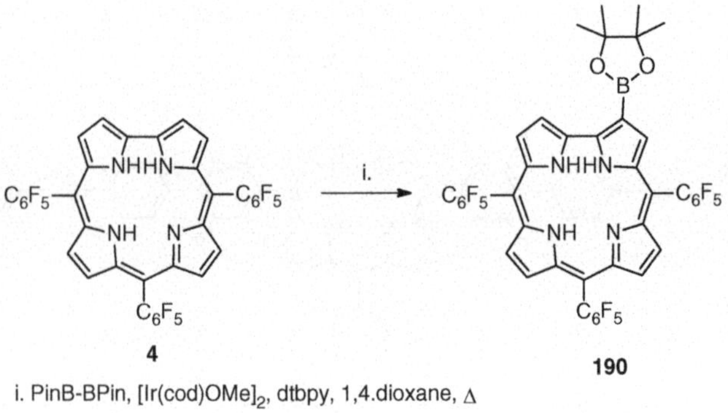

Scheme 58 Direct borylation of 5,10,15-tris(pentafluorophenyl)corrole

The use of the Sonogashira methodology to functionalize corroles has attracted also the interest of some research groups. For instance, Gryko and co-workers [120] explored this methodology as an important step to prepare the multichromophoric system **187** via functionalization of the *trans*-A_2B-corrole **186** with the *N*-(4-ethynylphenyl)-*N*,*N*-diphenylamine (Scheme 56). From the electrochemical and spectroscopic data authors were able to conclude that there was a weak coupling between the different structural entities of the triad **187**. The photophysical measurements showed that an efficient energy transfer takes place from the triphenylamine and perylene bis-imide functions to the corrole unit, which is followed by a subsequent electron transfer from the corrole to the perylene bis-imide.

Zhan et al. [121] were able to prepare the new alkynyl corroles **189** by coupling ethynylbenzene with the adequate iodocorrole **188** under different catalytic conditions (temperature, Pd salts, presence and absence of CuI) (Scheme 57). It has been established that the metal corroles **Cu188** and **Mn188** give better yields than the free base **188** at room temperature. It is also known that the efficiency of

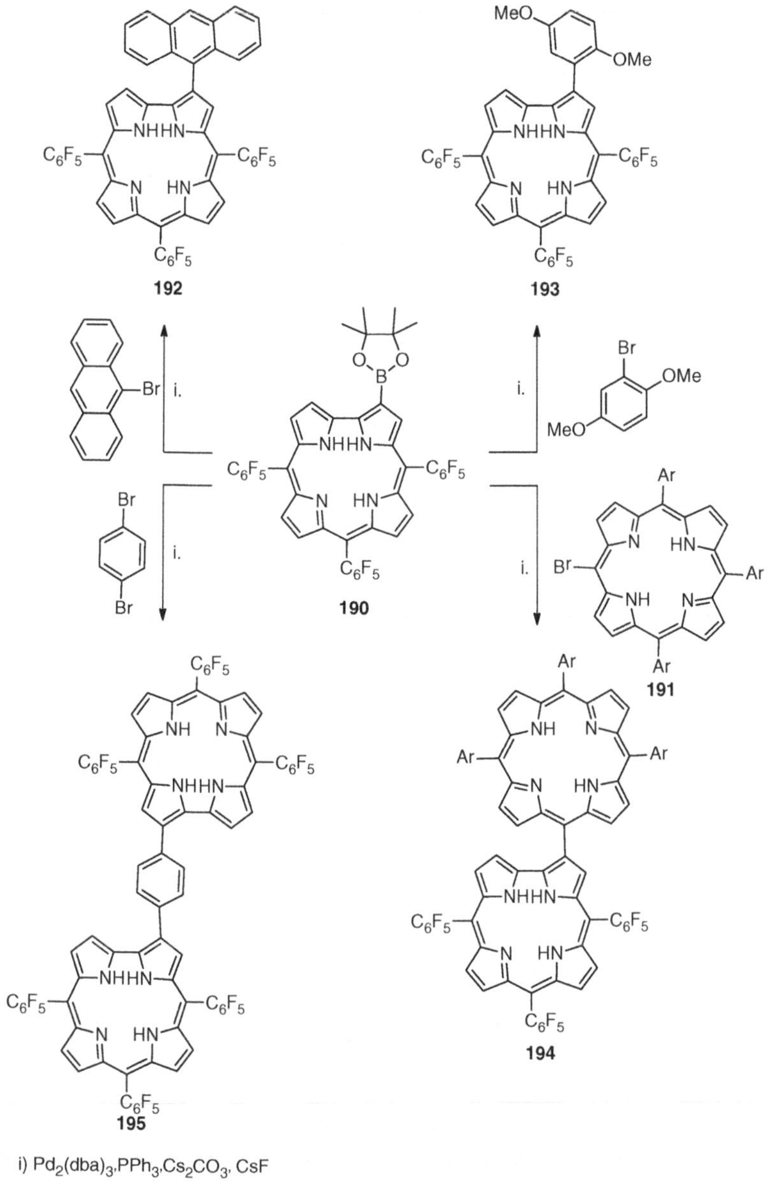

Scheme 59 Suzuki-Miyaura cross coupling reactions of monoborylated corrole **190**

the process was not affected by the source of palladium or by the presence or absence of CuI. However, the free base **188**, in the presence of copper(I) iodide and triethylamine, afforded the desired alkynyl adduct, which was accompanied by the corresponding alkynyl copper(III) complex **Cu189** as a minor product. Authors

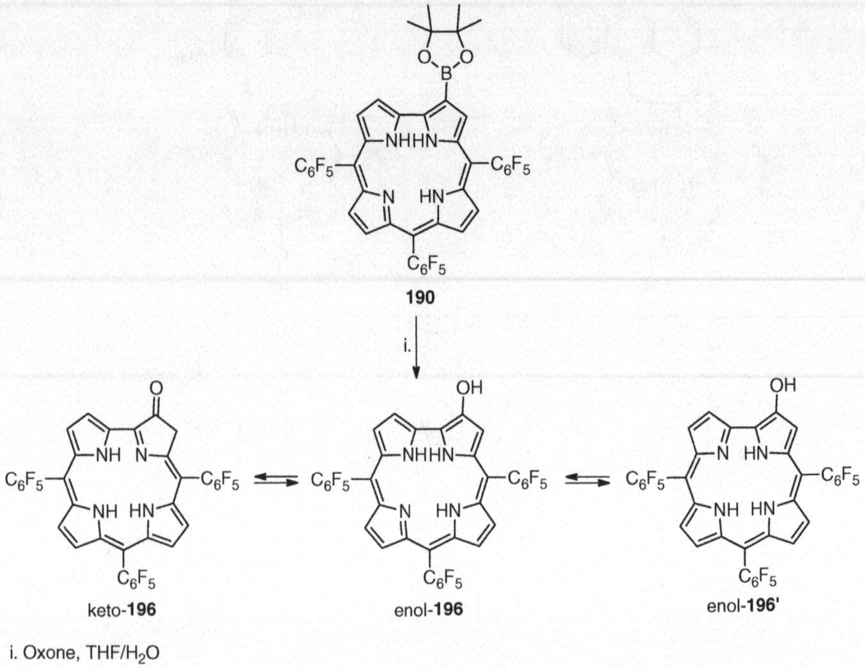

Scheme 60 Treatment of monoborylated corrole **190** with oxone

confirm that the copper complex can be obtained by the reaction of free base corrole **189** and CuI in CH_2Cl_2 at room temperature with the addition of Et_3N.

The possibility of using corroles as the organoboron component in Suzuki–Miyaura cross-coupling reactions was considered by Osuka and co-workers in the construction of other systems based on corroles. As an extension of their studies involving porphyrins, the boryl group was introduced via direct borylation of the corrole macrocycle through C–H cleavage by iridium(I) catalyst [122]. The studies carried out with 5,10,15-tris(pentafluorophenyl)corrole **4** using bis(pinacolato) diborane (PinB–BPin), catalytic amounts of $[Ir(cod)OMe]_2$ (cod = cycloocta-1,8-diene) and 4,4′-ditert-butyl-2,2′-bipyridyl (dtbpy) afforded the monoborylated corrole **190** in 91% yield (Scheme 58).

The reaction of the monoborylated corrole **190** with 5-bromoanthracene, 1-bromo-2,5-dimethoxybenzene, 1,4-dibromobenzene, and *meso*-bromoporphyrin **191**, under Suzuki–Miyaura conditions, afforded the expected adducts **192–195** in excellent yields (Scheme 59).

In the same publication the authors reported that treatment of the monoborylated corrole **190** with oxone gives access to the oxidized product **196** in 62% yield (Scheme 60). The predominance of the keto or enol forms is dependent on the solvent used.

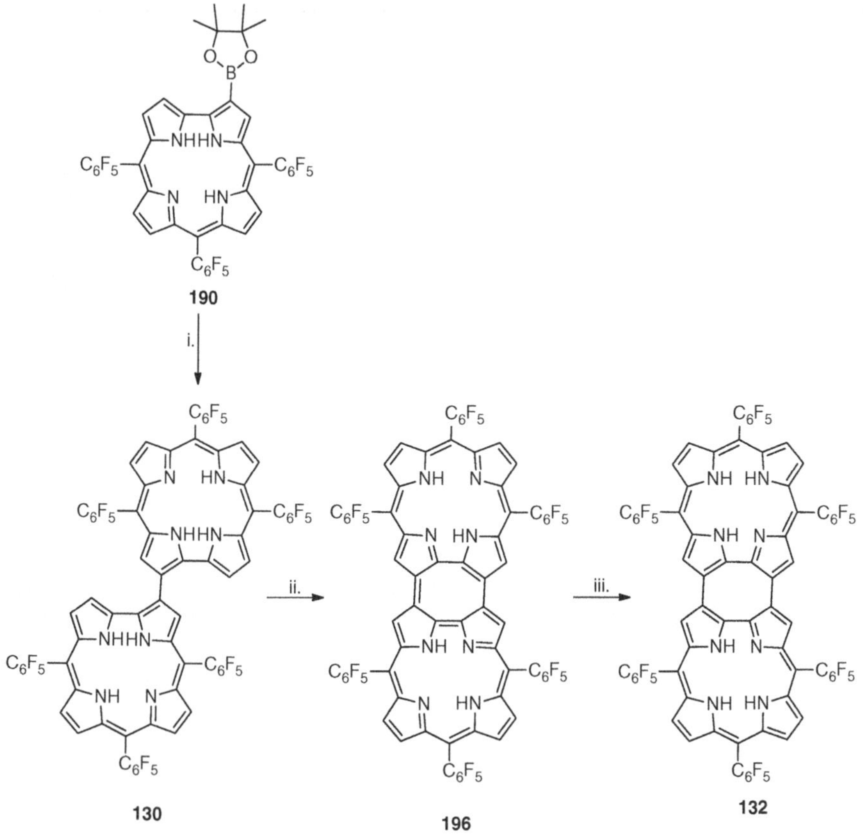

i. $PdCl_2$(dppb), chloroacetone, THF, H_2O, 70 °C
ii. DDQ, toluene, 50 °C
iii. $NaBH_4$, THF, MeOH

Scheme 61 Synthesis of doubly linked corrole dimers

Soon afterwards, the same group used the 2-borylcorrole **190** in the synthesis of the doubly linked corrole dimer **132** (Scheme 61) [106]. The synthetic strategy involved a palladium-catalyzed oxidative coupling using chloroacetone as an oxidant, affording the 2,2′-linked corrole dimer **130** in excellent yield. Further oxidation of **130** with DDQ in toluene provided product **196**, as an air- and moisture-stable solid in a moderate yield. The electronic absorption spectrum of **196** showed a broad spectrum reaching the near-infrared region. Reduction of **196** with $NaBH_4$ resulted in a color change in solution from brown to green and **132** was obtained quantitatively.

The possibility of using the Heck coupling conditions between the octa-brominated corroles **Cu197** and **Cu198** and methyl acrylate as a synthetic strategy to afford the related tetrabenzocorroles **Cu199** and **Cu200** was recently reported

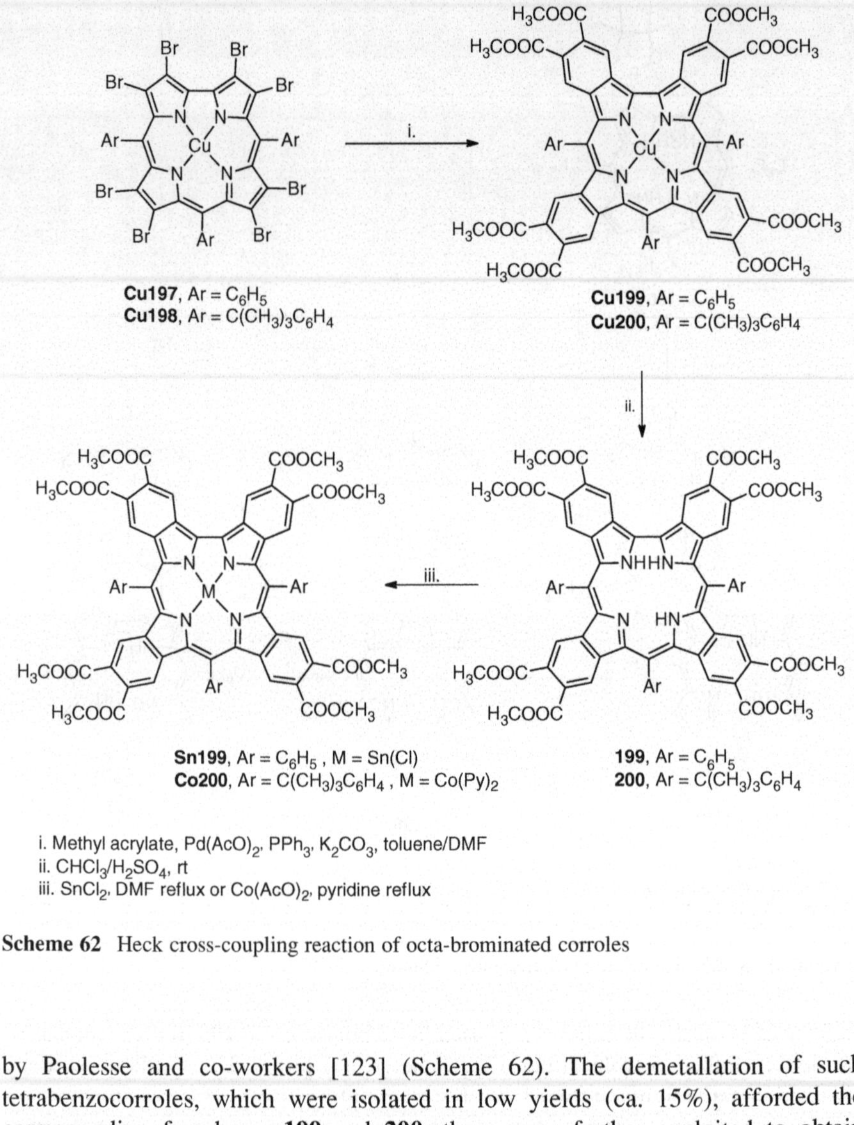

Scheme 62 Heck cross-coupling reaction of octa-brominated corroles

by Paolesse and co-workers [123] (Scheme 62). The demetallation of such tetrabenzocorroles, which were isolated in low yields (ca. 15%), afforded the corresponding free bases **199** and **200**; these were further exploited to obtain the Sn and Co complexes in good yields. It has also been reported that the tetrabenzocorroles and their corrole precursors demonstrate to have similar stabilities. In the same publication another synthetic route leading to tetrabenzo-corroles, *via* the condensation of benzaldehyde and a tetrahydroisoindole, was also considered.

6 Final Remarks

In comparison with porphyrins, corroles can be considered as a young group of tetrapyrrolic macrocycles. This review considers a literature update on the work published about the functionalization of the corrole macrocycle. This includes the reactivity of such tetrapyrroles under electrophilic or nucleophilic substitutions, in cycloaddition transformations, and also in a new chemical avenue catalyzed by transition metal complexes. Several types of substituents can be introduced into the corrole macrocycle. This means that researchers dealing with studies on the potential applications of corroles can have access to a wide variety of such compounds.

Acknowledgments Authors are grateful to "Fundação para a Ciência e a Tecnologia – FCT, QREN, FEDER and COMPETE" Lisbon, for funding QOPNA (the Research Unit in Organic Chemistry), CICECO, and research projects. Thanks are also due to FCT for grants awarded to PhD and Post-Doc students. Thanks are also due to the University of Aveiro, to all students involved in the work carried out in Aveiro.

References

1. Johnson AW, Price R (1960) Synthesis of derivatives of corrole (pentahydrocorrin). J Chem Soc 1649
2. Johnson AW, Kay IT (1965) Corroles. Part I. Synthesis. J Chem Soc 1620
3. Gross Z, Galili N, Saltsman I (1999) The first direct synthesis of corroles from pyrrole. Angew Chem Int Ed 38:1427
4. Paolesse R, Jaquinod L, Nurco DJ, Mini S, Sagone F, Boschi T, Smith KM (1999) 5,10,15-Triphenylcorrole: a product from a modified Rothemund reaction. Chem Commun 1307
5. Gryko DT, Koszarna B (2003) Refined methods for the synthesis of *meso*-substituted A_3- and *trans*-A_2B-corroles. Org Biomol Chem 1:350
6. Lindsey JS (2000) Synthesis of *meso*-substituted porphyrins. In: Kadish KM, Smith KM, Guilard R (eds) The porphyrin handbook. Academic, Boston
7. Koszarna B, Gryko DT (2006) Efficient synthesis of *meso*-substituted corroles in a H_2O-MeOH mixture. J Org Chem 71:3707
8. Collman JP, Decréau RA (2003) Microwave-assisted synthesis of corroles. Tetrahedron Lett 44:1207–1210
9. Kumari P, Chauhan SMS (2008) Efficient synthesis of 5,10,15-triarylcorroles using Amberlyst 15 under solvent-free conditions. J Heterocycl Chem 45:779
10. Zhan HY, Liu HY, Chen HJ, Jiang HF (2009) Preparation of *meso*-substituted *trans*-A_2B-corroles in ionic liquids. Tetrahedron Lett 50:2196
11. Dogutan DK, Stoian SA, McGuire R, Schwalbe M, Teets TS, Nocera DG (2011) Hangman corroles: efficient synthesis and oxygen reaction chemistry. J Am Chem Soc 133:131
12. Nardis S, Monti D, Paolesse R (2005) Novel aspects of corrole chemistry. Mini Rev Org Chem 2:355
13. Schwalbe M, Dogutan DK, Stoian SA, Teets TS, Nocera DG (2011) Xanthene-modified and Hangman iron corroles. Inorg Chem 50:1368

14. Tasior M, Gryko DT, Pielachinska DJ, Zanelli A, Flamigni L (2010) *Trans*-A_2B-corroles bearing a Coumarin moiety – from synthesis to photophysics. Chem Asian J 5:130
15. Flamigni L, Wyrostek D, Voloshchuk R, Gryko DT (2010) Solvent polarity effect on intramolecular electron transfer in a corrole-naphthalene bisimide dyad. Phys Chem Chem Phys 12:474
16. Paolesse R (2008) Corrole: the little big porphyrinoid. Synlett 15:2215
17. Gryko DT (2008) Adventures in the synthesis of *meso*-substituted corroles. J Porphyrins Phthalocyanines 12:906
18. Bröring M, Milsmann C, Ruck S, Köhler S (2009) Bis-picket-fence corroles. J Organomet Chem 694:1011
19. El Ojaimi M, Gros CP, Barbe J-M (2008) Efficient two-step synthesis of face-to-face *meso*-substituted bis(corrole) dyads. Eur J Org Chem 1181
20. Rohand T, Dolusic E, Ngo TH, Maes W, Dehaen W (2007) Efficient synthesis of aryldipyrromethanes in water and their application in the synthesis of corroles and dipyrromethenes. Arkivoc: 307
21. Guilard R, Gryko DT, Canard G, Barbe J-M, Koszarna B, Brandes S, Tasior M (2002) Synthesis of corroles bearing up to three different *meso* substituents. Org Lett 4:4491
22. Gryko DT, Jadach K (2001) A simple and versatile one-pot synthesis of *meso*-substituted *trans*-A_2B-corroles. J Org Chem 66:4267
23. Aviv-Harel I, Gross Z (2009) Aura of corroles. Chem Eur J 15:8382
24. Flamigni L, Gryko DT (2009) Photoactive corrole-based arrays. Chem Soc Rev 38:1635
25. Aviv I, Gross Z (2007) Corrole-based applications. Chem Commun 1987
26. Aviv-Harel I, Gross Z (2011) Coordination chemistry of corroles with focus on main group elements. Coord Chem Rev 255:717
27. Harvey PD, Stern C, Gros CP, Guilard R (2007) The photophysics and photochemistry of cofacial free base and metallated bisporphyrins held together by covalent architectures. Coord Chem Rev 251:401
28. Palmer JH (2012) Transitions metals corroles coordination chemistry. In: Mingos DMP, Day P, Dahl JP (eds) Molecular electronic structures of transition metal complexes I. Springer, Heidelberg
29. Paolesse R, Nardis S, Venanzi M, Mastroianni M, Russo M, Fronczek FR, Vicente MGH (2003) Vilsmeier formylation of 5,10,15-triphenylcorrole: expected and unusual products. Chem Eur J 9:1192
30. Saltsman I, Goldberg I, Gross Z (2003) One-step conversions of a simple corrole into chiral and amphiphilic derivatives. Tetrahedron Lett 44:5669
31. Gross Z, Galili N (1999) *N*-Substituted corroles: a novel class of chiral ligands. Angew Chem Int Ed 38:2366
32. Simkhovich L, Mahammed A, Goldberg I, Gross Z (2001) Synthesis and characterization of germanium, tin, phosphorus, iron, and rhodium complexes of tris(pentafluorophenyl)corrole, and the utilization of the iron and rhodium corroles as cyclopropanation catalysts. Chem Eur J 7:1041
33. Simkhovich L, Iyer P, Goldberg I, Gross Z (2002) Structure and chemistry of *N*-substituted corroles and their rhodium(I) and zinc(II) metal-ion complexes. Chem Eur J 8:2595
34. Tardieux C, Gros CP, Guilard R (1998) On corrole chemistry. An isomerization study and oxidative cleavage of the corrole macroring to a biliverdin structure. J Heterocycl Chem 35:965
35. Paolesse R, Sagone F, Macagnano A, Boschi T, Prodi L, Montalti M, Zaccheroni N, Bolletta F, Smith KM (1999) Photophysical behaviour of corrole and its symmetrical and unsymmetrical dyads. J Porphyrins Phthalocyanines 3:364
36. Gros CP, Barbe JM, Espinosa E, Guilard R (2006) Room-temperature autoconversion of free-base corrole into free-base porphyrin. Angew Chem Int Ed 45:564

37. Swider P, Nowak-Krol A, Voloshchuk R, Lewtak JP, Gryko DT, Danikiewicz W (2010) Mass spectrometry studies on *meso*-substituted corroles and their photochemical decomposition products. J Mass Spectrom 45:1443
38. Barata JFB, Neves MGPMS, Tomé AC, Faustino MAF, Silva AMS, Cavaleiro JAS (2010) How light affects 5,10,15-tris(pentafluorophenyl)corrole. Tetrahedron Lett 51:1537
39. Hirabayashi S, Omote M, Aratani N, Osuka A (2012) Directly linked corrole oligomers via facile oxidative3–3 coupling reaction. Bull Chem Soc Jpn 85:558
40. Stefanelli M, Shen J, Zhu WH, Mastroianni M, Mandoj F, Nardis S, Ou ZP, Kadish KM, Fronczek FR, Smith KM, Paolesse R (2009) Demetalation of silver(III) corrolates. Inorg Chem 48:6879
41. Manoj F, Nardis S, Pomarico G, Paolesse R (2008) Demetalation of corrole complexes: an old dream turning into reality. J Porphyrins Phthalocyanines 12:19
42. Ngo TH, Van Rossom W, Dehaen W, Maes W (2009) Reductive demetallation of Cu-corroles – a new protective strategy towards functional free-base corroles. Org Biomol Chem 7:439
43. Paolesse R, Nardis S, Sagone F, Khoury RG (2001) Synthesis and functionalization of *meso*-aryl-substituted corroles. J Org Chem 66:550
44. Du R-B, Liu C, Shen D-M, Chen Q-Y (2009) Partial bromination and fluoroalkylation of 5,10,15-tris(pentafluorophenyl)corrole. Synlett 16:2701
45. Nardis S, Pomarico G, Mandoj F, Fronczek FR, Smith KM, Paolesse R (2010) One-pot synthesis of *meso*-alkyl substituted isocorroles: the reaction of a triarylcorrole with Grignard reagent. J Porphyrins Phthalocyanines 14:752
46. Nardis S, Mandoj F, Paolesse R, Fronczek FR, Smith KM, Prodi L, Montalti M, Battistini G (2007) Synthesis and functionalization of germanium triphenylcorrolate: the first example of a partially brominated corrole. Eur J Inorg Chem 2345
47. Palmer JH, Durrell AC, Gross Z, Winkler JR, Gray HB (2010) Near-IR phosphorescence of iridium(III) corroles at ambient temperature. J Am Chem Soc 132:9230
48. Wagnert L, Rubin R, Berg A, Mahammed A, Gross Z, Levanon H (2010) Photoexcited triplet state properties of brominated and nonbrominated Ga(III)-corroles as studied by time-resolved electron paramagnetic resonance. J Phys Chem B 114:14303
49. Wagnert L, Berg A, Stavitski E, Berthold T, Kothe G, Goldberg I, Mahammed A, Simkhovich L, Gross Z, Levanon H (2006) Exploring the photoexcited triplet states of aluminum and tin corroles by time-resolved Q-band EPR. Appl Magn Reson 30:591
50. Palmer JH, Day MW, Wilson AD, Henling LM, Gross Z, Gray HB (2008) Iridium corroles. J Am Chem Soc 130:7786
51. Mahammed A, Tumanskii B, Gross Z (2011) Effect of bromination on the electrochemistry, frontier orbitals and spectroscopy of metallocorroles. J Porphyrins Phthalocyanines 15:1275
52. Wasbotten IH, Wondimagegn T, Ghosh A (2002) Electronic absorption, resonance Raman, and electrochemical studies of planar and saddled copper(III) *meso*-triarylcorroles. Highly substituent-sensitive Soret bands as a distinctive feature of high-valent transition metal corroles. J Am Chem Soc 124:8104
53. Golubkov G, Bendix J, Gray HB, Mahammed A, Goldberg I, DiBilio AJ, Gross Z (2001) High-valent manganese corroles and the first perhalogenated metallocorrole catalyst. Angew Chem Int Ed 40:2132
54. Schechter A, Stanevsky M, Mahammed A, Gross Z (2012) Four-electron oxygen reduction by brominated cobalt corrole. Inorg Chem 51:22
55. Mahammed A, Gray HB, Meier-Callahan AE, Gross Z (2003) Aerobic oxidations catalyzed by chromium corroles. J Am Chem Soc 125:1162
56. Ngo TH, Puntoriero F, Nastasi F, Robeyns K, Van Meervelt L, Campagna S, Dehaen W, Maes W (2010) Synthetic, structural, and photophysical exploration of *meso*-pyrimidinyl-substituted AB_2-corroles. Chem Eur J 16:5691
57. Mahammed A, Botoshansky M, Gross Z (2012) Chlorinated corroles. Dalton Trans 41:10938

58. Vestfrid J, Botoshansky M, Palmer JH, Durrell AC, Gray HB, Gross Z (2011) Iodinated aluminum(III) corroles with long-lived triplet excited states. J Am Chem Soc 133:12899
59. Mahammed A, Goldberg I, Gross Z (2001) Highly selective chlorosulfonation of tris (pentafluorophenyl)corrole as a synthetic tool for the preparation of amphiphilic corroles and metal complexes of planar chirality. Org Lett 3:3443
60. Mahammed A, Gross Z (2010) Chlorosulfonated corrole: a versatile synthon for advanced materials. J Porphyrins Phthalocyanines 14:911
61. Gross Z, Mahammed A (2002) Selective sulfonation and deuteration of free-base corroles. J Porphyrins Phthalocyanines 6:553
62. Saltsman I, Mahammed A, Goldberg I, Tkachenko E, Botoshansky M, Gross Z (2002) Selective substitution of corroles: nitration, hydroformylation, and chlorosulfonation. J Am Chem Soc 124:7411
63. Agadjanian H, Weaver JJ, Mahammed A, Rentsendorj A, Bass S, Kim J, Dmochowski IJ, Margalit R, Gray HB, Gross Z, Medina-Kauwe LK (2006) Specific delivery of corroles to cells via noncovalent conjugates with viral proteins. Pharm Res 23:367
64. Lim P, Mahammed A, Okun Z, Saltsman I, Gross Z, Gray HB, Termini J (2012) Differential cytostatic and cytotoxic action of metallocorroles against human cancer cells: potential platforms for anticancer drug development. Chem Res Toxicol 25:400
65. Hwang JY, Lubow DJ, Sims JD, Gray HB, Mahammed A, Gross Z, Medina-Kauwe LK, Farkas DL (2012) Investigating photoexcitation-induced mitochondrial damage by chemo-therapeutic corroles using multimode optical imaging. J Biomed Opt 17:015003-1
66. Agadjanian H, Ma J, Rentsendorj A, Valluripalli V, Hwang JY, Mahammed A, Farkas DL, Gray HB, Gross Z, Medina-Kauwe LK (2009) Tumor detection and elimination by a targeted gallium corrole. Proc Natl Acad Sci USA 106:6105
67. Haber A, Aviram M, Gross Z (2012) Variables that influence cellular uptake and cytotoxic/ cytoprotective effects of macrocyclic iron complexes. Inorg Chem 51:28
68. Catrinescu M-M, Chan W, Mahammed A, Gross Z, Levin LA (2012) Superoxide signaling and cell death in retinal ganglion cell axotomy: effects of metallocorroles. Exp Eye Res 97:31
69. Hwang JY, Lubow J, Chu D, Ma J, Agadjanian H, Sims J, Gray HB, Gross Z, Farkas DL, Medina-Kauwe LK (2011) A mechanistic study of tumor-targeted corrole toxicity. Mol Pharm 8:2233
70. Hwang JY, Lubow J, Chu D, Sims J, Alonso-Valenteen F, Gray HB, Gross Z, Farkas DL, Medina-Kauwe LK (2012) Photoexcitation of tumor-targeted corroles induces singlet oxygen-mediated augmentation of cytotoxicity. J Control Release 163:368
71. Sudhakar K, Velkannan V, Giribabu L (2012) Synthesis, electrochemical and photophysical properties of β-carboxy triarylcorroles. Tetrahedron Lett 53:991
72. Vale LSHP, Barata JFB, Santos CIM, Neves MGPMS, Faustino MAF, Tomé AC, Silva AMS, Paz FAA, Cavaleiro JAS (2009) Corroles in 1,3-dipolar cycloaddition reactions. J Porphyrins Phthalocyanines 13:358
73. Sorasaenee K, Taqavi P, Henling LM, Gray HB, Tkachenko E, Mahammed A, Gross Z (2007) Amphiphilic aluminium(III) and gallium(III) corroles. J Porphyrins Phthalocyanines 11:189
74. Santos CIM, Oliveira E, Barata JFB, Faustino MAF, Cavaleiro JAS, Neves MGPMS, Lodeiro C (2012) Corroles as anion chemosensors: exploiting their fluorescence behaviour from solution to solid-supported devices. J Mater Chem 22:13811
75. Mastroianni M, Zhu W, Stefanelli M, Nardis S, Fronczek FR, Smith KM, Ou Z, Kadish KM, Paolesse R (2008) β-Nitro derivatives of germanium(IV) corrolates. Inorg Chem 47:11680
76. Stefanelli M, Mandoj F, Mastroianni M, Nardis S, Mohite P, Fronczek FR, Smith KM, Kadish KM, Xiao X, Ou Z, Chen P, Paolesse R (2011) Amination reaction on copper and germanium β-nitrocorrolate. Inorg Chem 50:8281
77. Pomarico G, Fronczek FR, Nardis S, Smith KM, Paolesse R (2011) Synthetic protocols for the nitration of corroles. J Porphyrins Phthalocyanines 15:1085

78. Pomarico G, Xiao X, Nardis S, Paolesse R, Fronczek FR, Smith KM, Fang Y, Ou Z, Kadish KM (2010) Synthesis and characterization of free-base, copper, and nickel isocorroles. Inorg Chem 49:5766
79. Stefanelli M, Mastroianni M, Nardis S, Licoccia S, Fronczek FR, Smith KM, Zhu W, Ou Z, Kadish KM, Paolesse R (2007) Functionalization of corroles: the nitration. Inorg Chem 46:10791
80. Smith KM, Barnett GH, Evans B, Martynenko Z (1979) Novel *meso*-substitution reactions of metalloporphyrins. J Am Chem Soc 101:5953
81. Stefanelli M, Pomarico G, Tortora L, Nardis S, Fronczek FR, McCandless GT, Smith KM, Manowong M, Fang YC, Kadish KM, Rosa A, Ricciardi G, Paolesse R (2012) β-Nitro-5,10,15-tritolylcorroles. Inorg Chem 51:6928
82. Nardis S, Stefanelli M, Mohite P, Pomarico G, Tortora L, Manowong M, Chen P, Kadish KM, Fronczek FR, McCandless GT, Smith KM, Paolesse R (2012) β-Nitro derivatives of iron corrolates. Inorg Chem 51:3910
83. Vaz Serra VI, Pires SMG, Alonso CMA, Neves MGPMS, Tomé AC, Cavaleiro JAS (2013) *meso*-Tetraarylporphyrins bearing nitro or amino groups: synthetic strategies and reactivity profiles. Top Heterocycl Chem 33. doi:10.1007/7081_2013_101
84. Collman JP, Decréau RA (2005) 5,10,15-Tris(*o*-aminophenyl) corrole (TAPC) as a versatile synthon for the preparation of corrole-based hemoprotein analogs. Org Lett 7:975
85. Collman JP, Kaplun M, Decréau RA (2006) Metal corroles as electrocatalysts for oxygen reduction. Dalton Trans 554
86. Barbe J-M, Canard G, Brandès S, Guilard R (2005) Organic–inorganic hybrid sol–gel materials incorporating functionalized cobalt(III) corroles for the selective detection of CO. Angew Chem Int Ed 44:3103
87. Barbe J-M, Canard G, Brandès S, Guilard R (2005) Synthesis and physicochemical characterization of *meso*-functionalized corroles: precursors of organic–inorganic hybrid materials. Eur J Org Chem 4601
88. Barbe J-M, Canard G, Brandès S, Guilard R (2007) Selective chemisorption of carbon monoxide by organic–inorganic hybrid materials incorporating cobalt(III) corroles as sensing components. Chem Eur J 13:2118
89. Gao Y, Liu J, Jiang W, Xia M, Zhang W, Li M, Åkermark B, Sun L (2007) Synthesis and photophysical and electrochemical properties of a binuclear Ru(bpy)3-Cu(III) corrole complex. J Porphyrins Phthalocyanines 11:463
90. Richeter S, Jeandon C, Ruppert R, Callot HJ (2002) A modular approach to porphyrin oligomers using metal ions as connectors. Chem Commun 266
91. Richeter S, Hadj-Aïssa A, Taffin C, van der Lee KA, Leclercq D (2007) Synthesis and structural characterisation of the first *N*-heterocyclic carbene ligand fused to a porphyrin. Chem Commun 2148
92. Costa JIT, Tomé AC, Neves MGPMS, Cavaleiro JAS (2011) 5,10,15,20-Tetrakis (pentafluorophenyl)porphyrin: a versatile platform to novel porphyrinic materials. J Porphyrins Phthalocyanines 15:1117
93. Aviezer D, Cotton S, David M, Segev A, Khaselev N, Galili N, Gross Z, Yayon A (2000) Porphyrin analogues as novel antagonists of fibroblast growth factor and vascular endothelial growth factor receptor binding that inhibit endothelial cell proliferation, tumor progression, and metastasis. Cancer Res 60:2973
94. Hori T, Osuka A (2010) Nucleophilic substitution reactions of *meso*-5,10,15-tris (pentafluorophenyl)-corrole; synthesis of ABC-type corroles and corrole-based organogels. Eur J Org Chem 2379
95. Cardote TAF, Barata JFB, Faustino MAF, Preuß A, Neves MGPMS, Cavaleiro JAS, Ramos CIV, Santana-Marques MGO, Röder B (2012) Pentafluorophenylcorrole-D-galactose conjugates. Tetrahedron Lett 53:6388

96. Barata JFB, Daniel-da-Silva AL, Neves MGPMS, Cavaleiro JAS, Trindade T (2013) Corrole-silica hybrid particles: synthesis and effects on singlet oxygen generation. RSC Adv 3:274
97. Maes W, Ngo TH, Vanderhaeghen J, Dehaen W (2007) *Meso*-pyrimidinyl-substituted A_2B-corroles. Org Lett 9:3165
98. Ngo TH, Nastasi F, Puntoriero F, Campagna S, Dehaen W, Maes W (2010) *Meso*-pyrimidinyl-substituted A_2B- and A_3-corroles. J Org Chem 75:2127
99. Cavaleiro JAS, Tomé AC, Neves MGPMS (2010) *Meso*-tetraarylporphyrin derivatives: new synthetic methodologies. In: Kadish KM, Smith KM, Guilard R (eds) Handbook of porphyrin science. World Scientific, Singapore
100. Cavaleiro JAS, Neves MGPMS, Tomé AC (2003) Cycloaddition reactions of porphyrins. Arkivoc: 107
101. Tomé AC, Neves MGPMS, Cavaleiro JAS (2009) Porphyrins and other pyrrolic macrocycles in cycloaddition reactions. J Porphyrins Phthalocyanines 13:408
102. Silva AMG, Cavaleiro JAS (2008) Porphyrins in Diels–Alder and 1,3-dipolar cycloaddition reactions. In: Gribble GW, Joule JA (eds) Progress in heterocyclic chemistry, vol 19. Elsevier, Amsterdam, p 44
103. Barata JFB, Silva AMG, Faustino MAF, Neves MGPMS, Tomé AC, Silva AMS, Cavaleiro JAS (2004) Novel Diels–Alder and thermal [4+4] cycloadditions of corroles. Synlett 7:1291
104. Barata JFB (2009) Synthesis and reactivity studies of new corrole derivatives. Department of Chemistry, University of Aveiro, Aveiro, p 248
105. Barata JFB, Silva AMG, Neves MGPMS, Tomé AC, Silva AMS, Cavaleiro JAS (2006) β,β'-Corrole dimers. Tetrahedron Lett 47:8171
106. Hiroto S, Furukawa K, Shinokubo H, Osuka A (2006) Synthesis and biradicaloid character of doubly linked corrole dimers. J Am Chem Soc 128:12380
107. Li CJ, Fechtel M, Feng YQ, Krautler B (2012) Corroles programmed for regioselective cycloaddition chemistry – synthesis of a bisadduct with C-60-fullerene. J Porphyrins Phthalocyanines 16:556
108. Vale LSHP, Barata JFB, Neves MGPMS, Faustino MAF, Tomé AC, Silva AMS, Paz FAA, Cavaleiro JAS (2007) Novel quinone-fused corroles. Tetrahedron Lett 48:8904
109. Ramos CIV, Santana-Marques MG, Ferrer-Correia AJ, Barata JFB, Tomé AC, Neves MGPMS, Cavaleiro JAS, Abreu PE, Pereira MM, Pais AACC (2012) Differentiation of aminomethyl corrole isomers by mass spectrometry. J Mass Spectrom 47:516
110. D'Souza F, Chitta R, Ohkubo K, Tasior M, Subbaiyan NK, Zandler ME, Rogacki MK, Gryko DT, Fukuzumi S (2008) Corrole-fullerene dyads: formation of long-lived charge-separated states in nonpolar solvents. J Am Chem Soc 130:14263
111. Lewandowska K, Barszcz B, Wolak J, Graja A, Grzybowski M, Gryko DT (2013) Vibrational properties of new corrole-fullerene dyad and its components. Dyes Pigm 96:249
112. Scrivanti A, Beghetto V, Matteoli U, Antonaroli S, Marini A, Mandoj F, Paolesse R, Crociani B (2004) Iminophosphine–palladium(0) complexes as highly active catalysts in the Suzuki reaction. Synthesis of undecaaryl substituted corroles. Tetrahedron Lett 45:5861
113. Berg S, Thomas KE, Beavers CM, Ghosh A (2012) Undecaphenylcorroles. Inorg Chem 51:9911
114. König M, Reith LM, Monkowius U, Knör G, Bretterbauer K, Schoefberger W (2011) Suzuki–Miyaura cross-coupling reaction on copper-*trans*-A_2B-corroles with excellent functional group tolerance. Tetrahedron 67:4243
115. Kinzel T, Zhang Y, Buchwald SL (2010) A new palladium precatalyst allows for the fast Suzuki–Miyaura coupling reactions of unstable polyfluorophenyl and 2-heteroaryl boronic acids. J Am Chem Soc 132:14073
116. Reith LM, Koenig M, Schwarzinger C, Schoefberger W (2012) BiIII-corroles: a versatile platform for the synthesis of functionalized corroles. Eur J Inorg Chem 4342
117. Reith LM, Stiftinger M, Monkowius U, Knor G, Schoefberger W (2011) Synthesis and characterization of a stable bismuth(III) A_3-corrole. Inorg Chem 50:6788

118. Bröring M, Funk M, Milsmann C (2009) Investigating different strategies towards the preparation of chiral and achiral bis-picket fence corroles. J Porphyrins Phthalocyanines 13:107
119. Nigel-Etinger I, Mahammed A, Gross Z (2011) Covalent versus non-covalent (biocatalytic) approaches for enantioselective sulfoxidation catalyzed by corrole metal complexes. Catal Sci Technol 1:578
120. Tasior M, Gryko DT, Shen J, Kadish KM, Becherer T, Langhals H, Ventura B, Flamigni L (2008) Energy- and electron-transfer processes in corrole-perylenebisimide-triphenylamine array. J Phys Chem C 112:19699
121. Zhan H-Y, Liu H-Y, Lu J, Wang A-Z, You L-L, Wang H, Ji L-N, Jiang H-F (2010) Alkynyl corroles: synthesis by Sonogashira coupling reaction and the physicochemical properties. J Porphyrins Phthalocyanines 14:150
122. Hiroto S, Hisaki I, Shinokubo H, Osuka A (2005) Synthesis of corrole derivatives through regioselective Ir-catalyzed direct borylation. Angew Chem Int Ed 44:6763
123. Pomarico G, Nardis S, Paolesse R, Ongayi OC, Courtney BH, Fronczek FR, Vicente MGH (2011) Synthetic routes to 5,10,15-triaryl-tetrabenzocorroles. J Org Chem 76:3765

Top Heterocycl Chem (2014) 33: 143–202
DOI: 10.1007/7081_2013_99

Published online: 19 April 2013

Degradation Pathways for Porphyrinoids

Jacek Wojaczyński

Abstract Porphyrin, a tetrapyrrolic aromatic macrocycle, is relatively resistant to degradation. However, certain strong oxidants (e.g. chromic acid) cause its decomposition to monopyrrolic units. More often, ring opening caused by attack of oxidant on a *meso*-position has been observed. Such degradation by metal salts (thallium(III), cerium(IV)), nitric acid, and other reagents has been studied. Light-driven macrocycle opening by dioxygen has also been noted. Coupled oxidation of metalloporphyrins has been investigated mainly as a mimics of heme degradation observed in vivo.

Modifications of parent porphyrin macrocycle can cause a prominent change of its reactivity toward oxidants. In particular, inversion of one of the pyrrole rings (in N-confused porphyrin) or removal of one of the methine bridges (in corrole) increases macrocycle susceptibility to oxidative ring opening.

Keywords Biliverdin · Coupled oxidation · Degradation · Photooxidation · Tetrapyrrole

Contents

J. Wojaczyński (✉)
Department of Chemistry, University of Wrocław, 14 F. Joliot-Curie St., 50383 Wrocław, Poland
e-mail: jacek.wojaczynski@chem.uni.wroc.pl

Abbreviations

CAN	Cerium(IV) ammonium nitrate
DDQ	2,3-Dichloro-5,6-dicyanobenzoquinone
FCC	Fluorescent chlorophyll catabolite
HO	Heme oxygenase
NBS	*N*-Bromosuccinimide
NCC	Nonfluorescent chlorophyll catabolite
$OEBH_3$	2,3,7,8,12,13,17,18-Octaethylbilindione
$OEPH_2$	2,3,7,8,12,13,17,18-Octaethylporphyrin
$OEPOH_3$	2,3,7,8,12,13,17,18-Octaethyloxophlorin (2,3,7,8,12,13,17,18-octaethyl-5-hydroxyporphyrin)
PDT	Photodynamic therapy
$TPPH_2$	5,10,15,20-Tetraphenylporphyrin
TTFA	Thallium(III) trifluoroacetate
TTN	Thallium(III) nitrate

1 Scope and Limitations

This review is focused on degradation of tetrapyrrolic macrocycles: porphyrins, their N-confused isomers, and corroles (**1**–**3**, Fig. 1). "Degradation" is understood here as a disruption of a macrocyclic system. For this reason, reactions leading only to the lowering of number of rings of the starting pentacyclic system are not included, although formation of secochlorins **4** [1–3] or vacataporphyrins **5** [4, 5] (Fig. 2) also results in a qualitative change of the macrocycle properties. Similarly, processes connected with the loss of the macrocyclic aromaticity without ring opening (e.g. formation of phlorins) will not be discussed unless they serve as a preliminary stage of the actual degradation. Ring-opening reactions of phthalocyanines, porphyrazines, and similar macrocycles as well as systems containing less or more than four pyrrolic

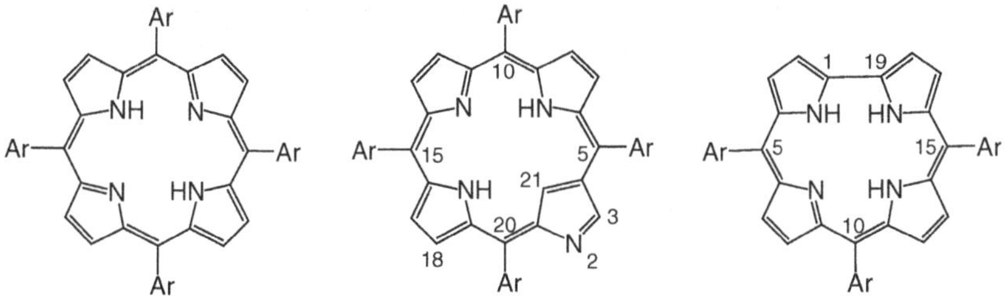

1 2 3

Fig. 1 Porphyrin, its N-confused isomer and corrole (*meso*-aryl derivatives are shown)

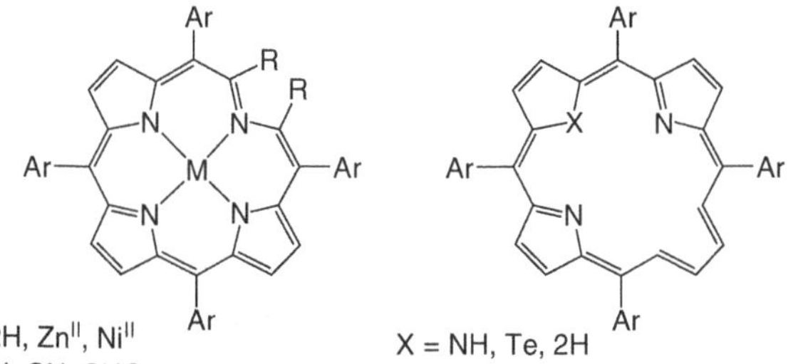

M = 2H, Zn^{II}, Ni^{II}
R = 2H, CN, CHO
4

X = NH, Te, 2H
5

Fig. 2 Examples of secochlorins (**4**) and vacataporphyrins (**5**) [1–5]

rings are not presented. The emphasis is laid on the literature published in the years 2000–2012, but for the sake of comparison, older achievements are also briefly described.

The porphyrin macrocycle containing a conjugated 18 π-electron system is known to be highly stable toward destruction. This fact inspired search for methods of ring opening. The interest in degradation of cyclic tetrapyrroles is connected with several aspects: analytical (structure determination), biochemical (heme and chlorophyll metabolism, formation of algae biliproteins), catalytic (stability of porphyrin derivatives used as catalysts and photosensitizers), and synthetic (preparation of linear oligopyrrolic systems exhibiting interesting properties: helical chirality [6–9], conformational flexibility connected with possible *E–Z* isomerization [10], specific and sometimes unpredictable coordination modes [11–14]).

A direct opening of porphyrin macroring is achieved when one of the C(α)–C (*meso*) bonds is cleft. Reactions at the macrocycle periphery occur preferentially on *meso* positions unless sterical reasons preclude access to this part of molecule [15]. In general, degradation is caused by various oxidants (reduction with hydriodic acid in acetic acid being a notable exception) and is thus preceded by their attack on one of the methine bridges. On the other hand, numerous examples of pyrrole- and metal-centered oxidations have been also described, which can also constitute a preliminary step of further macrocycle decomposition.

This chapter is divided into eight sections. Section 2 is devoted to traditional methods of structure determination based on destruction of tetrapyrrolic systems. In Sect. 3, macrocycle opening by oxidants is discussed, excluding light-driven reactions with dioxygen (Sect. 4) and coupled oxidation of metalloporphyrins (Sect. 5). Biodegradation is shortly presented in Sect. 6, followed by concluding remarks (Sect. 7) and reference list.

2 Degradation Used as Analytical Tool

Classical methods used for structure elucidation of tetrapyrrolic compounds (both cyclic ones and linear derivatives) utilized oxidative degradation with chromic acid, potassium permanganate [16, 17], and ozone [18] or hydriodic acid reduction [19]. Analysis of the resulting monopyrrolic units (maleimides, succinimides) which could be identified, allowed recognition of β-substitution pattern, and in certain cases also *meso* substituents [15]. Among those methods, chromic acid (CrO_3/H_2SO_4) oxidation used in combination with gas chromatography and mass spectrometry has been most widely applied, particularly for identification of chlorophyll derivatives, bilins, and geoporphyrins [20–27]. More recently, this method was used in the analysis of hematoporphyrin derivative used in photodynamic therapy [28, 29]. A new method was described allowing quantitative determination of chlorophyll derivatives by analysis of amount of ethylmethylmaleimide formed during degradation with chromic acid [30].

Formally, part of analytical methods commonly used for the characterization of newly synthesized tetrapyrrolic macrocycles also involves destruction of the molecule. Elementary (combustion) analysis is widely performed, though the results are sometimes not quite satisfactory due to the ease of incorporation of various guest molecules, including solvents, in the crystal lattice of porphyrins [31]. Also a conventional method of carbon isotopic composition of geoporphyrins relies on combustion to CO_2 which is examined by mass spectrometry [32, 33]. Fragmentation observed in certain techniques of mass spectrometry serves as a source of a valuable structural information [34–36]. Analytical data based on other methods involving sample decomposition, such as combustion calorimetry experiments [37, 38], differential scanning calorimetry, and thermogravimetry [39–41] are less frequently reported.

3 Ring Opening by Oxidants

Ring-centered reactions of porphyrin derivatives with various oxidants can lead to opening of the macrocycle without its complete disintegration. Systematic research on oxidation of tetrapyrrolic macrocycles was performed in the 1960–1970s;

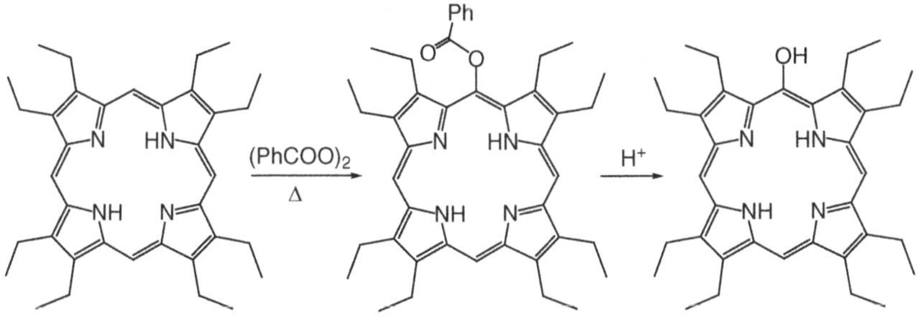

6 7 8

Scheme 1 Synthesis of octaethyloxophlorin [43, 44, 46]

in most of the recent contributions specifically modified systems or reactions conducted under modified conditions have been discussed.

In Sect. 3.1, reactions of porphyrins and their complexes with redox innocent metals are described. Degradation of iron and manganese porphyrin complexes by reagents which are typically used in metalloporphyrin-catalyzed oxidations is discussed in Sect. 3.2. The section is concluded by description of reactivity of N-confused porphyrins and corroles.

3.1 Oxidation of Porphyrins and Their Complexes

Reactions of porphyrins and their complexes with oxidants were extensively studied by Bonnett and coworkers [42–46] and Smith et al. [47–53]. Special attention was devoted to *meso* oxidation leading to oxophlorin (5-hydroxyporphyrin) derivatives due to importance of iron oxophlorins as intermediates in the process of heme degradation. Octaethyloxophlorin (OEPOH$_3$, **8**) was obtained from the reaction of 2,3,7,8,12,13,17,18-octaephylporphyrin (OEPH$_2$, **6**) with benzoyl peroxide [43, 44, 46]. A radical attack at *meso* position gave 5-benzoyloxyporphyrin **7** at ca. 30% yield, and its hydrolysis led to the desired product **8** (Scheme 1). This compound was also prepared by ring synthesis and by coupled oxidation (see 5.1) [43, 44].

Bonnett et al. prepared octaethyloxophlorin **8** by treatment of (OEP)Fe^{II}(py)$_2$ dissolved in pyridine with hydrogen peroxide [43, 45]. Later it was found that reaction did not occur with zinc(II), nickel(II), copper(II), iron(III), and cobalt(III) complexes, while oxophlorins were obtained for Fe(II), Co(II) and Mn(II) or Mn(III) (i.e. metal ions with an easily accessible higher oxidation state) [45]. Conversion of iron(III) oxophlorin into verdoheme analog and its further conversion to biliverdin **9** (Fig. 3) was also described [45].

Kalish et al. demonstrated that treatment of deuteroheme, mesoheme, or protoheme with hydrogen peroxide in pyridine solution yielded all four isomeric oxophlorin complexes in comparable yields [54]. In contrast, oxidation of iron(II)

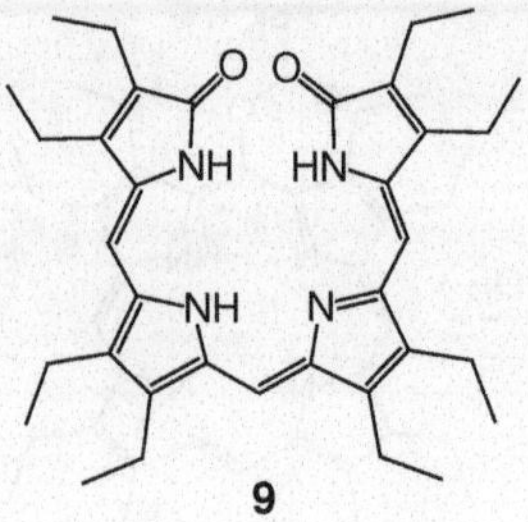

9

Fig. 3 Octaethylbilindione – a synthetic biliverdin analogue

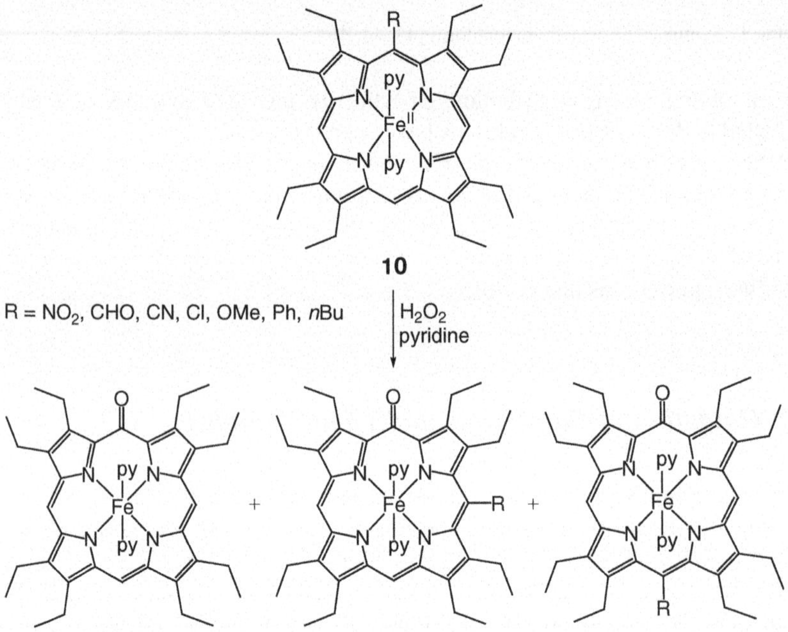

10

11 12 13

Scheme 2 Oxidation of 5-substituted iron(II) octaethylporphyrins [55]

5-substituted-octaethylporphyrins (5-R-OEP)$Fe^{III}(py)_2$ (R = NO_2, CHO, CN, Cl, OMe, Ph, *n*-Bu) exhibited a strong dependence on the nature of the substituent: yields of (OEPO)$Fe(py)_2$, a product of replacement of R group with oxygen function, varied from 0% (R = Ph, *n*-Bu) to 100% (R = NO_2), while ratio of *cis* to *trans*-oxygenated products (**12** and **13**, Scheme 2) changed from 5.0 (R = CN) to 1.4 (R = Ph) [55].

Treatment of zinc or magnesium complexes of octaethylporphyrin **14** with thallium(III) trifluoroacetate (TTFA) followed by demetallation gave high yields (55–79%) of oxophlorin **8** [49, 50]. 5-Trifluoroacetoxyporphyrins **15** were isolated as stable intermediates of this process (Scheme 3). Similar reactivity was observed when lead(IV) or mercury(II) trifluoroacetates were used, but yields of oxophlorins

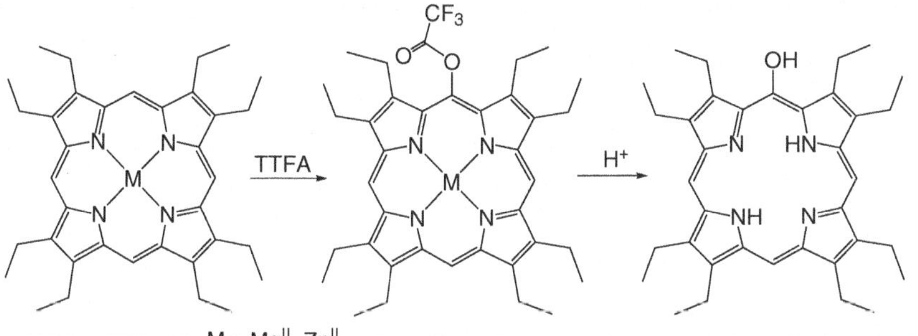

14 15 8

Scheme 3 Oxidation of OEP complexes with TTFA leading to $OEOPOH_3$ [49, 50]

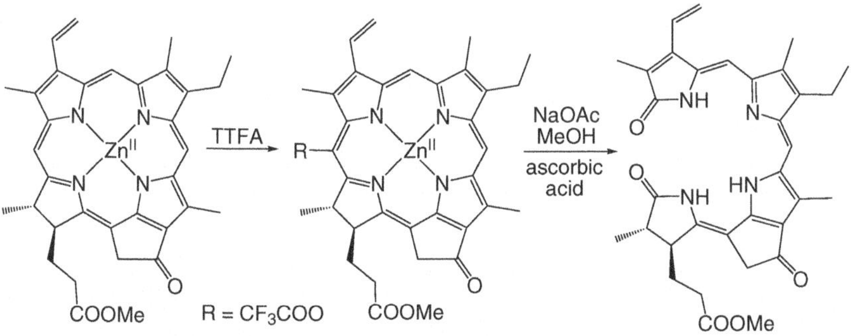

16 17 18

Scheme 4 TTFA oxidation of zinc(II) methyl pyropheophorbide *a* [53]

were significantly lower (19–37%) [50]. Iron(III), copper(II), and nickel(II) complexes of OEP were found resistant to the TTFA attack. An analogous reaction of zinc(II) methyl pyropheophorbide *a* **16** with TTFA, followed by hydrolysis in the presence of ascorbic acid and air proceeded regioselectively to give dihydrobiliverdin **18** (Scheme 4) [53].

In contrast to OEP complexes, zinc tetraphenylporphyrin ((TPP)Zn^{II} **19**) was converted by TTFA, thallium(III) nitrate (TTN) or cerium(IV) ammonium nitrate (CAN) into a ring-opened tetrapyrrole **20** along with 5,15-disubstituted products **21**, **22** (Scheme 5) [51, 52]. These compounds were obtained after acidic workup and chromatography on alumina column. The proper structure of compound **20**, formed by addition of water molecule to the demetallated primary product, was established in the course of studies on photooxidation of TPP complexes (Sect. 4.1).

Interestingly, when zinc(II) 5,10,15-triarylporphyrins were reacted with thallium(III) trifluoroacetate, an oxidative dimerization was observed leading to *meso–meso* linked diporphyrins (Scheme 6) [56]. A similar reactivity of zinc di- and triarylporphyrins with silver(I) salts was reported by Osuka and coworkers [57–59].

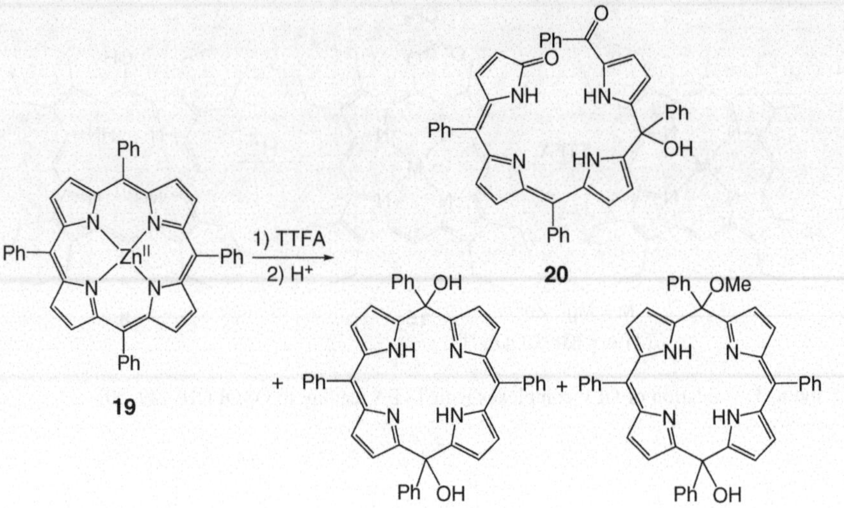

Scheme 5 TTFA oxidation of (TPP)Zn^{II} [51, 52]

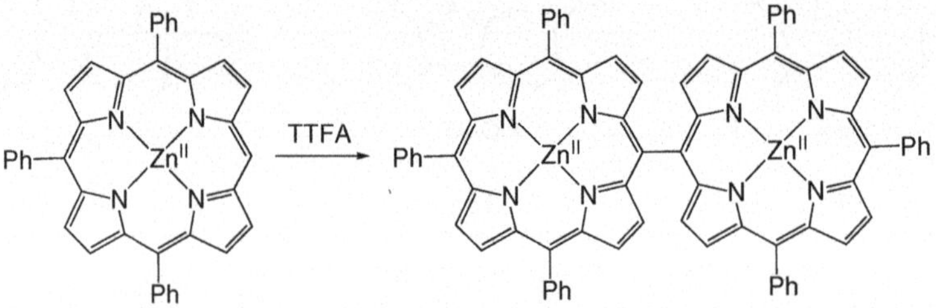

23 24

Scheme 6 Dehydrodimerization of zinc(II) triphenylporphyrin [56]

In case of TTN and CAN oxidation of (TPP)Zn^{II}, β-nitrated product **25** (Fig. 4) was also isolated [51, 52]. *Meso*-nitration of octaethylporphyrin was reported by Bonnett and Dimsdale, who used fuming nitric acid–acetic acid mixture for this reaction; ring opening was not observed under these conditions [42]. Catalano et al. established the dependence of the site of reaction with nitrogen dioxide on the metal coordinated to tetraphenylporphyrin [60]. Nickel(II), copper(II), and palladium(II) complexes were exclusively converted to 2-nitro derivatives, while for more electropositive zinc(II) and magnesium(II) ions ring opening resulting from the reaction at *meso* position was noted. This observation was rationalized by a different symmetry of π-cation radicals formed by oxidation of metalloporphyrin with NO_2. Also reaction of (TPP$^{+\bullet}$)Zn^{II}(ClO_4) with various nucleophiles yielded mainly 2-substituted derivatives, but in the particular case of nitrite anion,

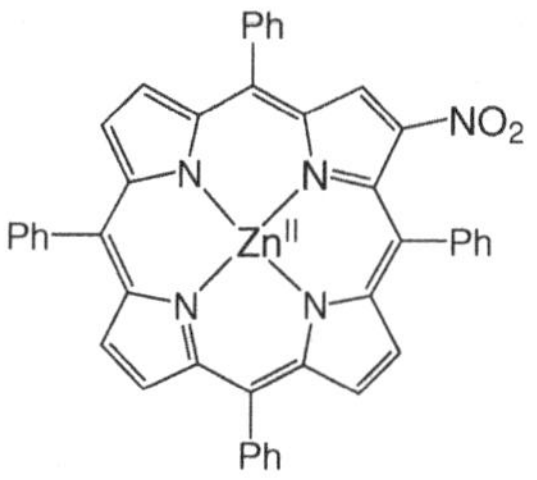

25

Fig. 4 Zinc 2-nitro-5,10,15,20-tetraphenylporphyrin

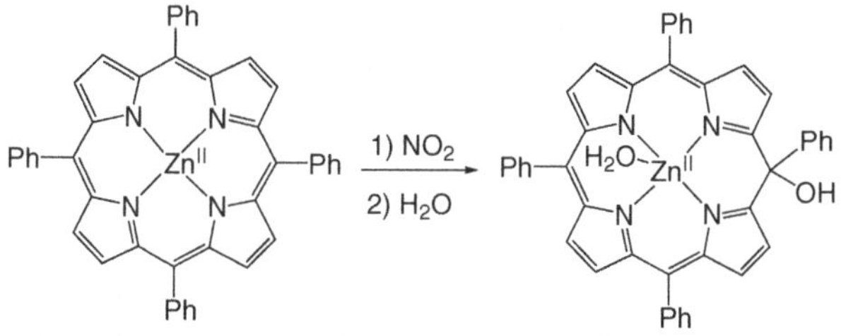

19 **26**

Scheme 7 Formation of zinc(II) isoporphyrin [63]

β-nitrated porphyrin product was accompanied with an open-chain compound **20** [61]. More recently, Sarkar et al. described a formation of *meso*-hydroxylated isoporphyrin **26** upon treatment of *meso*-tetrakis(3,4,5-trimethoxyphenyl)porphyrin iron(III) or zinc complex with NO_2 (O_2 and NO, Scheme 7) [62, 63]. Further degradation of iron isoporphyrin in solution was observed, and formation of verdoheme- and biliverdin-type products was postulated on the basis of UV–vis spectra. In contrast, zinc derivative remained stable in presence of air and light.

Oxidation of macrocycle can be facilitated by an appropriate modification of the porphyrin ring (both sterical aspects and generation of specific reactivity by substitution are of importance). Ring opening of sterically hindered, dodecasubstituted porphyrins **27** via $NaNO_2$ treatment in the presence of trifluoroacetic acid and air was studied by Ongayi et al. [64–66]. Authors attributed the ease of degradation of porphyrinic substrates **27** to the tendency to relieve steric strain. The proposed reaction pathway involved oxidation of macrocycle by NO^+ to a π-cation radical followed by ring opening by dioxygen. A primary bilitrienone product **29** was isolated in 70% yield (Scheme 8), but only for nonyl-substituted system, while in case of *meso*-tetraphenyl derivative the unstable compound **29** was converted to a biladienone **31** by addition of water. Two isomers of hydrated benzoylbiliverdin **31** were separated, presumably differing in the configuration of C(4)–C(5) bond. Hydration of nonyl derivative **29** was observed as well, but it could be inverted by heating the product **31** above 40°C [66].

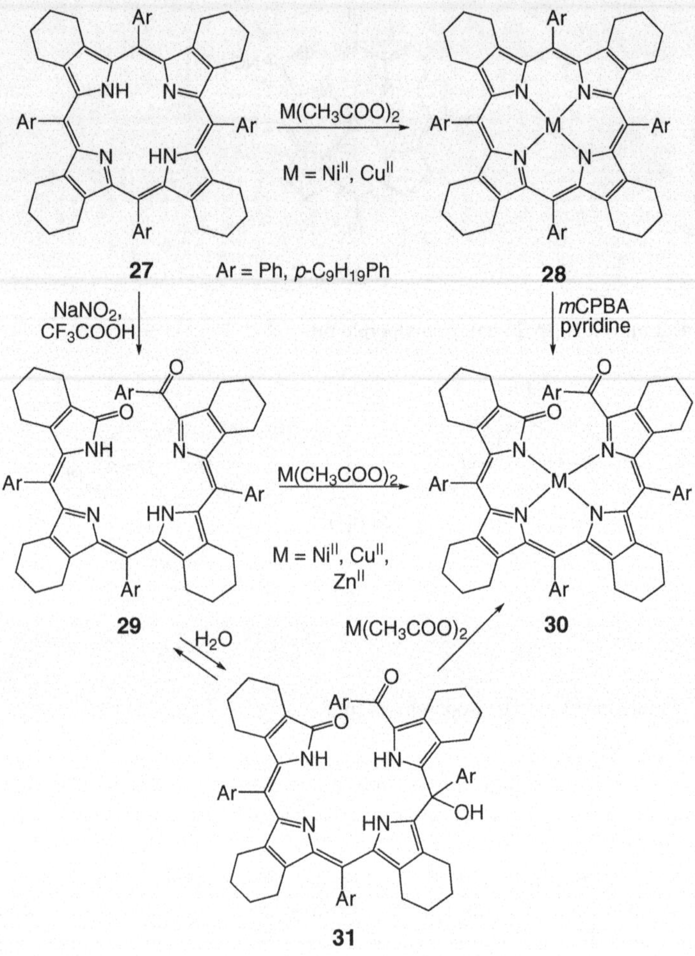

Scheme 8 Degradation of dodecasubstituted porphyrins [64–66]

Metallation of **31** with Ni(II), Cu(II), and Zn(II) ions led to formation of 4N chelates **30** in which a dehydrated form of tetrapyrrole was found [65]. Nickel(II) and copper(II) complexes were also prepared by an alternative route from the corresponding metalloporphyrins **28** which were oxidized using *meta*-chloroperoxybenzoic acid in pyridine in the presence of air (Scheme 8) [65].

Yashunsky, Morozova, and Ponomarev described a conversion of nickel complexes of 5-formylporphyrin oximes **32** in a mixed water-organic solvent system into brown-yellow products [67, 68]. These products were identified as open-chain tripyrroryloxazoles **33** and were isolated by column chromatography in ca. 50% yield (Scheme 9) [68]. A mechanism was proposed involving conversion of oxime substituent into 1,2-oxazine ring and oxidation of formed intermediates by dioxygen leading to fission of pyrrolic β,β′ bond and elimination of α-carbon.

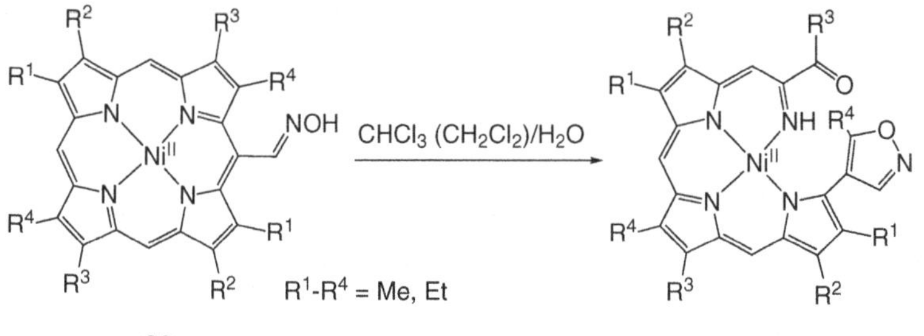

Scheme 9 Conversion of 5-formylporphyrin oximes to tripyrroryloxazoles [67, 68]

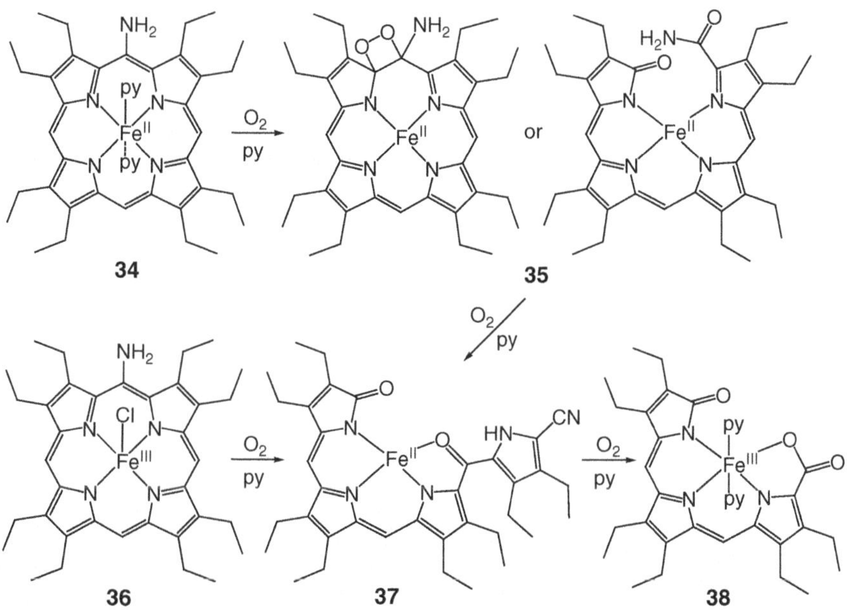

Scheme 10 Ring opening of iron *meso*-aminoporphyrin complexes [69, 70]

A remarkable ease of ring opening was observed for *meso*-amino-substituted octaethylporphyrin complexes, $(H_2N\text{-OEP})Fe^{II}(py)_2$ **34** and $(H_2N\text{-OEP})Fe^{III}Cl$ **36** [69, 70]. The exposure of their pyridine solutions to dioxygen resulted in its regioselective attack at the substituted carbon; ring opening was followed by a second oxidation step introducing another *meso*-oxygen atom; at the same time the terminal amide fragment was dehydrated to cyano group (Scheme 10). A resulting (3N + O) complex **37** and its analog with an axial ethanol ligand were characterized by X-ray crystallography. In the case of **34** oxidation, a green

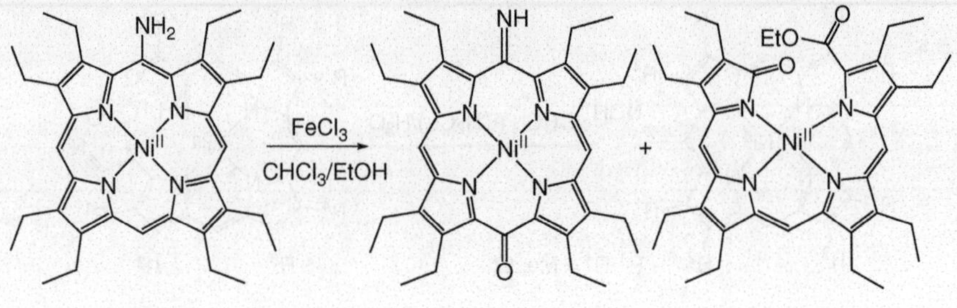

39 **40** **41**

Scheme 11 Oxidation of nickel(II) *meso*-aminoporphyrin

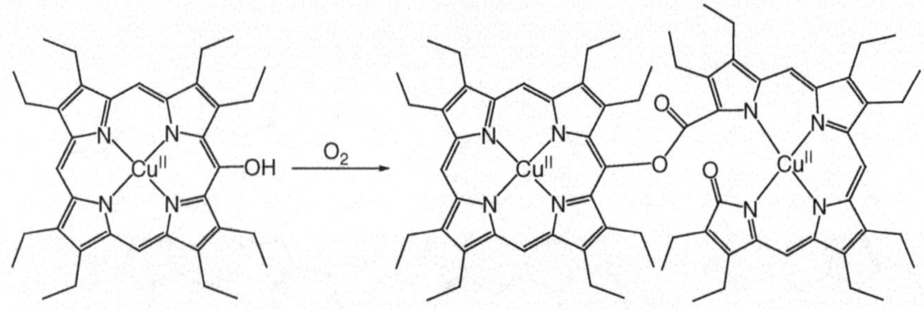

42 **43**

Scheme 12 Formation of dinuclear copper complex [14]

intermediate was detected [69]. Its ^{1}H NMR spectrum indicated a significant degree of ligand radical character and symmetry lowering with respect to the starting iron (II) complex **34**, which was attributed to the formation of dioxygen adduct or iron biliverdin derivative **35**. A prolonged contact with dioxygen resulted in a slow conversion of compound **37** to a mixture of tripyrrole complex **38** and small amounts of another unidentified product [70].

A pyridine solution of nickel(II) complex of 5-aminooctaethylporphyrin **39** remained unchanged upon exposure to dioxygen [71]. A slow reaction was observed, however, when iron(III) chloride was used as oxidant (Scheme 11), yielding a biliverdin derivative **41** as a minor isolated product (10% yield).

Phillips et al. reported an oxidative ring opening of copper oxophlorin complex **42** yielding an ester-linked, dinuclear copper complex **43** (Scheme 12) [14]. A proposed mechanism included oxidation of macrocycle by dioxygen leading to (OEPO$^{\bullet}$)Cu^{II} complex, its reaction with the starting (OEPOH)Cu^{II} to produce a C–O link, ring opening by addition of dioxygen and termination of the process by superoxide anion.

A formation of verdoheme analog **45**, which was further hydrolyzed to octaalkylbiliverdin **46**, was observed by Chang et al. upon oxygenation of cobalt(II) porphyrin substituted with naphthoic acid **44** (Scheme 13) [72]. The substituent was believed to support the activation of molecular oxygen by the metal center and was finally cleft

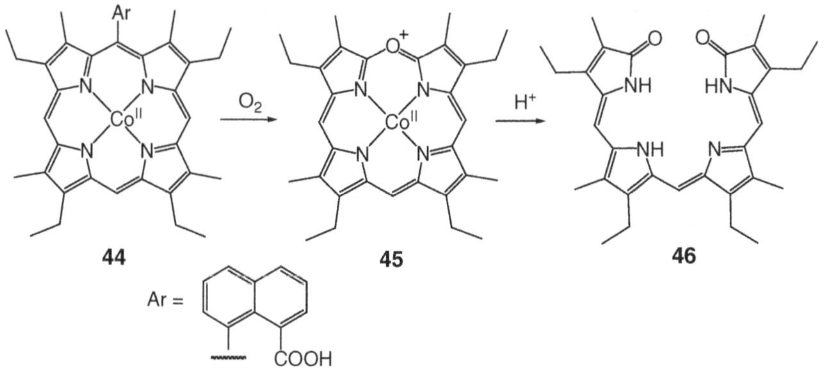

Scheme 13 Oxygenation of cobalt(II) porphyrin substituted with naphthoic acid [72]

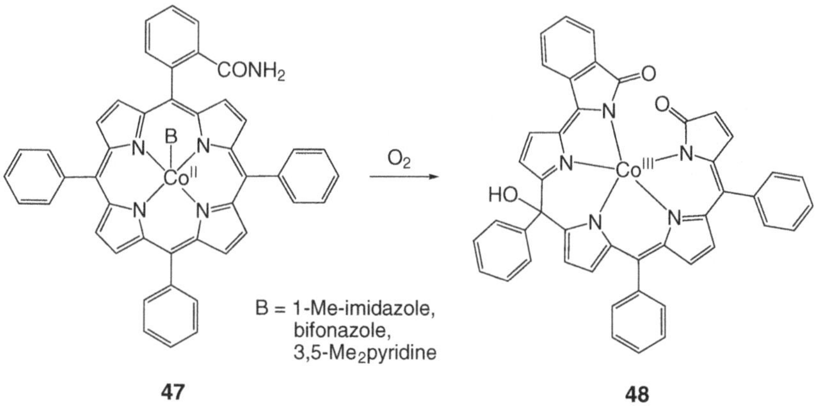

Scheme 14 Formation of Co(III) complex of an acyclic penatpyrrole [73]

as 8-formyl-1-naphthalenecarboxylic acid. A helical cobalt(III) complex of acyclic pentapyrrole **48** was obtained by Yamanishi et al. by treatment of cobalt(II) 5-(2-carbamoylphenyl)-10,15,20-triphenylporphyrin **47** with 1-methylimidazole and air (Scheme 14) [73]. An amide substituent and axial base (imidazole and pyridine derivatives were tested) was found essential for dioxygen activation, which resulted in breaking in C(4)–C(5) bond, followed by formation of oxoisoindole ring and addition of hydroxyl group to a *meso* position. Chiral HPLC separation of racemic **48** was performed. The application of chiral axial ligands bearing (*S*) configuration: nicotine, cotinine, or bifonazole led to the preferential formation of (*M*)-helical form of pentapyrrolic product.

An unexpected ring opening upon bromination of tetraphenylporphyrin with 20 equivalents of *N*-bromosuccinimide (NBS) in chloroform–methanol solution was described by Liu et al. [74]. From a mixture of reaction products which was treated with zinc acetate, crystals of compound **50** were isolated (Scheme 15). An X-ray

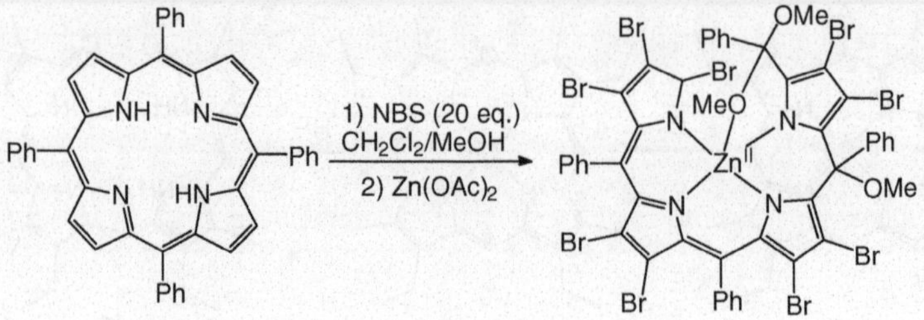

49 **50**

Scheme 15 Ring opening upon bromination of $TPPH_2$ [74]

structure of this zinc complex revealed the presence of nine bromo substituents at pyrrole rings and three methoxy groups attached to *meso* positions. Various *para*-phenyl-substituted tetraarylporphyrins could also be converted to the corresponding ring-opened products formed in 11–46% yield; also zinc tetraphenylporphyrin underwent a similar reaction, while the use of copper(II) and nickel(II) as central ions resulted only in β-bromination. A mechanism of the transformation was proposed involving MeOBr (formed from NBS and methanol) as an active species responsible for perbromination of pyrrole rings to form a highly congested dodecasubstituted macrocycle. The steric hindrance could be released by addition of another MeOBr molecule to C(*meso*)–C(α) bond followed by nucleophillic addition of methoxide to the *meso* positions of ring-opened product.

3.2 *Degradation of Metalloporphyrin Catalysts*

In this part, we shall discuss reactions of iron and manganese complexes with reagents which are typically used in metalloporphyrin-catalyzed oxidations (hydroxylations, epoxydations): peroxides, peroxyacids, and molecular oxygen [75–78]. Since typically an organic substrate is used in an excess in these processes, the problem of catalyst stability under such conditions has been often neglected. If this has been taken into account, methods of increasing metalloporphyrin robustness have been sought, mainly via its appropriate modification [79–81]. It was achieved by a substitution of porphyrin ring increasing catalytical activity and/or providing steric protection not only against formation of μ-oxo dimer PFe^{III}–O–$Fe^{III}P$ but also against attack of oxidants on *meso* positions [75]. Possible inter- and intramolecular processes leading to degradation of metalloporphyrin have been addressed [82, 83], though papers devoted to the analysis of catalyst stability have been relatively rare [84, 85].

In the recent years, several groups concentrated their efforts on the analysis of oxidation of porphyrin complexes by different oxidants used for the metalloporphyrin-catalyzed oxidations of organic substrates. Starting from simple, rather qualitative observations of possible decomposition of macrocycle as indicated by intensity lowering of Soret band in the UV–vis spectra, the studies have been typically extended to the analysis of reaction kinetics and attempts of determination of possible reaction mechanisms. However, in most cases the fate of catalyst and structures of degradation products have not been considered.

Stephenson and Bell investigated mechanism and kinetics of iron porphyrin-catalyzed epoxidation of olefins by hydrogen peroxide [86, 87]. Among factors affecting the activity of catalyst, oxidative degradation of porphyrin ring and μ-oxo dimer formation were discussed. The authors attributed the macrocycle decomposition to the attack of hydroxyl radicals (generated from of coordinated hydrogen peroxide). This hypothesis was in agreement with the observation that factors increasing the rate of hydroxyl radical generation contributed also to porphyrin degradation. The efficiency of iron porphyrin epoxidation catalysts was also studied by Cunningham and coworkers [88–90]. They connected the observed bleaching of the catalyst with its direct oxidation in the resting state (Fe(III)) rather than the high-valent intermediates.

Rocha Gonsalves and coworkers analyzed the epoxidation of alkenes by peroxides catalyzed by manganese porphyrins [91]. Two mechanisms of degradation of catalysts were found, depending on their structure and reaction conditions: an intramolecular pathway predominated when a metallo-oxo species was an active intermediate, while a metalloacylperoxo derivative favored an intermolecular one.

Ungvarai-Nagy and coworkers reacted iron(III) complexes of protoporphyrin IX and tetra(4-sulfonatophenyl)porphyrin with bromate and observed macrocycle degradation in acidic solutions [92–94]. Türk et al. investigated the stability of water-soluble porphyrins and their manganese(III) complexes toward peroxides and sodium hypochlorite [95–98]. The degradation rate constants were found dependent on the structure of porphyrin substrate, nature of oxidant, and pH of the solution. However, possible degradation pathways and structures of products formed were not discussed. Lente and Fábián studied kinetics and mechanism of oxidation of water-soluble porphyrin **51** with hydrogen peroxide and peroxomonosulfate anion [99]. The analysis of ESI mass spectra of the reaction mixture revealed the presence of iron complex of biliverdin-type tetrapyrrole **52** and a sulfonated benzoic acid **53** as dominant products of porphyrin decomposition (Scheme 16). Hopefully, this precedent will prompt further works on structural characterization of ring-opened oligopyrroles produced in the course of degradation of metalloporphyrin catalysts.

3.3 Oxidation of N-Confused Porphyrins

Though N-confused porphyrins have been known for almost two decades [100, 101], relative little studies have been devoted to their degradation. However, the instability of these macrocycles during metallation performed under aerobic conditions has been frequently observed. This led Furuta et al. to investigate the nature of the degradation product [102]. They found that in the course of reaction with copper(II) acetate in the

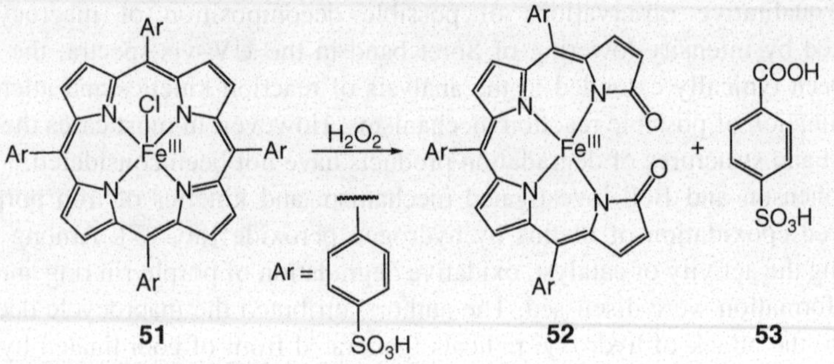

Scheme 16 Degradation of water-soluble iron porphyrin catalyst [99]

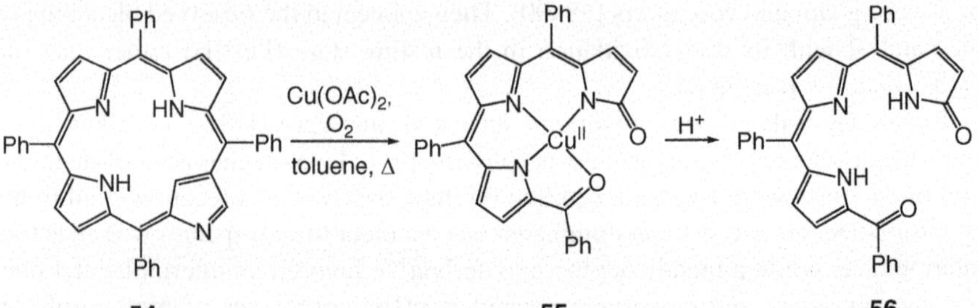

Scheme 17 Degradation of N-confused porphyrin [102]

presence of air N-confused tetraphenylporphyrin **54** underwent an oxidative transformation. Copper(II) complex of a linear tripyrrole **55** was isolated from the reaction mixture in 34% yield (Scheme 17). No other products were identified. Free tripyrrinone **56** and its zinc(II), nickel(II), palladium(II), platinum(II), and cobalt(II) derivatives were obtained [102]; crystal structures of Cu(II) and Pd(II) complexes showed a square-planar, N_3O-coordination mode [103].

A suggested mechanism of the degradation involved two successive reactions with molecular oxygen, activated by coordinated Cu(II) ion, leading to scission of two C(*meso*)–C(α) bonds. Further studies on the regioselectivity of the process, performed on 5-(2-pyridyl) derivative, showed that the N-confused pyrrole was cleft together with 5-aryl substituent, which proved the primary attack of dioxygen at C(1)–C(20) bond [102]. In contrast to this observation, Pawlicki et al. found that copper(II) complex of pyrrole-appended O-confused tetraaryloxaporphyrin **57** reacted with dioxygen yielding both possible tripyrrolic degradation products **58**, **59** (resulting from breaking of either C(1)–C(20) or C(4)–C(5) bond) formed in 7:3 ratio, along with and the product of oxygen atom insertion into a copper–carbon bond **60** (Scheme 18) [104]. Apparently, *meso*- and pyrrole substitution can direct the attack of dioxygen molecule; a discussion on the regioselectivity of oxidative ring opening of N-confused porphyrin can be found in the part devoted to photooxidation of tetrapyrroles (Sect. 4.1).

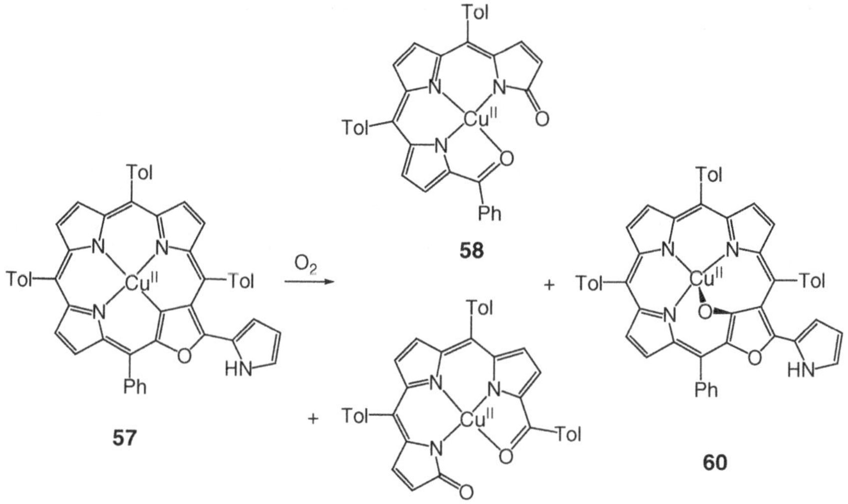

Scheme 18 Oxygenation of copper(II) complex of pyrrole-appended O-confused oxaporphyrin

3.4 Oxidation of Corroles

Despite general similarity to porphyrins, corroles exhibit a specific and sometimes unpredictable reactivity [105]. Both macrocycle families share a common 18-π-electron system, but lack of one *meso* bridge in corroles leads to increase of electron density and, as a consequence, a susceptibility to oxidative ring opening. Interestingly, all reports on such reactions concern *meso*-substituted systems [105], though any systematic and comprehensive research on factors influencing corrole stability has not been performed. Most work in the field concentrated on photooxidation of corroles (see Sect. 4.2). Macrocycle opening by certain oxidants has been also described, though typically formation of biliverdin-type compounds only accompanied the reaction of major interest.

A fully brominated open-chain tetrapyrrole **61** (Fig. 5) was identified as a reaction by-product resulting from breaking of C(4)–C(5) bond of germanium(IV) 5,10,15-triphenylcorrole treated with bromine [106]. A linear tetrapyrrole **62** was formed in minor quantities when triarylcorroles were reacted with 4-amino-4*H*-1,2,4-triazole [107]. This time, C(5)–C(6) bond of the original macrocycle was cut (Fig. 5). Ring opening at C(10) was observed upon conversion of triarylcorrole **63** to a corresponding porphyrin (Scheme 19) [108]. A proposed mechanism of the transformation involved a [2 + 2] cycloaddition of two corroles and cleaving of a spirocyclobutane intermediate by dioxygen connected with an extrusion of *meso*-carbon bearing *para*-nitrophenyl substituent.

Other pathways of corrole oxidation were reported, including isocorrole formation by 2,3-dichloro-5,6-dicyanobenzoquinone (DDQ) treatment [109, 110] or demetallation [111, 112] and oxidative dimerization of 5,10,15-tris(pentafluorophenyl)corrole with formation of β-β′ bond(s) upon heating in 1,2,4-trichlorobenzene [113].

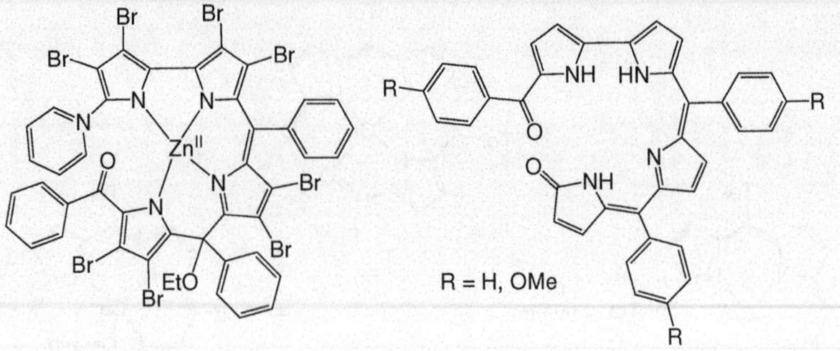

61 **62**

Fig. 5 Ring-opened products of corrole oxidation [106, 107]

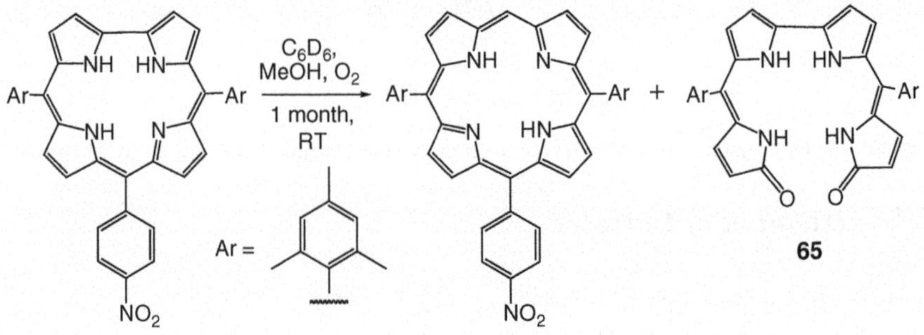

63 **64** **65**

Scheme 19 Conversion of triarylcorrole to poprhyrin and a linear tetrapyrrole [108]

4 Photooxidation of Tetrapyrroles

Photooxidation of tetrapyrrolic macrocycles and their complexes is considered as the most important process responsible for the frequently observed photobleaching of these compounds [114]. This phenomenon is connected with the ability of porphyrin derivatives to activate molecular oxygen in the presence of light. Energy transfer from the excited state of the macrocycle to the ground state of the dioxygen molecule results in the generation of singlet oxygen. As a practical consequence, tetrapyrroles are used as photosensitizers for degradation of various organic substrates [115–117] and in photodynamic therapy (PDT) for treatment of cancer, macular degeneration, chronic skin diseases, and other conditions [118–121]. Under certain conditions, also tetrapyrrole itself can be attacked by singlet oxygen, which may eventually lead to ring opening.

In the context of not only photosensitizer stability but also other applications of tetrapyrroles, light-driven reactivity of porphyrin derivatives toward O_2 is of particular interest. Photobleaching of photosensitizers used in photodynamic

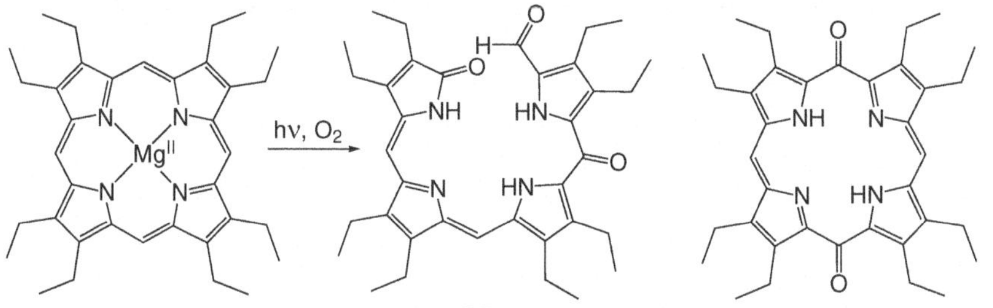

Scheme 20 Photooxidation of (OEP)Mg^{II} [122]. Compounds **67** and **68** were found among products of light-driven oxidation of oxophlorin [123]

therapy was thoroughly reviewed by Bonnett and Martínez [114]. Thus, older contributions will be only briefly described in his chapter, and the attention will be focused on recent developments in the field.

4.1 *Photooxidation of Porphyrins, N-Confused Porphyrins and Phlorins*

Most metal-free porphyrins are not prone to photooxidative degradation due to the relative high value of oxidation potential. However, their deprotonation or conversion to complexes of electropositive metal ions (e.g. with Zn(II), Cd(II) or Mg(II)) lowers redox potential and therefore the robustness of the system toward oxidative degradation is also reduced.

Fuhrhop and Mauzerall reported the photooxidation of magnesium(II) octaethylporphyrin **66** and identified a linear tetrapyrrole **67** as the final product for this transformation (Scheme 20) [122]. This compound was also found by Bonnett et al. as one of the two main products of photooxidation of octaethyloxophlorin **8** in neutral solution (the other being 5,15-dioxoderivative **68**, Scheme 20) [123]. Light-driven ring opening of zinc, magnesium, cadmium, thallium(I) complexes of tetraphenylporphyrin **69** as well as the porphyrin dianion (TPP^{2-}) was examined by several groups [124–128]. A proper structure of the final product **20** or **70** was finally established by Cavaleiro and coworkers [128]. A bilindione derivative bearing –OR substituent in 15-position resulted from dioxygen attack on the C(*meso*)–C(α) bond, followed by demetallation and addition of water or alcohol (ROH, Scheme 21). As proved by isotope labeling studies, both carbonyl oxygen atoms are derived from the single molecule of O_2 [125, 126]. Silva et al. studied effects of substitution of tetraarylporphyrin on the degradation of cadmium(II) complexes and showed that the reaction was governed by steric factors rather than electronic ones [129]. The presence of substituents in *ortho* positions of phenyl rings prevented the macrocycle from the dioxygen attack (Scheme 22).

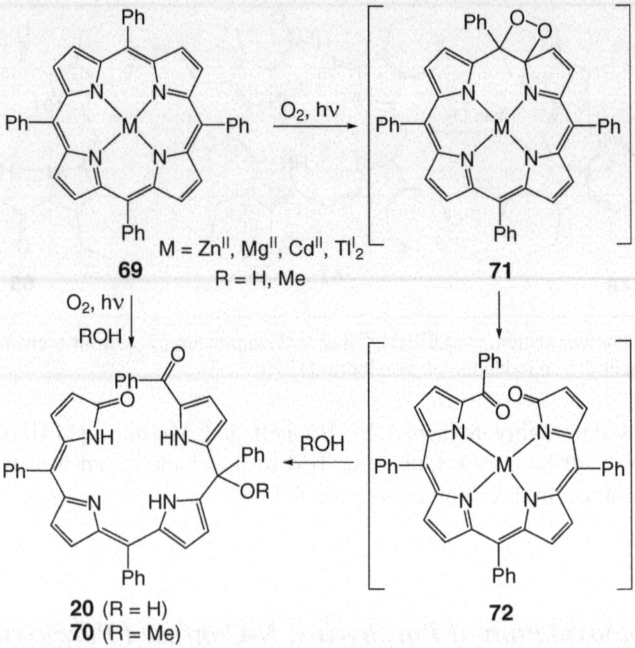

Scheme 21 Photooxidation of TPP complexes [128]

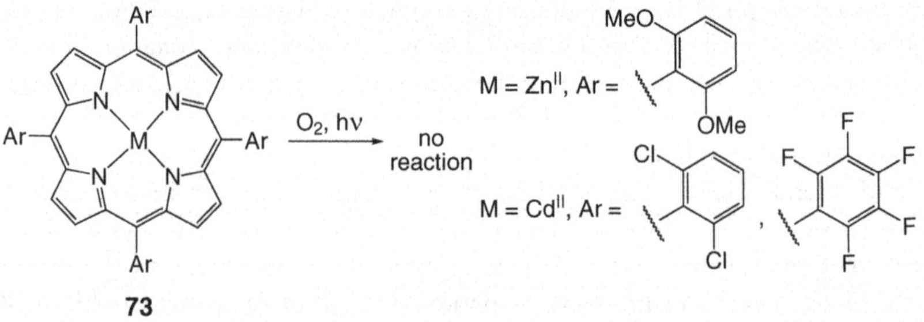

Scheme 22 Zinc(II) and cadmium(II) tetraarylporphyrins not prone to photooxidation [129]

Both cadmium(II) tetra(3,4,5-trimethoxyphenyl)porphyrin **74** and zinc(II) 2,3,12,13-tetrabromoporphyrin **77**, however, were converted to the corresponding open-chain products **75**, **76** (two forms were observed) and **78**, respectively (Schemes 23, 24).

Zinc(II) complexes of linear tetrapyrrole **20** were obtained [130]. Depending on metallation conditions, 3N + O or 4N coordination was found in these chelates, in the latter the loss of methanol or water led to a fully conjugated structure (Scheme 25). Copper(II) complex, formed by transmetallation of photooxidation product of magnesium(II) tetraphenylporphyrin, heated with excess of copper(II) acetate yielded a dinuclear species **82** (Fig. 6); the additional *meso*-oxygen bridging two copper ions originated probably from water since compound **82** was obtained also under dioxygen-free conditions [12].

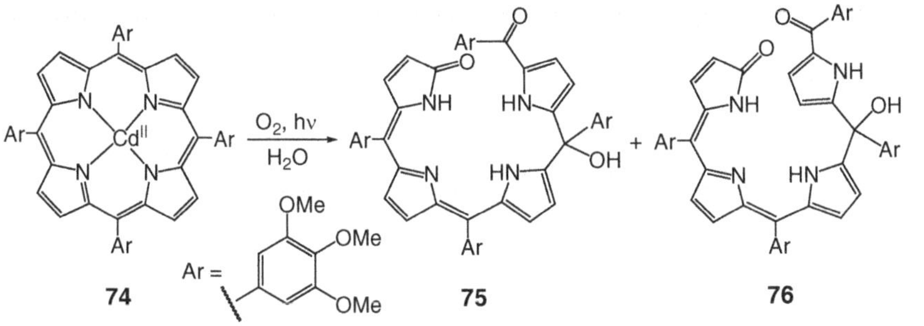

Scheme 23 Photooxidation of cadmium(II) tetra(3,4,5-trimethoxyphenyl)porphyrin [129]

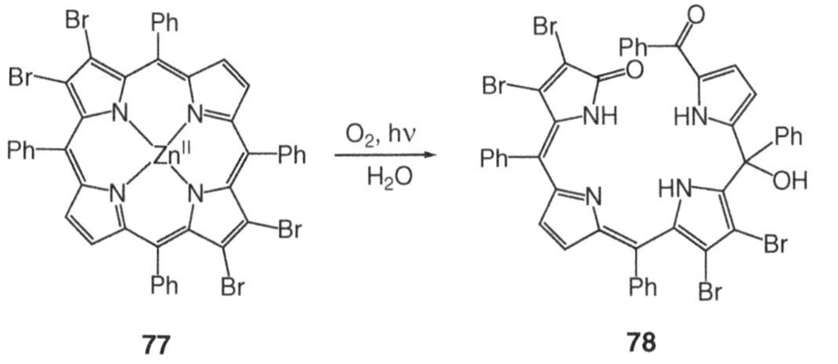

Scheme 24 Photooxidation of zinc(II) β-tetrabromoporphyrin [129]

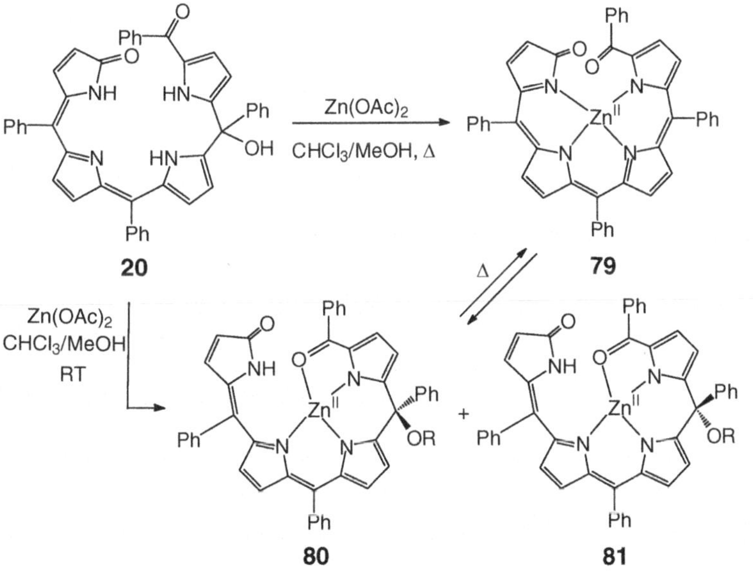

Scheme 25 Metallation of bilindione **20** [130]

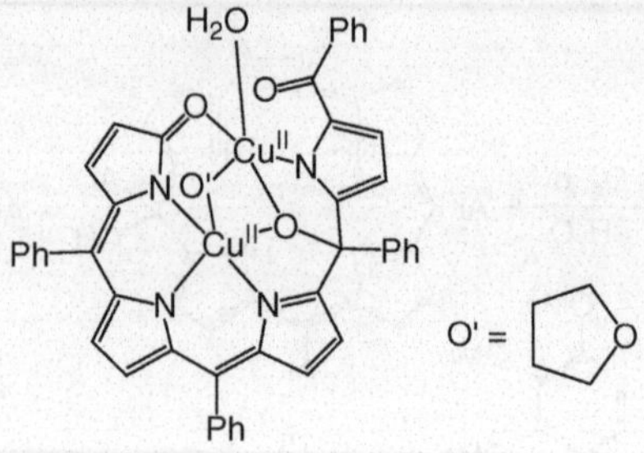

82

Fig. 6 Dinuclear Cu(II) complex

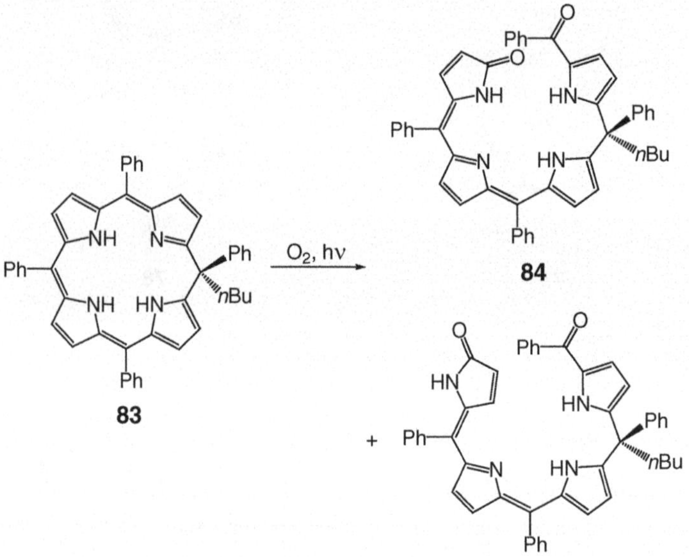

83

84

85

Scheme 26 Photooxidation of *meso*-substituted phlorin [131]

Mixed 3N + O copper(II), nickel(II), and zinc(II) complexes were formed from ligands **84** and **85**, obtained by the photooxidation of a *meso*-substituted phlorin **83** [131]. Two isomers of bilindione and its complexes were described, with a different orientation of the terminal pyrrolone ring (Scheme 26). Their interconversion upon irradiation which caused *E*–*Z* isomerization was demonstrated. LeSaulnier et al. investigated photodegradation of phlorins bearing different number of mesityl substituents **86** (Fig. 7) [132]. As expected, the incorporation of bulky mesityl substituents enhanced phlorin stability.

Photobleaching of certain metal-free porphyrins was also observed, not necessarily connected with ring-opening reactions. Water-soluble, cationic 5,10,15,20-tetrakis(1-pentyl-4-pyridyl)porphyrin underwent fast photodegradation in aqueous media [133].

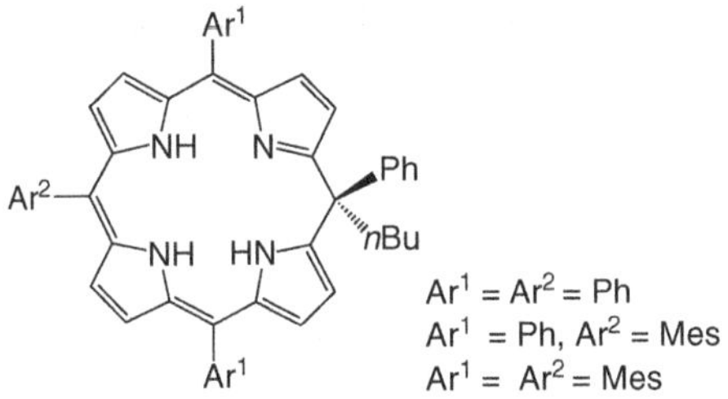

86

Fig. 7 Mesityl-substituted phlorins

Niziolek and coworkers observed that lipid peroxidation in membranes, mediated by protoporphyrin IX as a singlet oxygen photosensitizer, can be prolonged in the presence of nitric oxide [134]. NO was found to protect the macrocycle against oxidative destruction. Cavaleiro et al. carried out photochemical studies on stability of porphyrins and their copper(II) complexes and showed that the latter had shorter triplet lifetimes and were more stable with respect to photodegradation than the respective free bases [135]. Similarly, perfluorination of phenyl substituents of tetraphenylporphyrin had a beneficial effect on the macrocycle robustness.

When 2-aza-21-carba-5,10,15,20-tetraphenylporphyrin (inverted porphyrin **54**) was dissolved in dichloromethane and irradiated with visible light in the presence of air, only traces of degradation products could be detected. Instead, photooxidation of the dianion of N-confused tetraphenylporphyrin **87** was performed which led to a mixture of linear oligopyrroles within 1 h [136]. Chromatographic separation yielded fractions containing tripyrrinone **56** (33% of reacted substrate), its dimethyl acetal **88** (24%) and N-confused tetrapyrrole **89** (31%, Scheme 27). Upon metallation with palladium(II), compound **89** converted into complex **90** containing a conjugated N-confused biliverdin analog acting as a binucleating ligand with two types of coordination surroundings: (NNNO) and (CNOO) (Scheme 27) [136]. Further exploration of photooxidation products led to detection of the additional, unexpected tetrapyrrolic compound **70** (present in ca. 6–9% yield), typically formed in the course of TPP^{2-} degradation [137]. This observation led to a conclusion that two different mechanisms operate in one molecule. Apart from 1,2-dioxygen addition, which is common for tetrapyrrolic macrocycles, the rare 1,3-addition was also found (Scheme 28).

Compound **89**, the major isolated tetrapyrrolic product of photooxidation of N-confused porphyrin dianion resulted from cleavage of C(10)–C(11) bond of the original macrocycle. However, changing of reaction conditions (metallation with zinc or replacing of methoxide with ethoxide for the conversion of N-confused porphyrin to its dianion) allowed us to detect other tetrapyrrolic degradation products (Wojaczyński J, Popiel M, Gońka E, Latos-Grażyński L, unpublished results). DFT calculations performed on inverted porphyrin dianion did not show any significant differences among *meso* positions which could be responsible for any preference of dioxygen attack. Apparently, the observed product distribution reflects not only the regioselectivity of O_2 addition but also the relative stability of

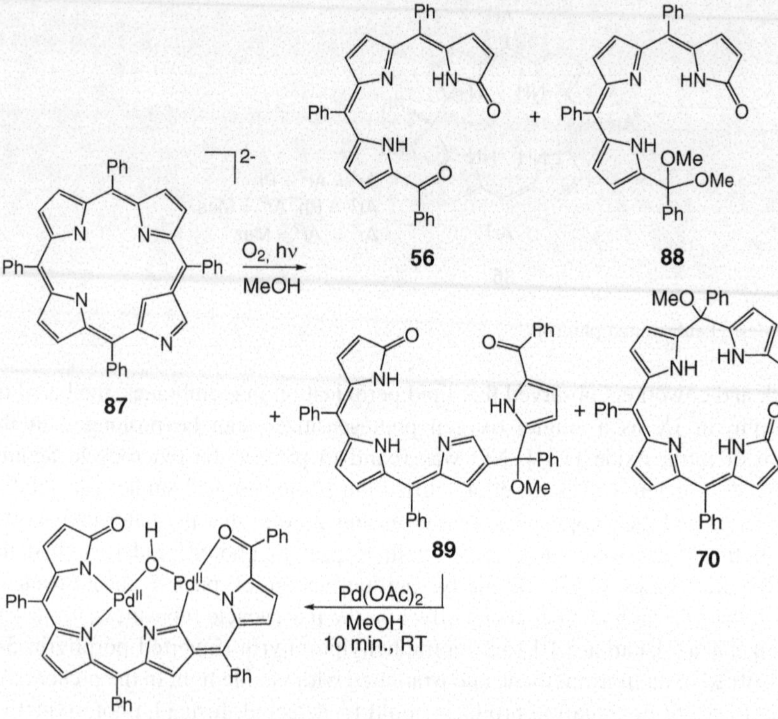

Scheme 27 Photooxiadtion of dianion of N-confused porphyrin and formation of dinuclear palladium complex of N-confused biliverdin derivative [136]

degradation products under given conditions since part of them can undergo further reactions (as proved by the observation of tripyrrinone products **56**, **88** which could be formed from primary ring opening at C(5) or C(20) followed by loss of inverted pyrrole in the second oxidation step).

4.2 *Photooxidation of Corroles*

The question of photochemical stability of corroles is particularly important in context of their possible application in photoactive devices, chromophores for light energy conversion and singlet oxygen generation [138–140]. Early observations indicated a stepwise degradation of corroles in solution in the presence of light and air. The process was monitored by UV–vis spectroscopy since a systematic lowering of Soret band intensity was observed [141, 142]. The presence of electron-withdrawing substituents in corrole ring or complexation with metal ion was shown to increase the macrocycle robustness. The first proposal of a structure of

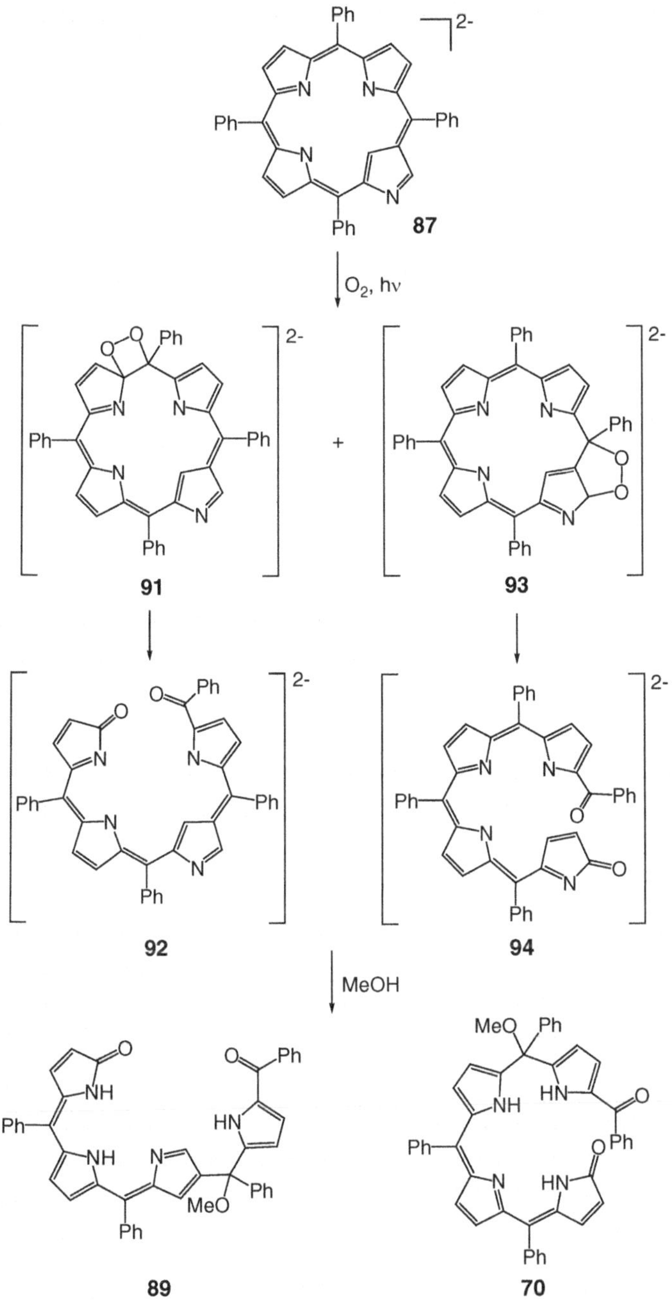

Scheme 28 Mechanism of 1,2- and 1,3-dioxygen addition to dianion **87** [137]

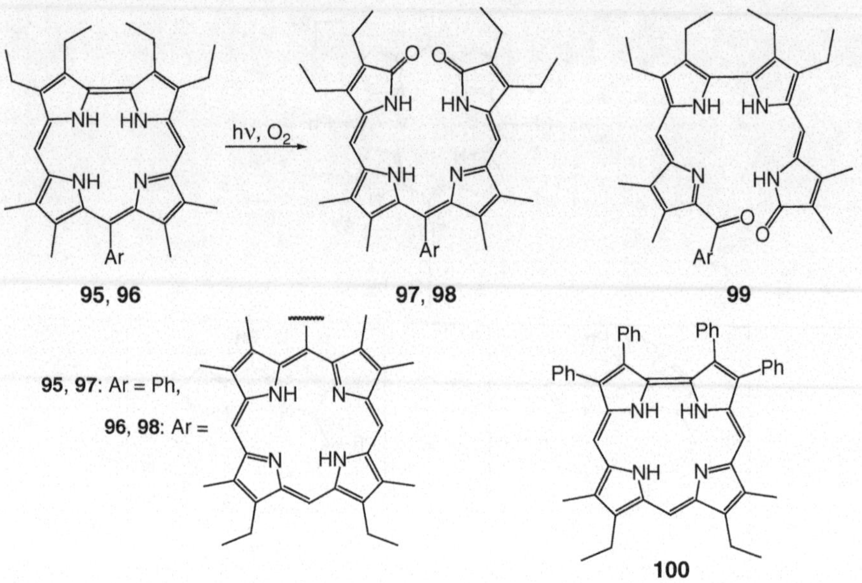

Scheme 29 Photooxidation of *meso*-aryl-substituted corroles **95**, **96** [143, 144]. An alternative structure of degradation product **99** and a tetraphenyl analogue of compound **95** are also shown

degradation product was made by Guilard and coworkers who investigated photooxidation of 2,3,17,18-tetraethyl-7,8,12,13-tetramethyl-10-phenylcorrole **95** (Scheme 29) [143]. A biliverdin derivative **97** was obtained in 24% yield and characterized by ^{1}H NMR, IR, MS and elemental analysis which were in general agreement with an intuitive assumption that pyrrole–pyrrole (C(1)–C(19)) bond was attacked by dioxygen molecule. No other reaction products were isolated. Opening of corrole ring by breaking of C(α)–C(α) bond was also postulated by Paolesse et al. for photooxidation of β-octaalkylcorrole with a porphyrin attached to a 10-position **96** [144]. In both cases the symmetry of resulting ^{1}H NMR spectrum was lower than expected for the proposed structure (an analogous triarylbilindione obtained by Yamauchi et al. by coupled oxidation of iron porphyrin exhibited a simple ^{1}H NMR pattern [145]). The difference was attributed to isomerization of biliverdin moiety to (*E*,*Z*,*Z*) configuration; however, certain spectral features (e.g. a doublet at ca. 8 ppm which could be assigned to *ortho*-aryl protons) suggest that a structure resulting from opening at aryl-substituted *meso* position **99** could be considered as well. On the other hand, the observation that 2,3,17,18-tetraphenyl analog **100** (*meso*-unsubstituted!) was found far more stable than **95** and a similar behavior of corresponding cofacial bis(corroles) connected with a 10-anthracene bridge suggested efficiency of a steric protection of bipyrrole fragment limiting the access of dioxygen molecule to C(1)–C(19) bond [146].

Degradation of *meso*-triarylcorroles has received a considerable attention [141, 142], but only a systematic mass spectrometry study on decomposition pathways of these compounds by Świder et al. led to identification of isocorroles and biliverdin

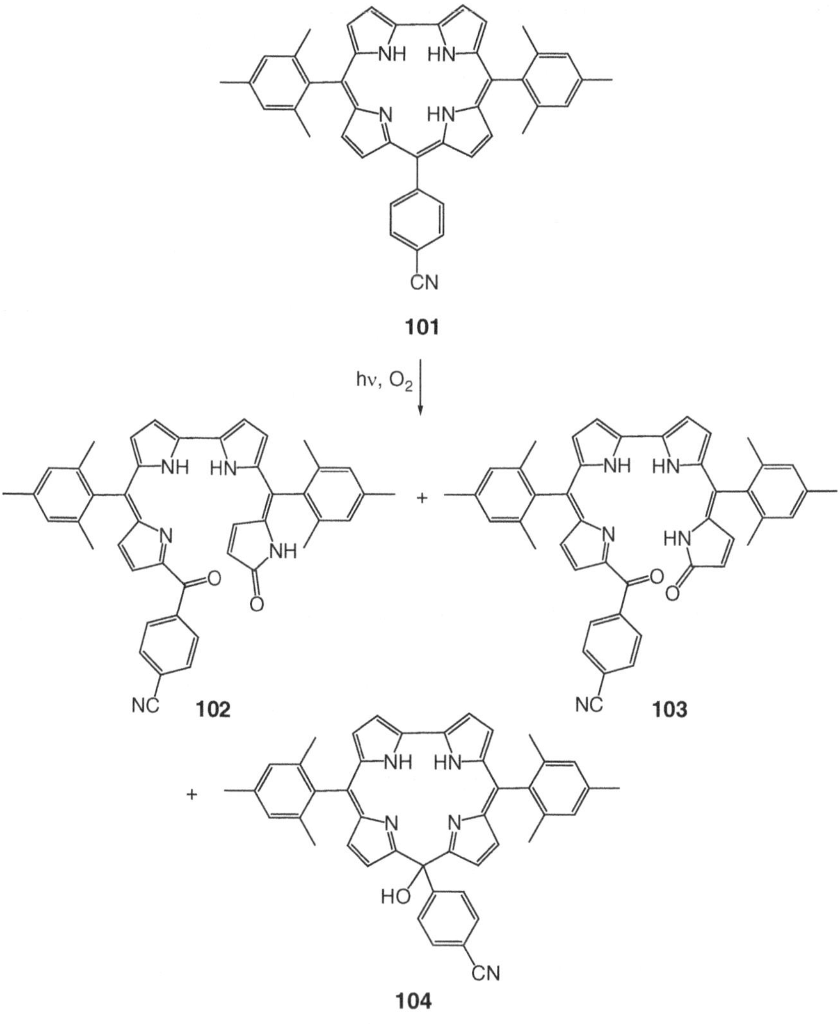

Scheme 30 Photooxidation of triarylcorrole [36]

derivatives as photooxidation products [36]. Preparative degradation experiment was conducted with corrole **101** with 5 and 15 positions protected by bulky substituents, which was dissolved in acetonitrile and exposed to sunlight for 60 h. Three major compounds **102–104** were isolated from the reaction mixture (Scheme 30), indicating dioxygen attack on *meso*-C(10) carbon atom. In our studies on photooxidation of triphenylcorrole and tris(*p*-methoxyphenyl)corrole, scission of C(9)–C(10), but also of C(4)–C(5) bond of symmetrical, non-hindered substrate was noted [147]. As can be seen, any product resulting from breaking of a direct pyrrole–pyrrole bond has not been detected from photodegradation of triarylcorroles. One couldn't exclude, however, that the presence of β-alkyl substituents in compounds **95**, **96** directs dioxygen attack to the C(1)–C(19) bond. A strong dependence of reaction outcome on substitution of macrocycle is illustrated by reactivity of 5,10,15-tris(pentafluorophenyl) corrole which stirred at room temperature under ambient light and air slowly converted to 3,3′-linked dimer and 3,3′,17′,3″-trimer [148].

5 Coupled Oxidation

Heme oxygenase, responsible for the oxidative destruction of unwanted heme, requires molecular oxygen but also the source of electrons for its function (see Sect. 6 of this contribution). Oxidation of iron porphyrin in pyridine in the presence of reducing agent (hydrazine or ascorbic acid) has been used as a model for the enzymatic reaction [149, 150]. Pioneering studies by Lemberg (who described coupled oxidation of iron protoporphyrin IX with H_2O_2-ascorbic acid), Fischer and Libowitzky were performed on natural heme derivatives [151, 152]. Later on, higher symmetry synthetic model compounds such as complexes of octaethylporphyrin or ethioporphyrins have been used. A thorough analysis of coupled oxidation process was presented in a series of papers published in the years 1992–2008 by Balch, Latos-Grażyński, and coworkers. They isolated and characterized two main products of degradation of (OEP)$Fe^{II}(py)_2$ **105** caused by air in the presence of ascorbic acid: a diamagnetic verdoheme **106** (50%) and a paramagnetic dimeric iron biliverdin complex **107** (38%, Scheme 31) [153, 154]. In situ monitoring of the degradation of (OEP)$Fe^{II}(py)_2$ by dioxygen with hydrazine as sacrificial reductant identified iron oxophlorin, (OEPO)$Fe(py)_2$ **11** as a key intermediate of the process [155].

Oxidation of (OEP)$Fe^{III}Cl$ under pyridine-free conditions, but in the presence of cyanide ions as axial ligands, was also demonstrated [156]. Depending on cyanide concentration, iron oxophlorin or 5-oxaporphyrin complex (verdoheme) was formed. Coupled oxidation of Co(II) octaethylporphyrin leading to cobalt verdoheme and biliverdin analogs was also described [157]. In the recent years, degradation of iron complexes of β-unsubstituted, *meso*-arylporporphyrins under coupled oxidation conditions was investigated as well [145, 158–160].

5.1 *Oxophlorins*

The question of structure and reactivity of oxophlorins (hydroxyporphyrins) has been considered in numerous contributions. In addition to tautomeric equilibrium (Scheme 32), ocatethyloxophlorin was shown to undergo a facile one- and two-electron oxidation [161]. In consequence, it can serve as a trianionic, dianionic, and monoanionic ligand, and various electron distributions between metal ion and ligand are possible. Not surprisingly then, a rich coordination chemistry was observed for octaalkyloxophlorins: zinc(II), nickel(II), cobalt(II), copper(II), iron (III), and manganese(III) monomeric complexes with *meso*-hydroxyl groups

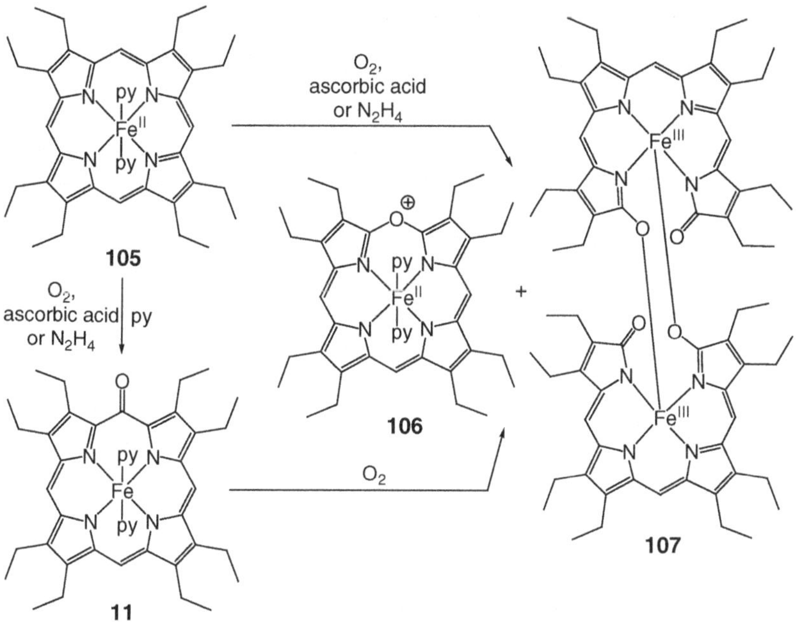

Scheme 31 Coupled oxidation of (OEP)$Fe^{II}(py)_2$ [153–155]

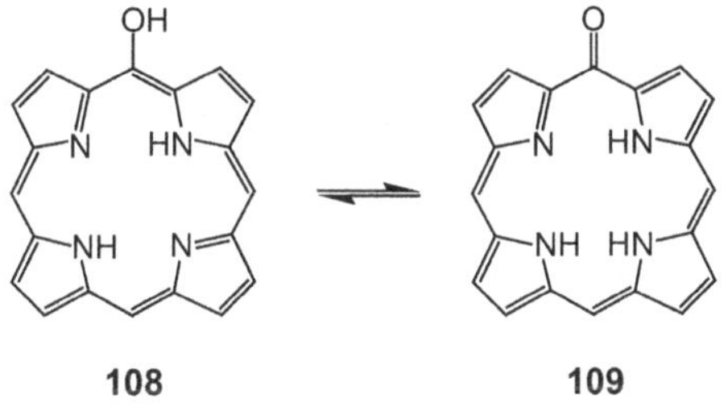

Scheme 32 Keto-enol tautomeric equilibrium of oxophlorin

[14, 162–166], dimeric complexes linked by *meso*-oxygen bridges with Fe(III), Mn(III), and In(III) [162, 165, 167–171], coordinated oxophlorin trianions [165, 166], coordinated radicals [163, 164, 168], and complexes of oxidized monoanionic form [168, 170] were reported. Variety of structures and their mutual interconversion is exemplified by iron(III) complexes shown in Scheme 33 [161]. The thorough overview of coordination chemistry of oxophlorins/*meso*-hydroxyporphyrins was published by Balch in 2000 [161].

Electronic structure of iron oxophlorin (OEPO)$Fe(py)_2$ and its analogs was a subject of a long-lasting debate [162, 167, 172–175]. Three possible electron distributions have been taken into account (Fig. 8). Patterns of paramagnetically shifted ^{1}H NMR signals observed for (OEPO)$Fe(py)_2$ and related species suggested a significant contribution of a ligand radical form $(OEPO^{\bullet})$–Fe^{II} [162, 172, 173]. A similar alteration of isotropic shifts was found for iron triphenyloxophlorin

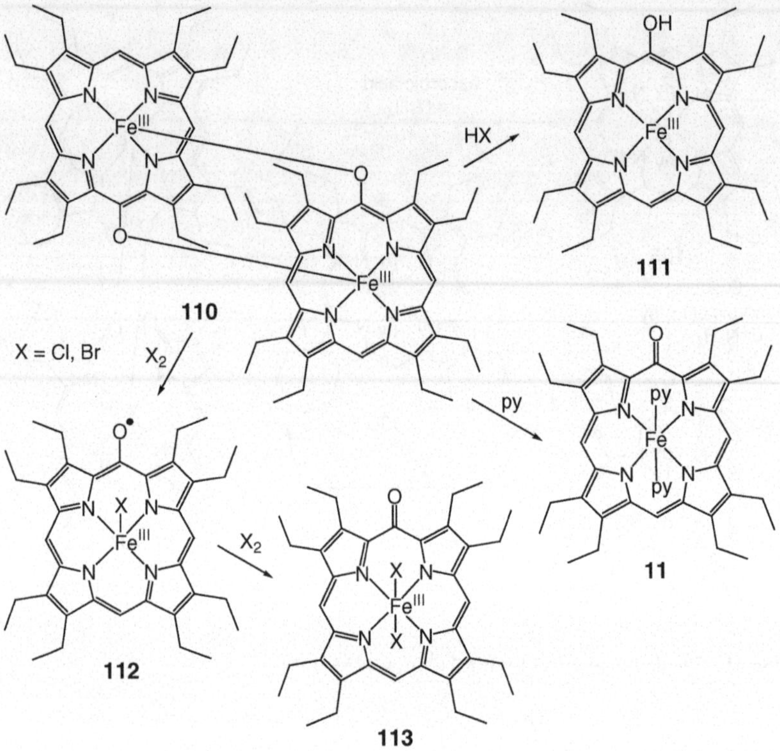

Scheme 33 Iron complexes of octaethyloxophlorin [161]

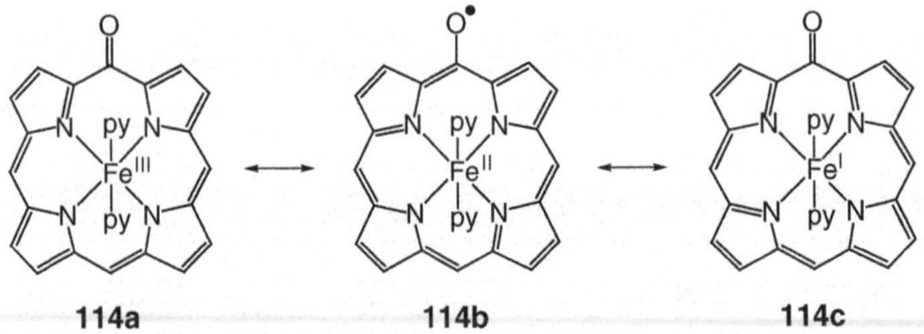

Fig. 8 Resonance forms of iron oxophlorin

complexes [158]. DFT calculated spin density maps for oxophlorin radicals allowed to reproduce the major observed spectroscopic features [176]. Later on, Rath et al. showed the dependence of electronic structure on the nature of axial ligands, with 2,6-xylyl isocyanide stabilizing the radical resonance structure [(OEPO$^{\bullet}$)Fe^{II}(CNR)$_2$] [177]. Recent crystallographic, magnetic, and spectroscopic measurements indicated the importance of Fe(III)/oxophlorin trianion form for bis-pyridine and bis-imidazole complexes [178]. DFT calculations of electronic structure of (OEPO)FeL$_2$ complexes

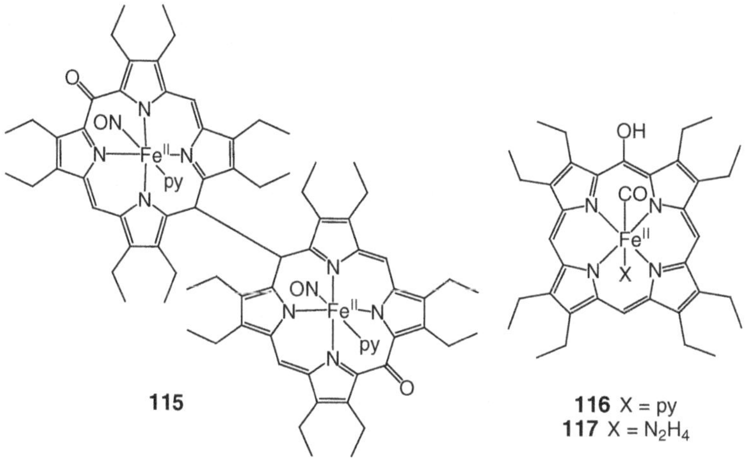

Fig. 9 Iron oxophlorin NO and CO complexes

performed by Gheidi et al. confirmed the dependence of electron distribution and iron spin state on the nature of axial ligands [179].

Reactivity of iron oxophlorin (OEPO)Fe(py)$_2$ (**11**) was extensively explored. Apart from coordination chemistry depicted in Scheme 33, interaction with small molecules was investigated [180, 181]. A reversible binding of NO to (OEPO)Fe (py)$_2$ connected with the formation of dimeric species **115** was reported (Fig. 9) [180]. A reduced form of oxophlorin, (OEPOH)Fe^{II}(py)$_2$, was converted to (OEPOH)Fe^{II}(CO)(py) (**116**) upon treatment with carbon monoxide, and pyridine could be replaced with hydrazine to form (OEPOH)Fe^{II}(CO)(N_2H_4) (**117**); both diamagnetic complexes were found extremely air sensitive and in the presence of dioxygen an immediate reaction leading to (OEPO)Fe(py)$_2$ **11** was observed [181].

Both redox processes preserving a basic skeleton of oxophlorin [168, 170] and coupled oxidation leading to verdoheme and biliverdin have been reported [155, 156]. Under certain conditions, oxidative degradation is not limited to macrocycle opening. Rath et al. observed that in the absence of reducing agent, addition of dioxygen to a pyridine solution of oxophlorin complex (OEPO)Fe(py)$_2$ (**11**) caused stepwise changes, resulting in formation of iron biliverdin **118**, and, finally, oxidative removal of pyrrole unit yielding a linear tripyrrole complex **38** (Scheme 34) [182]. This compound was also formed when compound **118** or verdoheme **106** was exposed to O_2.

5.2 Verdohemes

A green iron complex of 5-oxaporphyrin, called verdoheme, is an important intermediate in the process of heme oxidative cleavage by heme oxygenase [183]. It is also formed in the course of coupled oxidation of iron porphyrins but can be also obtained by dehydration of biliverdin in the presence or iron salts [184, 185].

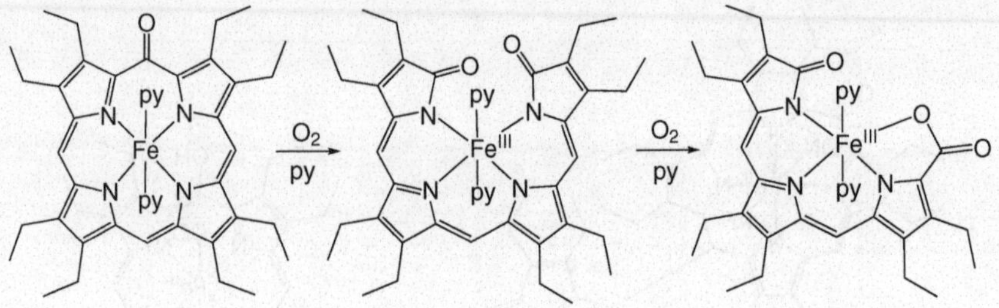

11 118 38

Scheme 34 Conversion of iron oxophlorin to a linear tripyrrole complex [182]

Metal-centered reactions have been reported, including changes of axial ligation, and metal oxidation and spin state, as demonstrated for iron (Scheme 35) and cobalt 5-oxaporphyrin complexes [153, 172, 186–192]. Coordination chemistry of verdohemes and biliverdin derivatives has been recently reviewed by Balch and Bowles [193].

Ligand transformations are particularly important for the study of macrocycle degradation since they can lead to linear tetrapyrrolic products. Two mechanisms of verdoheme ring opening leading to biliverdin have been described: an oxidative pathway [194, 195], resulting in release of Fe^{3+}, and a hydrolytic route (Scheme 36). The latter is generally believed to begin with addition of hydroxide to the macrocycle. To characterize this kind of reactivity of 5-oxapophyrin complexes, their conversions by anionic nucleophiles have been investigated [196–200]. Helical, ring-opened products resulting from the addition of alkoxide, thiolate, and amide ions to zinc(II) (**125**) and cobalt(II) verdoheme (**126**) were isolated and structurally characterized (Scheme 37) [197, 201]. More complex process was observed when cyanide ion was added to zinc 5-oxaporphyrin **125**, as macrocycle cleavage was accompanied with substitution at one or two *meso* positions (Scheme 38) [199]. A dimeric complex $[(OEBOMe)Fe^{II}]_2$ **130** was isolated from the reaction of iron(II) verdoheme with OMe^- ion (Scheme 39) [198]. Ring opening of Fe^{II} and Fe^{III} verdohemes with methoxide or hydroxide was monitored by ^{1}H NMR spectroscopy [200]. Characteristic alternating shift patterns indicating radical character of the particular intermediates and remarkable paramagnetic shifts of *meso* resonances of certain species were noted.

Utilizing O_2 as oxidant, Rath et al. demonstrated a conversion of Fe(II) verdoheme into a highly oxidized (Fe(IV) bound to bilindione ligand or Fe(III) coordinated to oxidized form of ligand) biliverdin complex (Scheme 40) [195]. Its reduction with zinc amalgam resulted in previously characterized dimeric $[(OEB)Fe^{III}]_2$ (**107**). Earlier, Saito and Itano reported that prolonged (1 month) exposure to air of verdoheme dissolved in ethylene glycol – pyridine solution led to several iron-free ring-opened products, including tripyrrolic ones [202]. Most of the starting material was recovered from the reaction.

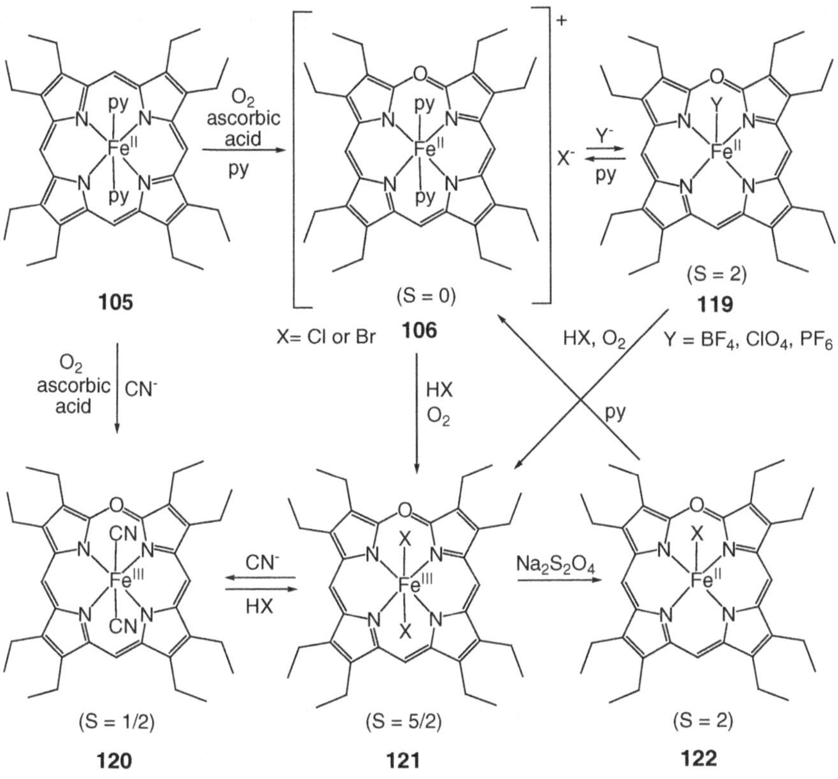

Scheme 35 Interconversion of iron 5-oxaporphyrin complexes [193]

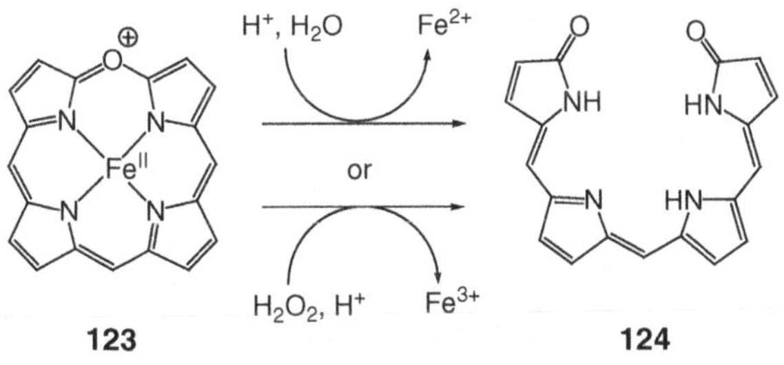

Scheme 36 Two pathways of verdoheme to biliverdin coversion

Theoretical study on factors determining verdoheme conversion to biliverdin was performed by Safari and coworkers. The role of axial ligands as well as coordinated metal ion was taken into account [203–206].

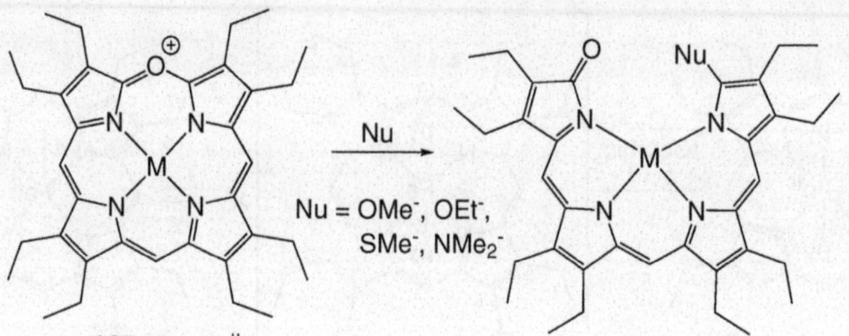

125 M = Zn^{II}
126 M = Co^{II}
127

Scheme 37 Opening of zinc(II) and cobalt(II) verdohemes by nucleophiles [197, 201]

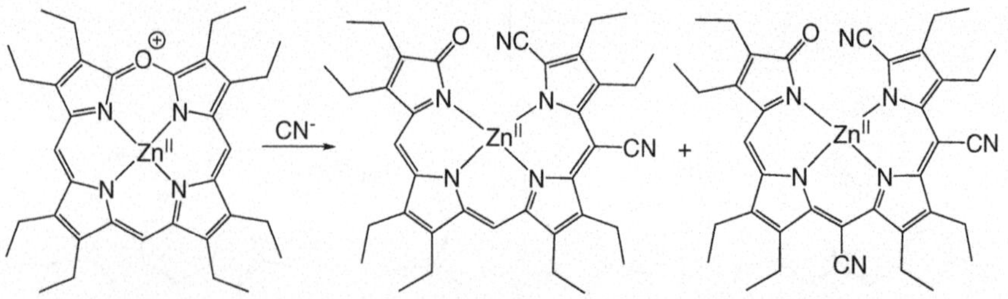

125 **128** **129**

Scheme 38 Zinc(II) verdoheme opening by cyanide [199]

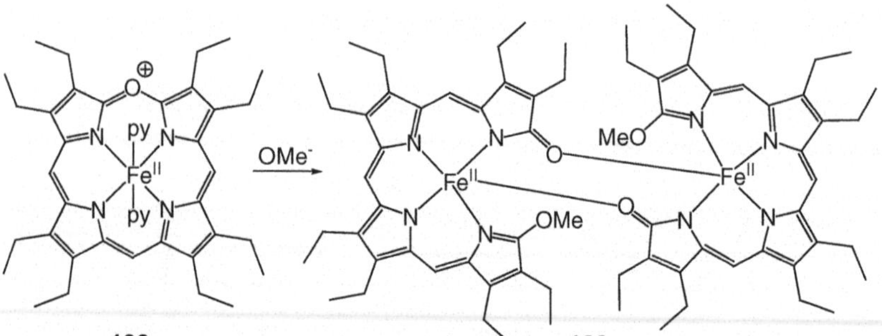

106 **130**

Scheme 39 Reaction of iron(II) verdoheme with methoxide [198]

5.3 *Biliverdins*

A dimeric helical iron(III) complex **107** of octaethylbilindione, a biliverdin analog, was obtained by Balch et al. along with verdoheme from the coupled oxidation of (OEP)Fe^{II}(py)$_2$ [154]. Its treatment with pyridine resulted in cleavage of Fe–O bonds and formation of monomeric (OEB)Fe^{III}(py)$_2$ **132** (Scheme 41). An easy

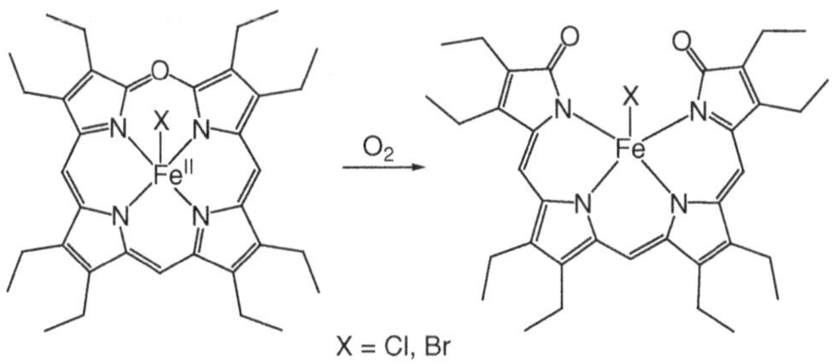

122 **131**

Scheme 40 Oxidation of iron(II) verdoheme [195]

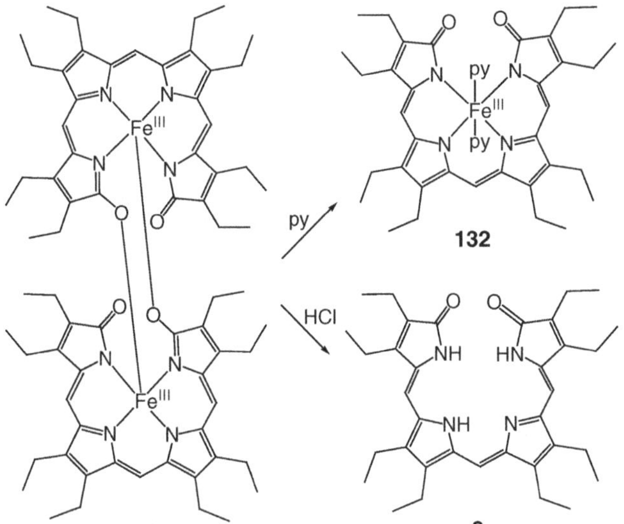

107 **132** **9**

Scheme 41 Splitting of dimeric iron(III) octaethylbiliverdin complex [154]

demetallation of $[(OEB)Fe^{III}]_2$ with hydrochloric acid released the blue bilindione $OEBH_3$ (**9**) [154]. Its complexes with other metal ions were investigated by Bonnett and coworkers [207, 208] and by Balch group [13, 209–215]. Interestingly, remetallation of $OEBH_3$ with iron has not been successful [193], while manganese, cobalt, nickel, copper, zinc, palladium, and boron complexes have been obtained. For Mn(III), a dimeric complex with oxygen bridges $[(OEB)Mn^{III}]_2$, which was cleft by pyridine to monomeric $(OEB)Mn^{III}(py)_2$ (in a full analogy with Fe(III) complexes) was described [210]. Spectroscopic investigations of monomeric, four-coordinate complexes of $OEBH_3$ with cobalt, nickel, copper, and palladium suggested their electronic structure consistent with the presence of M(II) ion and oxidized ligand radical $(OEB^{\bullet})M^{II}$ [208–210, 214] A significant degree of radical character was also postulated for iron complexes obtained by verdoheme ring

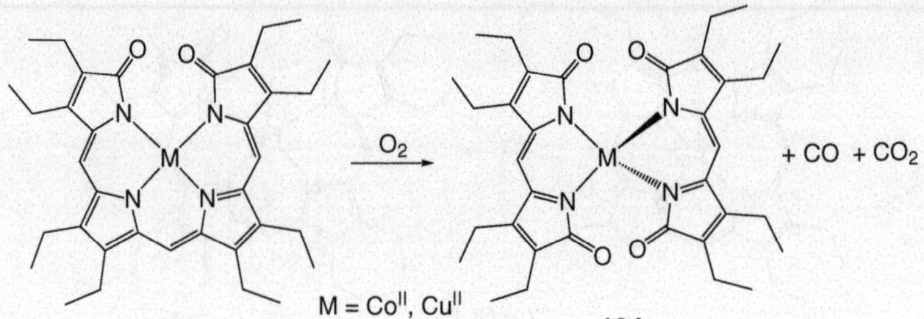

Scheme 42 Oxidation of cobalt and copper biliverdin complexes [209, 213]

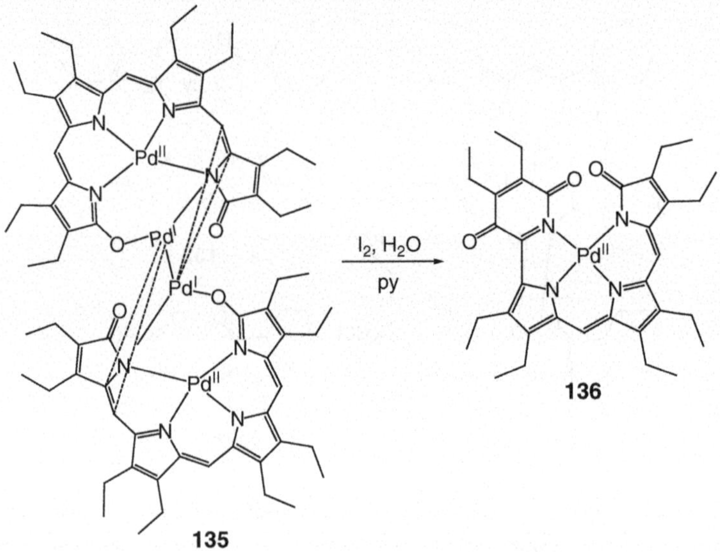

Scheme 43 Oxidation of tetranuclear palladium biliverdin complex by I_2 [215]

opening [200]. Cobalt biliverdins were alternatively obtained by a coupled oxidation of Co(II) octaethylporphyrin [157]. Oxidation with iodine converted (OEB$^•$) M^{II} complexes (M = Co, Ni, Pd) into ones containing an oxidized form of bilindione ligand [211, 214], while aerial oxidation of copper and cobalt complexes **133** resulted in cleavage of tetrapyrroles yielding complexes with two coordinated dipyrrolic units **134** (Scheme 42) [209, 213].

A unprecedented tetranuclear complex **135** consisting of two helical (OEB)Pd^{II} units bridged by $(Pd_2^{I})^{2+}$ fragment was isolated along with monomeric (OEB)Pd from the insertion of palladium into OEB ligand [13, 214, 215]. Reaction of this compound with iodine resulted in formation of rearranged monomeric complex **136**: an incorporation of oxidized *meso*-carbon into a terminal pyrrolone unit was observed (Scheme 43)[215].

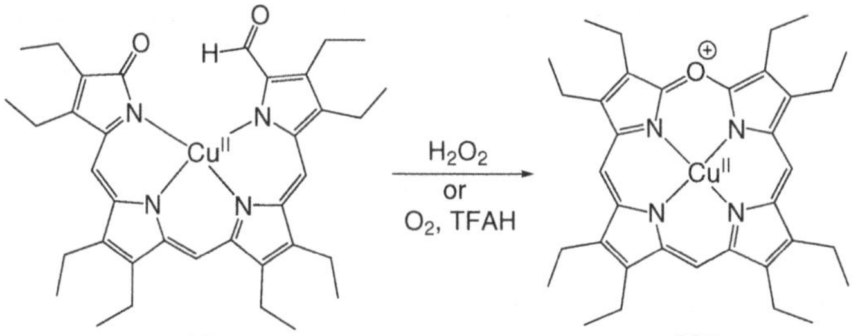

Scheme 44 Conversion of copper(II) formylbiliverdin to verdoheme [9]

Oxidative cyclization of biliverdin complexes leading to metalloverdohemes was also studied [216]. Nickel(II), cobalt(II), and copper(II) octaethylformylbiliverdins were converted to verdoheme analogues by treatment with hydrogen peroxide or (in case of Cu(II) species) by heating with trifuoroacetic acid under dioxygen (Scheme 44) [9]. Formation of carbon monooxide and dioxide was detected in the course of the reaction. Addition of trifluoroacetic acid to the dichloromethane solution of palladium octaethylbilindione also resulted in ring closure. Only 5 min of stirring at room temperature was found sufficient to cause the transformation [215] .

Formation of biliverdin derivatives in a process of coupled oxidation of iron porphyrins is not limited to β-octaalkyl derivatives. Mizutani's group worked out a high-yielding method of preparation of tetraphenylbiladienone **20** (a major product of degradation of TPP complexes by Tl(III), Ce(IV) or photooxidation, see Sects. 3.1 and 4.1) [145, 159, 160]. Iron *meso*-tetraphenylporphyrin subjected to coupled oxidation procedure in a chloroform solution yielded a mixture of isomeric biladienones **20** (63%) and **140** (15%; Scheme 45) [145]. Compound **140** could be photoisomerized to **20**, while the reverse transformation did not proceed. The additional bilindione products **141**, **142** were obtained when the reaction was carried out in refluxing chloroform; both compounds were converted to each other with visible light illumination. An X-ray structure of isomer **141** proved its *ZZZ* configuration and a helicoidal conformation. The procedure could be extended to other tetraarylporphyrins substituted in *para* positions with OCH_3, $COOCH_3$, CN, $OC_{12}H_{25}$, and $COOC_{12}H_{25}$ groups [159, 160]. The reaction was accelerated by electron-withdrawing substituents, which also favored the formation of triarylbilindiones (maximum yield of 19% was noted for *p*-$COOC_{12}H_{25}$ derivative) while electron-donating ones increased the amounts of biladienones (85% yield for methoxy-substituted substrate was found). Interestingly, the presence of one methoxy substituent in *ortho* position of each of phenyl groups did not prevent the macrocycle from oxidative degradation: both biladienone and bilindione were formed in 14% and 10% yield, respectively. Cyclization of bilindiones **141** was also described yielding the corresponding zinc triarylverdohemes, which were isolated as trfiluoroacetates [217].

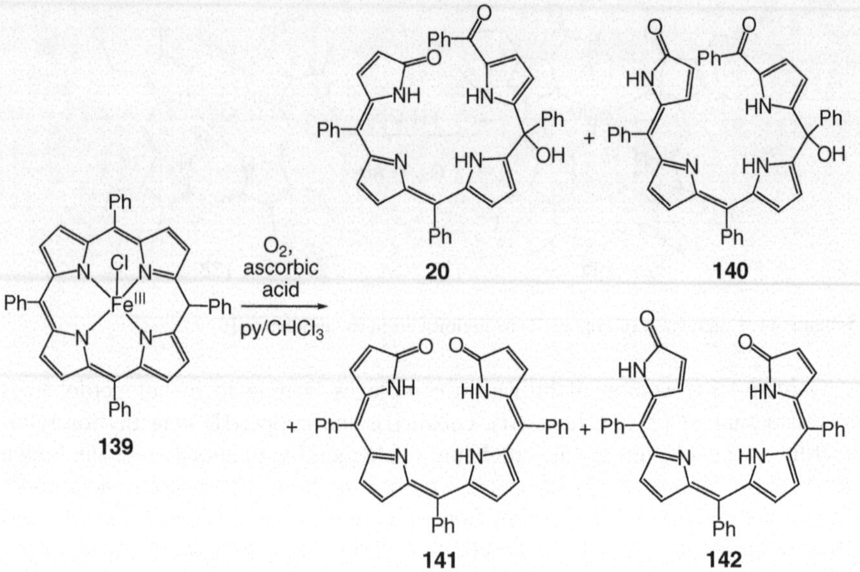

Scheme 45 Coupled oxidation of iron(III) tetraphenylporphyrin [145]

Theoretical studies on biliverdin and its complexes involved such aspects as molecular and electronic structure of its isomeric forms [218] and biliverdin-based metalloradicals [219], spin density distribution in metallobiliverdin radicals [220], energetics and dynamics of dimer formation by oxidized species [221], and mechanism of reduction to bilirubin [222].

5.4 *Regioselectivity of Coupled Oxidation*

Studies on regioselectivity of coupled oxidation of iron porphyrins were aimed to establish the influence of factors connected with a structure of macrocyclic substrate on the outcome of degradation process. Four isomeric biliverdins were isolated in comparable yields from coupled oxidation of iron(III) protoporphyrin IX, thus regioselectivity observed in natural systems (see Sect. 6) was lost [223, 224]. Later studies showed that replacement of 3-methyl group of mesoheme with CF_3 substituent had a great influence on product distribution: ring-opening occurred mainly at C(20) yielding δ isomer as a major product [225].

Coupled oxidation of 5- or 15-phenyl-substituted iron(III) protoporphyrin IX in pyridine solution yielded biliverdins opened only at three unsubstituted *meso* positions (as illustrated in Scheme 46 for 5-phenyl derivative) [35, 226]. Similarly, 5-aryl-mesohemes III were cleft at C(10), C(15) or C(20) yielding (due to symmetry of the starting complex) only two isomeric products [227]. The character of

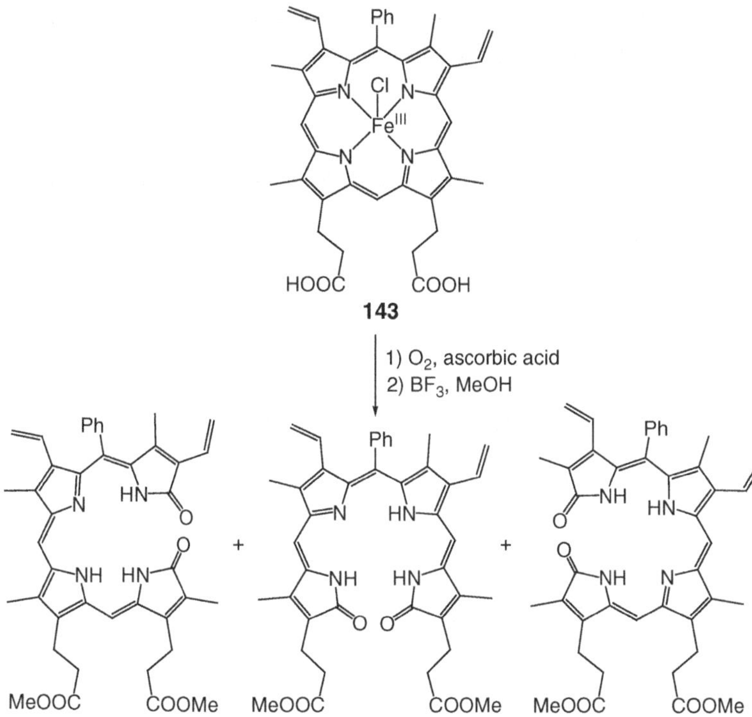

Scheme 46 Coupled oxidation of 5-phenylprotoheme [35]

aryl ring substituent influenced the reaction yield, but its impact on the product distribution was rather negligible. The identified biliverdin isomers served as references for studies on the regioselectivity of heme oxygenase (see Sect. 6 of this contribution) [228].

6 Biodegradation of Tetrapyrrolic Macrocycles

Degradation of tetrapyrrolic macrocycles is used by living organisms both as a method of removal of unwanted (redundant) heme or chlorophyll and as a way of synthesis of linear systems (bilins) which can also fulfill important physiological functions [229–232]. Mechanism of transformation of macrocycles to acyclic oligopyrroles has now become much more clear and better understood in the result of numerous studies on model reactions and determination of active intermediates and structures of key enzymes.

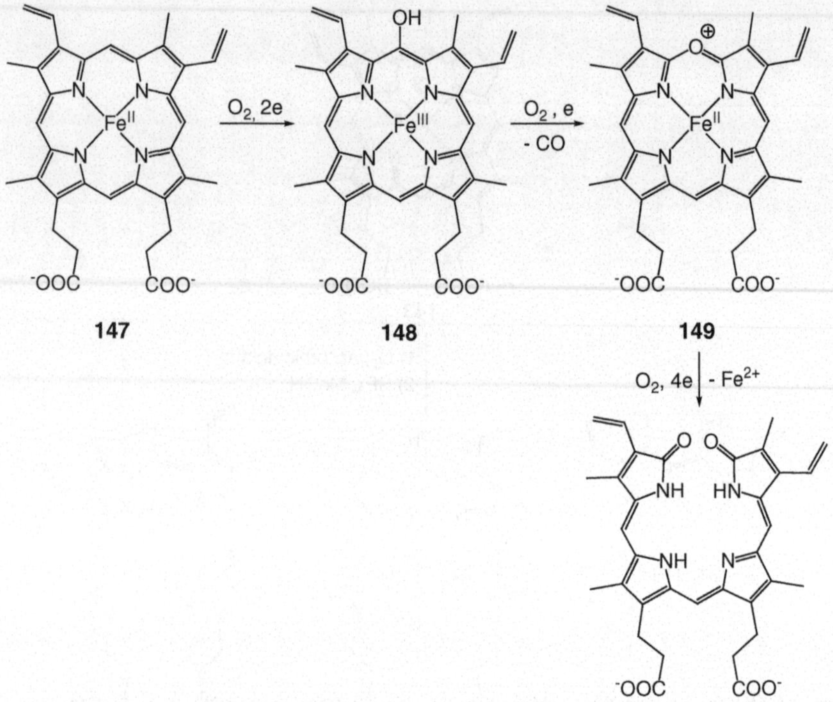

Scheme 47 Heme degradation catalyzed by heme oxygenase

6.1 Heme Oxygenase

Heme oxygenase (HO), an enzyme responsible for the oxidative conversion of heme to biliverdin, was discovered by Tenhunen et al. in 1968 [233]. Since that report, numerous studies have been devoted to understanding the mechanism of the enzymatic action [183, 234–239]. HO is unique among heme enzymes in that activation of dioxygen by prosthetic group is utilized for its own degradation. A regiospecific conversion of heme to biliverdin IXα, carbon monooxide and Fe^{2+} ions requires three molecules of O_2 and the total uptake of seven electrons, and proceeds in three successive steps (Scheme 47): *meso*-hydroxylation, followed by release of CO and verdoheme formation and ring opening connected with iron loss yielding free biliverdin. Formation of such metabolites implies other functions of heme oxygenase, involving iron homeostasis, cytoprotection against oxidative injury and cellular stress, and postulated role in cellular signaling.

In mammals three isoforms of HO have been identified; heme degradation enzymes can also be found in plants and some pathogenic bacteria [183, 240–242]. Many of these proteins have been structurally characterized, including cofactor-free enzymes and their complexes with heme and subsequent intermediates of its enzymatic conversion [240, 241, 243–248]. Since the structural aspects and mechanism of heme oxygenase have been thoroughly reviewed [183], only chosen aspects of recent investigations in the field will be presented in this contribution.

Several groups concentrated their efforts on detailed analysis of mechanism of heme degradation. A theoretical study on *meso*-hydroxylation step by Shaik and coworkers indicated a preference for homolytic dissociation of O–O bond in Fe–OOH intermediate and the crucial role of hydrogen bonding network of distal heme pocket in trapping of $^{\bullet}OH$ radical, in full agreement with the experimental data [249–251]. Verdoheme opening, the less understood third step of degradation process, was investigated by Ikeda-Saito and coworkers [239]. They prepared verdoheme complexes with various heme oxygenases and characterized them by various techniques [245, 252, 253]. A similarity of the final stage of heme oxidation to the first one was observed, including the participation of water cluster in the radical intermediate binding. Verdoheme-heme oxygenase complexes were also characterized by other groups [246, 254, 255].

Factors influencing regioselectivity of heme degradation have been also studied. The exclusive formation of α isomer of *meso*-hydroxyheme and, finally, of biliverdin α was substantiated by specific seating of heme in the protein and the construction of distal pocket limiting the access of coordinated dioxygen molecule to other *meso* positions [183, 236]. Mutant heme oxygenases were prepared with an altered regioselectivity which was attributed to various possible orientations of heme moiety [256, 257]; mutations can even change the typical function of enzyme to peroxidase activity [258]. Bacterial heme oxygenases were characterized exhibiting different preference of heme oxidation site as a result of specific seating of the heme [241, 259–261]. Part of regioselectivity studies utilized modified hemes to explore the impact of porphyrin ring substitution on the degradation process. Heme oxygenase was shown to accept various iron porphyrins as substrates, though the presence of propionate chains at C(13) and C(17) seemed to be an important feature required for enzymatic action [183, 234, 262]. Ikeda-Saito and coworkers showed that HO is capable of oxidizing of all isomers of *meso*-hydroxyhemin to the corresponding verdohemes, but only verdoheme α was further converted to biliverdin [263]. *Meso*-substitution effects were particularly important for the analysis of ring-opening mechanisms. Oxidation of mesoheme with methylated *meso*-position by human HO-1 was investigated by Torpey and Ortiz de Montellano [264]. Surprisingly, α-CH_3-derivative was converted to biliverdin α, while γ-CH_3-mesoheme yielded exclusively γ isomer (Scheme 48; in both cases the fate of extruded *meso* substituent remained unknown); β and δ-substitution resulted in a mixture of products (both methylated and *meso*-unsubstituted).

When protoheme substituted with 5- or 15-phenyl group was used as a substrate, biliverdin α was formed (Scheme 49; benzoic acid by-product was isolated in the first case) [228]. Mesobiliverdin α was identified as the major degradation product of various 5-aryl-mesohemes; isoporphyrin intermediate was detected in this reaction [265]. In contrast, 5-formylmesohemes were exclusively oxidized by heme oxygenase at non-substituted carbons (C(10) or C(20)) to give a formylated biliverdin derivative [266]. Generally, product distribution was found dependent mainly on the possible orientations of modified heme in the protein crevice, but electronic effects of substituents were also of importance.

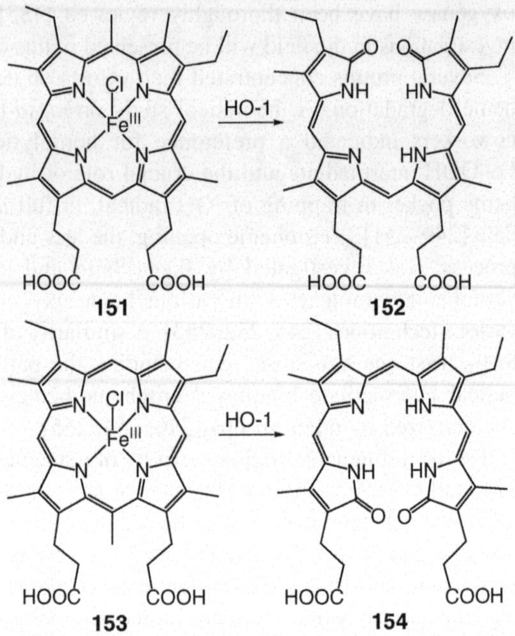

Scheme 48 Oxidation of methylated mesoheme by human HO-1 [264]

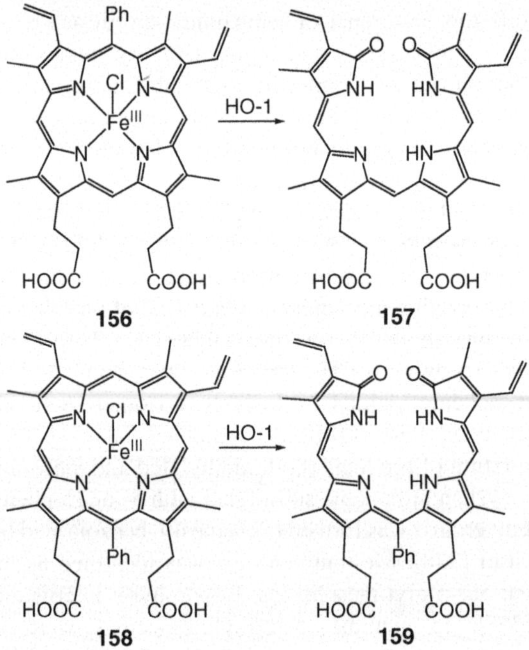

Scheme 49 Oxidation of phenyl-substituted protoheme by HO-1 [228]

The observation that under certain conditions various heme proteins also can exhibit oxygenase activity led to elaboration of protocol of coupled oxidation, which was used as a model of enzymatic heme degradation [267]. Though the detailed mechanism of *meso*-hydroxylation step is slightly different [237, 268], both processes share common intermediates: hydroxyheme, verdoheme, and, finally, iron- and metal-free biliverdin. In a typical experiment, these compounds are produced upon treatment of heme protein with an excess of ascorbate and dioxygen or H_2O_2; sometimes also the addition of pyridine was necessary to replace the protein axial ligands [231]. Coupled oxidation of hemoglobin (Hb) and myoglobin (Mb) has been most widely studied, leading mainly to α isomer of biliverdin, but in case of Hb a significant amount of β isomer is also produced [194, 269, 270]. This regioselectivity is changed for abnormal or mutant hemoglobins [270, 271] as well as for cobalt(II) porphyrins used as substrates [272]. Coupled oxidation of heme covalently attached to a variant of *Escherichia coli* cytochrome b_{562} yielded a verdoheme protein complex which could be converted with formic acid to protein-attached α-biliverdin [273]. One of axial ligand mutants of mitochondrial cytochrome b_5, H63V, also stopped at the verdoheme stage while H39V variant allowed to oxidize heme to biliverdin [274]. This different behavior was attributed to the presence of polar amino acid residues in H39V mutant able to interact with heme-bound iron.

6.2 Chlorophyll Degradation

The principal transformations and main intermediates of chlorophyll breakdown have been identified [275–278]. Chlorophyll *a* **160** and chlorophyll *b* **161** lose phytol side chain and magnesium ion and pheophorbide *a* **162** is formed (Scheme 50). Ring opening occurring exclusively at C(5) *meso* position yields a tetrapyrrole called red chlorophyll catabolite (RCC, **163**) which is further converted to fluorescent and nonfluorescent chlorophyll catabolites (FCCs and NCCs, respectively). A key ring-opening step is catalyzed by a specific enzyme, pheophorbide a oxygenase [279, 280]. Isotope labeling experiments showed that only one of newly introduced oxygen atoms is derived from O_2 molecule, while the second one probably originates from water.

Studies on photooxygenation of chlorophyll and bacteriochlorophyll derivatives were conducted in context of the catabolism of these compounds occurring in vivo. Typically, ring-opening reactions occurred by dioxygen attack on C(1)–C(20) bond [114, 281]. However, Iturraspe and Gossauer demonstrated the regioselectivity change by metal coordination: zinc(II) pyropheophorbide *a* methyl ester **16** led to C(20)-opened product **165** while cadmium complex **164** underwent cleavage of C (4)–C(5) bond yielding compound **166** (Scheme 51) [282]. Recent studies on the degradation of zinc chlorophyll derivatives substituted at 3- and 13-positions showed a systematic change of electronic absorption maxima (up to 919 nm) of the ring-opened products with the electron-withdrawing character of the substituent, demonstrating their attractiveness as near-infrared light absorbing pigments [283].

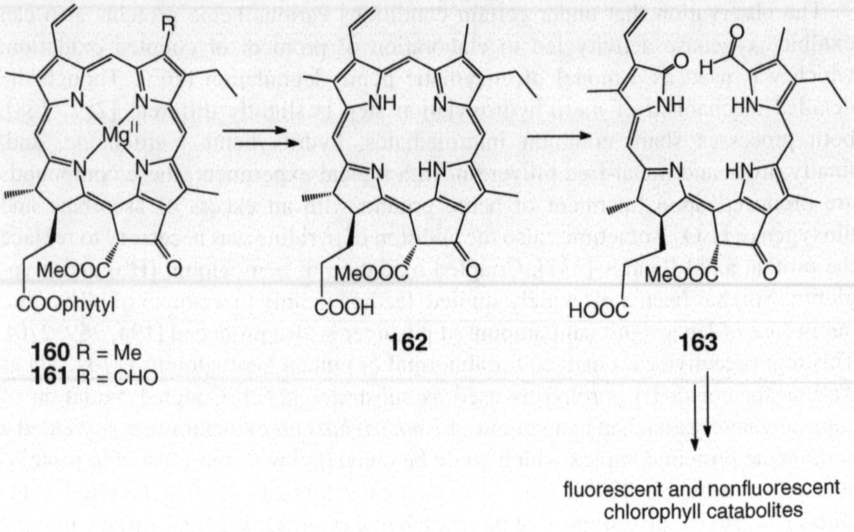

Scheme 50 Chlorophyll degradation

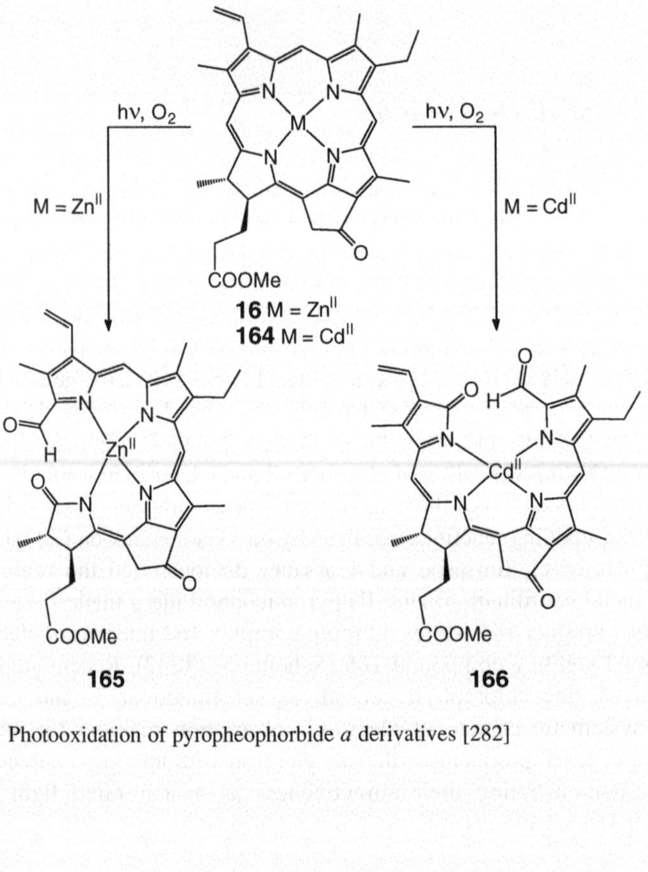

Scheme 51 Photooxidation of pyropheophorbide *a* derivatives [282]

7 Summary: Future Directions

The word "degradation" is commonly associated with the loss of quality, with a conversion of an object or a person to less attractive and less valuable state or form. These negative connotations, however, should not come to mind when porphyrin degradation is considered. Certainly, formed products lack many of properties of a parent compound, but at the same time they gained certain unique features, such as a conformational flexibility or an interesting coordination behavior. Ring opening of cyclic tetrapyrroles can be applied as the easiest method of preparation of these linear oligopyrroles.

On the other hand, many of degradation processes are not selective and are frequently accompanied by subsequent reactions (demetallation, *Z-E* isomerization, water/alcohol addition) which further increase the number of possible products. In many classical papers on porphyrin degradation, only major products were isolated and characterized, and the fate of the rest of starting material remains unknown. Perhaps the use of modern analytical techniques could lead to identification of minor decomposition products.

In general, a great progress has been made in deciphering of degradation processes of tetrapyrrolic macrocycles in nature and of their synthetic models. Still, some fields remain underexplored, including pathways of inactivation of metalloporphyrin catalysts. Since the ways of porphyrin ring modification are unlimited, new developments in the field can be expected because a specific reactivity can be generated connected with the particular substitution or/and metal ion insertion.

One can also imagine that wider synthetic availability of such members of porphyrinoid family, as expanded porphyrins, contracted ones, porphyrin isomers (N-confused, fused, porphycenes,...), and heteroporphyrins could result in investigations on their oxidative degradation. Ring opening of octaphyrins upon metallation with Cu(II) and interesting oxidative conversions of dithiaethyne-porphyrin and dioxaporphyrin which were described quite recently show a potential hidden in these porphyrin analogs [284–286].

References

1. Adams KR, Bonnett R, Burke PJ, Salgado A, Vallés MA (1993) The 2,3-secochlorin-2, 3-dione system. J Chem Soc Chem Commun 1860–1861
2. Brückner C, Rettig SJ, Dolphin D (1998) Formation of a *meso*-tetraphenylsecochlorin and a homoporphyrin with a twist. J Org Chem 63:2094–2098
3. Sessler JL, Shevchuk SV, Callaway W, Lynch V (2001) A one-step synthesis of a free base secochlorin from a 2,3-dimethoxy porphyrin. Chem Commun 968–969
4. Pacholska E, Latos-Grażyński L, Ciunik Z (2002) A direct link between annulene and porphyrin chemistry – 21-vacataporphyrin. Chem Eur J 8:5403–5406

5. Pacholska-Dudziak E, Szterenberg L, Latos-Grażyński L (2011) A flexible porphyrin–annulene hybrid: a nonporphyrin conformation for *meso*-tetraaryldivacataporphyrin. Chem Eur J 17:3500–3511
6. Mizutani T, Yagi S, Honmaru A, Murakami S, Furusyo M, Takagishi T, Ogoshi H (1998) Helical chirality induction in zinc bilindiones by amino acid esters and amines. J Org Chem 63:8769–8784
7. Mizutani T, Sakai N, Yagi S, Takagishi T, Kitagawa S, Ogoshi H (2000) Allosteric chirality amplification in zinc bilinone dimer. J Am Chem Soc 122:748–749
8. Hamakubo K, Yagi S, Nakazumi H, Mizutani T, Kitagawa S (2006) Homohelicity induction of propylene-linked zinc bilinone dimers by complexation with chiral amine and α-amino esters. Preorganization of structurally coupled homohelical subunits. Tetrahedron 62:3619–3628
9. Koerner R, Olmstead MM, Ozarowski A, Phillips S, Van Calcar PM, Winkler K, Balch AL (1998) Possible intermediates in biological metalloporphyrin oxidative degradation. Nickel, copper and cobalt complexes of octaethylformybiliverdin and their conversion to a verdoheme. J Am Chem Soc 120:1274–1284
10. Mizutani T, Yagi S (2004) Linear tetrapyrroles as functional pigments in chemistry and biology. J Porphyr Phthalocyanines 8:226–237
11. Bröring M (2010) Beyond dipyrrins: coordination interactions and templated macrocyclizations of open-chain oligopyrroles. In: Kadish KM, Smith KM, Guilard R (eds) Handbook of porphyrin science with applications to chemistry, physics, materials science, engineering biology and medicine, vol 8. World Scientific, Singapore, pp 343–501 (Chapter 41)
12. Koerner R, Olmstead MM, Ozarowski A, Balch AL (1999) A linear tetrapyrrole as a binucleating ligand with copper(II). Coordination beyond the usual $M–N_4$ bonding. Inorg Chem 38:3262–3263
13. Lord P, Olmstead MM, Balch AL (1999) Tetrapyrroles as π donors: a Pd_2^{2+} unit sandwiched between two helical bilindione–palladium moieties. Angew Chem Int Ed 38:2761–2763
14. Phillips S, Noll BC, Olmstead MM, Balch AL (2001) Oxidation of copper(II) hydroxyporphyrin (oxophlorin); oxidative ring opening and formation of an ester-linked dinuclear copper complex. Can J Chem 79:922–929
15. Fuhrhop J-H (1975) Irreversible reactions at the porphyrin periphery (excluding photochemistry). In: Smith KM (ed) Porphyrins and metalloporphyrins. Elsevier, Amsterdam, pp 625–666 (Chapter 15)
16. Nicolaus RA, Mangoni L, Caglioti L (1956) Pyrrole acids in the oxidation of the porphyrins. Ann Chim (Rome) 46:793–805
17. Gray CH, Nicholson DC, Nicolaus RA (1958) The IX-alpha structure of the common bile pigments. Nature 181:183–185
18. Battersby AR, Cardwell KS, Leeper FJ (1986) Stereochemical studies on porphyrin a: assignment of the absolute configuration of a model porphyrin by degradation. J Chem Soc Perkin Trans 1 1565–1580
19. Chapman RA, Roomi MW, Morton TC, Krajcarski DT, MacDonald SF (1971) The analytical reduction of porphyrins to pyrroles. Can J Chem 49:3544–3564
20. Ficken GE, Johns RB, Linstead RP (1956) Chlorophyll and related compounds. Part IV. The position of the extra hydrogens in chlorophyll. The oxidation of pyrophœophorbide-*a*. J Chem Soc 2272–2280
21. Morley HV, Holt AS (1961) Studies on chlorobium chlorophylls. II. The resolution of oxidation products of chlorobium pheophorbide (660) by gas–liquid partition chromatography. Can J Chem 39:755–760
22. Purdie JW, Holt AS (1965) Structures of chlorobium chlorophylls (650). Can J Chem 43:3347–3353

23. Ellsworth RK, Aronoff S (1968) Investigations on the biogenesis of chlorophyll *a*. I. Purification and mass spectra of maleimides from the oxidation of chlorophyll and related compounds. Arch Biochem Biophys 124:358–364
24. Rüdiger W (1969) Chromsäure- und Chromatabbau von Gallenfarbstoffen. Z Physiol Chem 350:1291–1300
25. Bonnett R, McDonagh AF (1969) Methylvinylmaleimide (nitrite body) from chromic acid oxidation of tetrapyrrolic pigments. Chem Ind 107–108
26. Ellsworth RK (1970) Gas chromatographic determination of some maleimides produced by the oxidation of heme and chlorophyll *a*. J Chromatogr A 50:131–134
27. Martin J, Quirke E, Shaw GJ, Soper PD, Maxwell JR (1980) Petroporphyrins—II. The presence of porphyrins with extended alkyl substituents. Tetrahedron 36:3261–3267
28. Gauler R, Hesse U, Risch N (1995) Derivatives of natural tetrapyrroles for photodynamic therapy, 3. Oxidative degradation studies on porphyrins with chromic acid. Liebigs Ann 2227–2230
29. Risch N, Hesse U, Josephs A, Gauler R (1996) Derivatives of natural tetrapyrroles for photodynamic therapy, 4. Oxidative degradation studies: direct analysis and determination of ether and ester linkages in porphyrin dimers and oligomers of hematoporphyrin derivative (HPD). Liebigs Ann 1871–1874
30. Satoh Y, Nomoto S, Hama T (2012) Comprehensive determination of chlorophyll derivatives by chromic acid oxidation. Chem Lett 41:571–573
31. Byrn MP, Curtis CJ, Hsiou Y, Khan SI, Sawin PA, Tendick SK, Terzis A, Strouse CE (1993) Porphyrin sponges: conservation of host structure in over 200 porphyrin-based lattice clathrates. J Am Chem Soc 115:9480–9497
32. Hayes JM, Freeman KH, Popp BN, Hoham CH (1990) Compound-specific isotopic analyses: a novel tool for reconstruction of ancient biogeochemical processes. Org Geochem 16:1115–1128
33. Yu Z, Sheng G, Fu J, Peng P (2000) Determination of porphyrin carbon isotopic composition using gas chromatography–isotope ratio monitoring mass spectrometry. J Chromatogr A 903:183–191
34. Van Berkel GJ, Glish GL, McLuckey SA, Tuinman AA (1989) Mechanism of porphyrin reduction and decomposition in a high-pressure chemical ionization plasma. J Am Chem Soc 111:6027–6035
35. Niemevz F, Buldain GY (2004) Phenyl biliverdin isomers obtained by chemical oxidation of iron(III) complex of 5-phenyl protoporphyrin IX. J Porphyr Phthalocyanines 8:989–995
36. Świder P, Nowak-Król A, Voloshchuk R, Lewtak JP, Gryko DT, Danikiewicz W (2010) Mass spectrometry studies on *meso*-substituted corroles and their photochemical decomposition products. J Mass Spectrom 45:1443–1451
37. Wu D, Xu G, Qu S, Xue R, Gu C, Zhang F (1989) Standard enthalpies of combustion and formation of porphyrin derivatives. Thermochim Acta 154:233–245
38. Patiño R, Campos M, Torres LA (2007) Strength of the Zn–N coordination bond in zinc porphyrins on the basis of experimental thermochemistry. Inorg Chem 46:9332–9336
39. Gamboa M, Campos M, Torres LA (2010) Study of the stability of 5,10,15,20-tetraphenylporphine (TPP) and metalloporphyrins NiTPP, CoTPP, CuTPP, and ZnTPP by differential scanning calorimetry and thermogravimetry. J Chem Thermodyn 42:666–674
40. Antina EV, Balantseva EV, Berezin MB (2011) Oxidative degradation of porphyrins and metalloporphyrins under polythermal conditions. Russ J Gen Chem 81:1222–1230
41. Gokakakar SD, Salker AV (2010) Thermal studies of cobalt, iron and tin metalloporphyrins. J Therm Anal Calorim 101:809–813
42. Bonnett R, Stephenson GF (1965) The *meso* reactivity of porphyrins and related compounds. I. Nitration. J Org Chem 30:2791–2798
43. Bonnett R, Dimsdale MJ (1968) The *meso*-hydroxylation and *meso*-benzoxylation of pyridine octaethylhaemochrome. Tetrahedron Lett 9:731–733

44. Bonnett R, Dimsdale MJ, Stephenson GF (1969) The *meso*-reactivity of porphyrins and related compounds. Part IV. Introduction of oxygen functions. J Chem Soc (C) 564–570
45. Bonnett R, Dimsdale MJ (1972) The *meso*-reactivity of porphyrins and related compounds. Part V. The *meso*-oxidation of metalloporphyrins. J Chem Soc Perkin Trans 1 2540–2548
46. Bonnett R, Cornell P, McDonagh AF (1976) The *meso*-reactivity of porphyrins and related compounds. Part VII. Benzoyloxylation of phenylpyrroles and of octaethylporphyrin. J Chem Soc Perkin Trans 1 794–800
47. Smith KM (1971) Reactions of porphyrins with thallium(III) trifluoroacetate. Chem Commun 540–541
48. Cavaleiro JAS, Smith KM (1971) Reactions of *trans*-octaethylchlorin with thallium(III) trifluoroacetate. Chem Commun 1384–1385
49. McCombie SW, Smith KM (1972) Oxophlorin (oxyporphyrin) synthesis. Tetrahedron Lett 13:2463–2464
50. Barnett GH, Hudson MF, McCombie SW, Smith KM (1973) Synthesis of oxophlorins (oxyporphyrins) from magnesium and zinc porphyrin chelates. J Chem Soc Perkin Trans 1 691–696
51. Evans B, Smith KM, Cavaleiro JAS (1976) Ring cleavage of *meso*-tetraphenylporphyrin. Tetrahedron Lett 17:4863–4866
52. Evans B, Smith KM, Cavaleiro JAS (1978) Bile pigment studies. Part 4. Some novel reactions of metalloporphyrins with thallium(III) and cerium(IV) salts. Ring cleavage of *meso*-tetraphenylporphyrin. J Chem Soc Perkin Trans 1 768–773
53. Huster MS, Smith KM (1988) Ring cleavage of chlorophyll derivatives: isolation of oxochlorin intermediates and ring opening via a two oxygen molecule mechanism. Tetrahedron Lett 29:5707–5710
54. Kalish HR, Latos-Grażyński L, Balch AL (2000) Heme/hydrogen peroxide reactivity: formation of paramagnetic iron oxophlorin isomers by treatment of iron porphyrins with hydrogen peroxide. J Am Chem Soc 122:12478–12486
55. Kalish H, Camp JE, Stępień M, Latos-Grażyński L, Balch AL (2001) Reactivity of mono-*meso*-substituted iron(II) octaethylporphyrin complexes with hydrogen peroxide in the absence of dioxygen. Evidence for nucleophilic attack on the heme. J Am Chem Soc 123:11719–11727
56. Wojaczyński J, Latos-Grażyński L, Chmielewski PJ, Van Calcar P, Balch AL (1999) ^{1}H NMR investigations of triphenylporphyrin metal complexes and electronic interactions in iron(III) complexes of *meso–meso*-linked 5,5′-bis(10,15,20-triphenylporphyrin). Inorg Chem 38:3040–3050
57. Osuka A, Shimidzu H (1997) meso, meso-Linked porphyrin arrays. Angew Chem Int Ed Engl 36:135–137
58. Yoshida N, Shimidzu K, Osuka A (1998) *meso-meso* Linked diporphyrins from 5,10,15-trisubstituted porphyrins. Chem Lett 27:55–56
59. Yoshida N, Aratani N, Osuka A (2000) Poly(zinc(II)-5,15-porphyrinylene) from silver(I)-promoted oxidation of zinc(II)-5,15-diarylporphyrins. Chem Commun 197–198
60. Catalano MM, Crossley MJ, Harding MM, King LG (1984) Control of reactivity at the porphyrin periphery by metal ion co-ordination: a general method for specific nitration at the β-pyrrolic position of 5,10,15,20-tetraarylporphyrins. J Chem Soc Chem Commun 1535–1536
61. Shine HJ, Padilla AG, Wu S-M (1979) Ion radicals. 45. Reactions of zinc tetraphenylporphyrin cation radical perchlorate with nucleophiles. J Org Chem 44:4069–4075
62. Abhilash GJ, Bhuyan J, Singh P, Maji S, Pal K, Sarkar S (2009) $^{\bullet}NO_2$-mediated *meso*-hydroxylation of iron(III) porphyrin. Inorg Chem 48:1790–1792
63. Bhuyan J, Sarkar S (2010) Oxidative degradation of zinc porphyrin in comparison with its iron analogue. Chem Eur J 16:10649–10652
64. Ongayi O, Fronczek FR, Vicente MGH (2003) Benzoylbiliverdins from chemical oxidation of dodeca-substituted porphyrins. Chem Commun 2298–2299

65. Ongayi O, Vicente MGH, Ou Z, Kadish KM, Kumar MR, Fronczek FR, Smith KM (2006) Synthesis and electrochemistry of undeca-substituted metallo-benzoylbiliverdins. Inorg Chem 45:1463–1470
66. Ongayi O, Vicente MGH, Ghosh B, Fronczek FR, Smith KM (2010) Bilitrienones from the chemical oxidation of dodecasubstituted porphyrins. Tetrahedron 66:63–67
67. Ponomarev GV, Morozova YV, Yashunsky DV (2001) Chemistry of oximes of *meso*-formylporphyrins. Opening of the porphyrin macrocycle into tripyrrolylisoxazoles. The revised structure of "isophlorins". Chem Heterocycl Compd 37:253–255
68. Morozova YV, Nesterov VV, Yashunsky DV, Antipin MY, Ponomarev GV (2005) Porphyrins. 40. Chemistry of oximes of metal complexes of *meso*-formyloctaalkylporphyrins. Synthesis, molecular and crystal structure of nickel complexes of "tripyrrolylisoxazoles". Chem Heterocycl Compd 41:598–605
69. Kalish H, Lee HM, Olmstead MM, Latos-Grażyński L, Rath SP, Balch AL (2003) Heme cleavage with remarkable ease: paramagnetic intermediates formed by aerobic oxidation of a *meso*-amino-substituted iron porphyrin. J Am Chem Soc 125:4674–4675
70. Rath SP, Kalish H, Latos-Grażyński L, Olmstead MM, Balch AL (2004) Facile ring opening of iron(III) and iron(II) complexes of *meso*-amino-octaethylporphyrin by dioxygen. J Am Chem Soc 126:646–654
71. Sprutta N, Rath SP, Olmstead MM, Balch AL (2005) Metal complexes of *meso*-amino-octaethylporphyrin and the oxidation of Ni^{II}(*meso*-amino-octaethylporphyrin). Inorg Chem 44:1452–1459
72. Chang CK, Avilés G, Bag N (1994) Verdoheme-like oxaporphyrin formation by oxygenation of a Co(II) porphyrinyl naphthoic acid. A new model of heme degradation. J Am Chem Soc 116:12127–12128
73. Yamanishi K, Miyazawa M, Yairi T, Sakai S, Nishina N, Kobori Y, Kondo M, Uchida F (2011) Conversion of cobalt(II) porphyrin into a helical cobalt(III) complex of acyclic pentapyrrole. Angew Chem Int Ed 50:6583–6586
74. Liu C, Shen D-M, Chen Q-Y (2006) Unexpected bromination ring-opening of tetraarylporphyrins. Chem Commun 770–772
75. Meunier B (1992) Metalloporphyrins as versatile catalysts for oxidation reactions and oxidative DNA cleavage. Chem Rev 92:1411–1456
76. Che C-M, Huang J-S (2009) Metalloporphyrin-based oxidation systems: from biomimetic reactions to application in organic synthesis. Chem Commun 3996–4015
77. Lu H, Zhang XP (2011) Catalytic C–H functionalization by metalloporphyrins: recent developments and future directions. Chem Soc Rev 40:1899–1909
78. Mansuy D (2007) A brief history of the contribution of metalloporphyrin models to cytochrome P450 chemistry and oxidation catalysis. C R Chim 10:392–413
79. Traylor PS, Dolphin D, Traylor TG (1984) Sterically protected hemins with electronegative substituents: efficient catalysts for hydroxylation and epoxidation. J Chem Soc Chem Commun 279–280
80. Banfi S, Montanari F, Quici S (1988) New manganese tetrakis(halogenoaryl)porphyrins featuring sterically hindering electronegative substituents: synthesis of highly stable catalysts in olefin epoxidation. J Org Chem 53:2863–2866
81. Moore KT, Horváth IT, Therien MJ (1997) High-pressure NMR studies of (porphinato)iron-catalyzed isobutane oxidation utilizing dioxygen as the stoichiometric oxidant. J Am Chem Soc 119:1791–1792
82. Leanord DR, Lindsay Smith JR (1991) Model systems for cytochrome P450 dependent monooxygenases. Part 8. A study of the epoxidation of (*Z*)-cyclooctene by iodosylbenzene catalysed by cationic iron(III) tetra(*N*-methylpyridyl)porphyrins adsorbed on Dowex MSC1. J Chem Soc Perkin Trans 2 25–30
83. Nappa MJ, Tolman CA (1985) Steric and electronic control of iron porphyrin catalyzed hydrocarbon oxidations. Inorg Chem 24:4711–4719

84. Pietzyk B, Fröhlich L, Göber B (1995) Characterization and stability of synthetic porphyrins. Pharmazie 50:747–750
85. Pietzyk B, Fröhlich L, Göber B (1996) Stability of synthetic Mn- and Fe-tetraphenylporphyrins in biomimetic systems of aromatic hydroxylation. Pharmazie 51:654–660
86. Stephenson NA, Bell AT (2005) A study of the mechanism and kinetics of cyclooctene epoxidation catalyzed by iron(III) tetrakispentafluorophenyl porphyrin. J Am Chem Soc 127:8635–8643
87. Stephenson NA, Bell AT (2007) Mechanistic insights into iron porphyrin-catalyzed olefin epoxidation by hydrogen peroxide: factors controlling activity and selectivity. J Mol Catal A Chem 275:54–62
88. Cunningham ID, Danks TN, O'Connell KTA, Scott PW (1999) Kinetics and mechanism of the hydrogen peroxide oxidation of a pentafluorophenyl-substituted iron(III) porphyrin. J Chem Soc Perkin Trans 2 2133–2139
89. Cunningham ID, Danks TN, Hay JN, Hamerton I, Gunathilagan S (2001) Evidence for parallel destructive, and competitive epoxidation and dismutation pathways in metalloporphyrin-catalysed alkene oxidation by hydrogen peroxide. Tetrahedron 57:6847–6853
90. Cunningham ID, Danks TN, Hay JN, Hamerton I, Gunathilagan S, Janczak C (2002) Stability of various metalloporphyrin catalysts during hydrogen peroxide epoxidation of alkene. J Mol Catal A Chem 185:25–31
91. Serra AC, Marçalo EC, Rocha Gonsalves AM'A (2004) A view on the mechanism of metalloporphyrin degradation in hydrogen peroxide epoxidation reactions. J Mol Catal A Chem 215:17–21
92. Rácz K, Burger M, Ungvarai-Nagy Z (2010) Comparison of the oxidation of two porphyrin complexes by bromate with respect to wave propagation. Physica D 239:752–756
93. Rácz K, Burger M, Nagy-Ungvarai Z (2006) Autocatalytic oxidation of hemin by acidic bromate. Int J Chem Kinet 38:503–509
94. Rácz K, Burger M, Lagzi I, Ungvarai-Nagy Z (2008) Oxidation of a water-soluble porphyrin complex by bromate. React Kinet Catal Lett 95:135–142
95. Türk H, Erdem M (2004) Structural stabilities of *N*-permethylated tetracations of *meso*-tetrakis(4-pyridyl)porphyrin, *meso*-tetrakis[4-(dimethylamino)phenyl]porphyrin and their manganese(III) complexes toward hydrogen peroxide, *tert*-butylhydroperoxide and sodium hypochlorite. J Porphyr Phthalocyanines 8:1196–1203
96. Türk H, Tay T, Berber H (2000) Structural stabilities of sulfonated manganese tetramesitylporphyrin and its β-brominated analogue toward NaOCl, H_2O_2 and $(CH_3)_3COOH$. J Mol Catal A Chem 160:323–330
97. Türk H, Berber H (2000) Structural stabilities of water-soluble MnTDCSPP, MnTSPP, and supported analogues toward hydrogen peroxide and sodium hypochlorite. Int J Chem Kinet 32:271–278
98. Türk H, Berber H (2001) Structural studies of water-soluble β-brominated manganese porphyrins: stabilities of $MnTDCSPPBr_8$ and $MnTSPPBr_8$ as homogeneous and supported reagents toward hydrogen peroxide and sodium hypochlorite. Turk J Chem 25:215–222
99. Lente G, Fábián I (2007) Kinetics and mechanism of the oxidation of water soluble porphyrin Fe^{III}TPPS with hydrogen peroxide and the peroxomonosulfate ion. Dalton Trans 4268–4275
100. Chmielewski PJ, Latos-Grażyński L, Rachlewicz K, Głowiak T (1994) Tetra-*p*-tolylporphyrin with an inverted pyrrole ring: a novel isomer of porphyrin. Angew Chem Int Ed Engl 33:779–781
101. Furuta H, Asano T, Ogawa T (1994) "N-confused porphyrin": a new isomer of tetraphenylporphyrin. J Am Chem Soc 11:767–768
102. Furuta H, Maeda H, Osuka A (2002) Regioselective oxidative liberation of aryl-substituted tripyrrinone metal complexes from N-confused porphyrin. Org Lett 4:181–184
103. Furuta H, Maeda H, Osuka A (2003) Crystal structures of palladium(II) and copper(II) complexes of *meso*-phenyl tripyrrinone. Inorg Chem Commun 6:162–164

104. Pawlicki M, Kańska I, Latos-Grażyński L (2007) Copper(II) and copper(III) complexes of pyrrole-appended oxacarbaporphyrin. Inorg Chem 46:6575–6584
105. Lemon CM, Brothers PJ (2011) The synthesis reactivity and peripheral functionalization of corroles. J Porphyr Phthalocyanines 15:809–834
106. Nardis S, Mandoj F, Paolesse R, Fronczek FR, Smith KM, Prodi L, Montalti M, Battistini G (2007) Synthesis and functionalization of germanium triphenylcorrolate: the first example of a partially brominated corrole. Eur J Inorg Chem 2345–2352
107. Mandoj F, Nardis S, Pomarico G, Stefanelli M, Schiaffino L, Ercolani G, Prodi L, Genovese D, Zaccheroni N, Fronczek FR, Smith KM, Xiao X, Shen J, Kadish KM, Paolesse R (2009) 6-Azahemiporphycene: a new member of the porphyrinoid family. Inorg Chem 8:10346–10357
108. Gros CP, Barbe J-M, Espinosa E, Guilard R (2006) Room-temperature autoconversion of free-base corrole into free-base porphyrin. Angew Chem Int Ed 45:5642–5645
109. Nardis S, Pomarico G, Fronczek FR, Vicente MGH, Paolesse R (2007) One-step synthesis of isocorroles. Tetrahedron Lett 48:8643–8646
110. Pomarico G, Xiao X, Nardis S, Paolesse R, Fronczek FR, Smith KM, Fang Y, Ou Z, Kadish KM (2010) Synthesis and characterization of free-base copper, and nickel isocorroles. Inorg Chem 49:5766–5774
111. Mandoj F, Nardis S, Pomarico G, Paolesse R (2008) Demetalation of corrole complexes: an old dream turning into reality. J Porphyr Phthalocyanines 12:19–26
112. Stefanelli M, Shen J, Zhu W, Mastroianni M, Mandoj F, Nardis S, Ou Z, Kadish KM, Fronczek FR, Smith KM, Paolesse R (2009) Demetalation of silver(III) corrolates. Inorg Chem 48:6879–6887
113. Barata JFB, Silva AMG, Neves MGPMS, Tomé AC, Silva AMS, Cavaleiro JAS (2006) β,β′-Corrole dimers. Tetrahedron Lett 47:8171–8174
114. Bonnett R, Martínez G (2001) Photobleaching of sensitisers used in photodynamic therapy. Tetrahedron 57:9513–9547
115. Wöhrle D, Wendt A, Weitmeyer A, Stark J, Spiller W, Schneider G, Müller S, Michelsen U, Kliesch H, Heuermann A, Ardeschirpur A (1994) Metal chelates of porphyrin derivatives as sensitizers in photooxidation processes of sulfur compounds and in photodynamic therapy of cancer. Russ Chem Bull 43:1953–1964
116. Silva M, Azenha ME, Pereira MM, Burrows HD, Sarakha M, Forano C, Ribeiro MF, Fernandes A (2010) Immobilization of halogenated porphyrins and their copper complexes in MCM-41: environmentally friendly photocatalysts for the degradation of pesticides. Appl Catal B 100:1–9
117. Kim H, Kim W, Mackeyev Y, Lee G-S, Kim H-J, Tachikawa T, Hong S, Lee S, Kim J, Wilson LJ, Majima T, Alvarez PJJ, Choi W, Lee J (2012) Selective oxidative degradation of organic pollutants by singlet oxygen-mediated photosensitization: tin porphyrin versus C_{60} aminofullerene systems. Environ Sci Technol 46:9606–9613
118. Sternberg ED, Dolphin D, Brückner C (1998) Porphyrin-based photosensitizers for use in photodynamic therapy. Tetrahedron 54:4151–4202
119. Ali H, van Lier JE (2010) Porphyrins and phthalocyanines as photosensitizers and radiosensitizers. In: Kadish KM, Smith KM, Guilard R (eds) Handbook of porphyrin science with applications to chemistry, physics, materials science, engineering biology and medicine, vol 4. World Scientific, Singapore, pp 1–119 (Chapter 16)
120. Arnaut LG (2011) Design of porphyrin-based photosensitizers for photodynamic therapy. Adv Inorg Chem 63:187–233
121. Ethirajan M, Chen Y, Joshi P, Pandey RK (2011) The role of porphyrin chemistry in tumor imaging and photodynamic therapy. Chem Soc Rev 40:340–362
122. Fuhrhop J-H, Mauzerall D (1971) The photooxygenation of magnesium-octaethylporphin. Photochem Photobiol 13:453–458
123. Bonnett R, Chaney BD (1987) *meso*-Reactivity of porphyrins and related compounds. Part 9. Photo-oxygenation of octaethyloxophlorin. J Chem Soc Perkin Trans 1 1063–1067

124. Matsuura T, Inoue K, Ranade AC, Saito I (1980) Photooxygenation of magnesium *meso*-tetraphenylporphyrin. Photochem Photobiol 31:23–26
125. Smith KM, Brown SB, Troxler RF, Lai J-J (1982) Photooxygenation of *meso*-tetraphenylporphyrin metal complexes. Photochem Photobiol 36:147–152
126. Smith KM, Brown SB, Troxler RF, Lai J-J (1980) Mechanism of photo-oxygenation of *meso*-tetraphenylporphyrin metal complexes. Tetrahedron Lett 21:2763–2766
127. Cavaleiro JAS, Hewlins MJE, Jackson AH, Neves GPMS (1986) Structures of the ring-opened oxidation products from *meso*-tetraphenylporphyrin. J Chem Soc Chem Commun 142–144
128. Cavaleiro JAS, Neves MGPS, Hewlins MJE, Jackson AH (1990) The photo-oxidation of meso-tetraphenylporphyrins. J Chem Soc Perkin Trans 1 1937–1943
129. Silva AMS, Neves MGPMS, Martins RRL, Cavaleiro JAS, Boschi T, Tagliatesta P (1998) Photo-oxygenation of *meso*-tetraphenylporphyrin derivatives: the influence of the substitution pattern and characterization of the reaction products. J Porphyr Phthalocyanines 2:45–51
130. Cavaleiro JAS, Hewlins MJE, Jackson AH, Neves MGPMS (1992) Structures of the zinc complexes of the bilinones formed by photo-oxidations of *meso*-tetraphenylporphyrins. Tetrahedron Lett 33:6871–6874
131. Jeandon C, Krattinger B, Ruppert R, Callot HJ (2001) Biladienones from the photooxidation of a *meso-gem*-disubstituted phlorin: crystal and molecular structures of the 3N + O coordinated nickel(II) and copper(II) complexes. Inorg Chem 40:3149–3153
132. LeSaulnier TD, Graham BW, Geier GR III (2005) Enhancement of phlorin stability by the incorporation of *meso*-mesityl substituents. Tetrahedron Lett 46:5633–5637
133. Herath HMA, Karunaratne V, Rajapakse RMG, Wickramasinghe A (2005) Synthesis, characterization and photochemistry of 5,10,15,20-tetrakis(4-*N*-pentylpyridyl)porphyrins, $[(TPePyP)H_2]^{4+}$ and $[(TPePyP)Zn^{II}]^{4+}$. J Porphyr Phthalocyanines 9:155–162
134. Niziolek M, Korytowski W, Girotti AW (2005) Self-sensitized photodegradation of membrane-bound protoporphyrin mediated by chain lipid peroxidation: inhibition by nitric oxide with sustained singlet oxygen damage. Photochem Photobiol 81:299–305
135. Cavaleiro JAS, Görner H, Lacerda PSS, MacDonald JG, Mark G, Neves MGPMS, Nohr RS, Schuchmann H-P, von Sonntag C, Tomé AC (2001) Singlet oxygen formation and photostability of *meso*-tetraarylporphyrin derivatives and their copper complexes. J Photochem Photobiol A Chem 144:131–140
136. Wojaczyński J, Latos-Grażyński L (2010) Photooxidation of N-confused porphyrin: a route to N-confused biliverdin analogues. Chem Eur J 16:2679–2682
137. Wojaczyński J, Popiel M, Szterenberg L, Latos-Grażyński L (2011) Common origin, common fate: regular porphyrin and N-confused porphyrin yield an identical tetrapyrrolic degradation product. J Org Chem 76:9956–9961
138. Aviv I, Gross Z (2007) Corrole-based applications. Chem Commun 1987–1999
139. Flamigni L, Gryko DT (2009) Photoactive corrole-based arrays. Chem Soc Rev 38:1635–1646
140. Ventura B, Degli Esposti A, Koszarna B, Gryko DT, Flamigni L (2005) Photophysical characterization of free-base corroles, promising chromophores for light energy conversion and singlet oxygen generation. New J Chem 9:1559–1566
141. Geier GR III, Chick JFB, Callinan JB, Reid CG, Auguscinski WP (2004) A survey of acid catalysis and oxidation conditions in the two-step one-flask synthesis of *meso*-substituted corroles via dipyrromethanedicarbinols and pyrrole. J Org Chem 69:4159–4169
142. Ding T, Alemán EA, Modarelli DA, Ziegler CJ (2005) Photophysical properties of a series of free-base corroles. J Phys Chem A 109:7411–7417
143. Tardieux C, Gros CP, Guilard R (1998) On corrole chemistry. An isomerization and oxidative cleavage of the corrole macroring to a biliverdin structure. J Heterocycl Chem 35:965–970
144. Paolesse R, Sagone F, Macagnano A, Boschi T, Prodi L, Montalti M, Zaccheroni N, Bolletta F, Smith KM (1999) Photophysical behaviour of corrole and its symmetrical and unsymmetrical dyads. J Porphyr Phthalocyanines 3:364–370

145. Yamauchi T, Mizutani T, Wada K, Horii S, Furukawa H, Masaoka S, Chang H-C, Kitagawa S (2005) A facile and versatile preparation of bilindiones and biladienones from tetraaryl-porphyrins. Chem Commun 1309–1311
146. Jérôme F, Gros CP, Tardieux C, Barbe J-M, Guilard R (1998) First synthesis of sterically hindered cofacial bis(corroles) and their bis(cobalt) complexes. Chem Commun 2007–2008
147. Wojaczyński J, Duszak M, Latos-Grażyński L (unpublished results)
148. Barata JFB, Neves MGPMS, Tomé AC, Faustino MAF, Silva AMS, Cavaleiro JAS (2010) How light affects 5,10,15-tris(pentafluorophenyl)corrole. Tetrahedron Lett 51:1537–1540
149. Warburg O, Negelein E (1930) Grünes Haemin aus Blut-Haemin. Chem Ber 63:1816–1818
150. Lemberg R (1956) Chemical mechanism of bile pigment formation. Rev Pure Appl Chem 6:1–23
151. Lemberg R, Cortis-Jones B, Norrie M (1937) Coupled oxidation of ascorbic acid and haemochromogens. Nature 139:1016–1017
152. Libowitzky H, Fischer H (1938) Bile pigments. XVIII. From coprohemin I to coproglaucobilin. Z Physiol Chem 255:209–233
153. Balch AL, Latos-Grażyński L, Noll BC, Olmstead MM, Szterenberg L, Safari N (1993) Structural characterization of verdoheme analogs. Iron complexes of octaethyloxoporphyrin. J Am Chem Soc 115:1422–1429
154. Balch AL, Latos-Grażyński L, Noll BC, Olmstead MM, Safari N (1993) Isolation and characterization of an iron biliverdin-type complex that is formed along with verdohemochrome during the coupled oxidation of iron(II) octaethylporphyrin. J Am Chem Soc 115:9056–9061
155. St Claire TN, Balch AL (1999) In situ monitoring of the degradation of iron porphyrins by dioxygen with hydrazine as sacrificial reductant. Detection of paramagnetic intermediates in the coupled oxidation process by ^{1}H NMR spectroscopy. Inorg Chem 38:684–691
156. Balch AL, Koerner R, Latos-Grażyński L, Lewis JE, St Claire TN, Zovinka EP (1997) Coupled oxidation of heme without pyridine. Formation of cyano complexes of iron oxophlorin and 5-oxaporphyrin (verdoheme) from octaethylheme. Inorg Chem 36:3892–3897
157. Balch AL, Mazzanti M, St Claire TN, Olmstead MM (1995) Production of oxaporphyrin and biliverdin derivatives by coupled oxidation of cobalt(II) octaethylporphyrin. Inorg Chem 34:2194–2200
158. Wojaczyński J, Stępień M, Latos-Grażyński L (2002) Monomeric and dimeric iron(III) complexes of 5-hydroxy-10,15,20-triphenylporphyrin: formation of cyano and pyridine complexes of (5-oxo-10,15,20-triphenylphlorin)iron. Eur J Inorg Chem 1806–1815
159. Asano N, Uemura S, Kinugawa T, Akasaka H, Mizutani T (2007) Synthesis of biladienone and bilatrienone by coupled oxidation of tetraarylporphyrins. J Org Chem 72:5320–5326
160. Nakamura R, Kakeya K, Furuta N, Muta E, Nishisaka H, Mizutani T (2011) Synthesis of *para*- or *ortho*-substituted triarylbilindiones and tetraarylbiladienones by coupled oxidation of tetraarylporphyrins. J Org Chem 76:6108–6115
161. Balch AL (2000) Coordination chemistry with *meso*-hydroxylated porphyrins (oxophlorins), intermediates in heme degradation. Coord Chem Rev 200–202:349–377
162. Balch AL, Latos-Grażyński L, Noll BC, Olmstead MM, Zovinka EP (1992) Chemistry of iron oxophlorins. 1. ^{1}H NMR and structural studies of five-coordinate iron(III) complexes. Inorg Chem 31:2248–2255
163. Balch AL, Noll BC, Zovinka EP (1992) Structural characterization of zinc(II) complexes of octaethyloxophlorin dianion and octaethyloxophlorin radical anion. J Am Chem Soc 114:3380–3385
164. Balch AL, Noll BC, Phillips SL, Reid SM, Zovinka EP (1993) Nickel(II) complexes of the octaethyloxophlorin dianion and octaethyloxophlorin radical dianion. Inorg Chem 32:4730–4736

165. Balch AL, Noll BC, Reid SM, Zovinka EP (1993) Coordination patterns for oxophlorin ligands. Pyridine-induced cleavage of dimeric manganese(III) and iron(III) octaethyloxophlorin complexes. Inorg Chem 32:2610–2611
166. Balch AL, Mazzanti M, Olmstead MM (1993) Cobalt complexes of octaethyloxophlorin. Metal-centered redox chemistry in the presence of a redox-active ligand. Inorg Chem 32:4737–4744
167. Masuoka N, Itano HA (1987) Radical intermediates in the oxidation of octaethylheme to octaethylverdoheme. Biochemistry 26:3672–3680
168. Balch AL, Latos-Grażyński L, Noll BC, Szterenberg L, Zovinka EP (1993) Chemistry of iron oxophlorins. 2. Oxidation of the iron(III) octaethyloxophlorin dimer and observation of stepwise two-electron oxidation of the oxophlorin macrocycle. J Am Chem Soc 115:11846–11854
169. Balch AL, Noll BC, Olmstead MM, Reid SM (1993) A cofacial dimeric, metallooxophlorin complex: [indium(III)(octaethyloxophlorin)]$_2$. J Chem Soc Chem Commun 1088–1090
170. Balch AL, Latos-Grażyński L, St Claire TN (1995) Chemistry of iron oxophlorins. 3. Reversible, one-electron oxidation of the iron(III) octaethyloxophlorin dimer. Inorg Chem 34:1395–1401
171. Garcia TY, Olmstead MM, Fettinger JC, Balch AL (2008) Cleavage of the indium(III) octaethyloxophlorin dimer, $\{In^{III}(OEPO)\}_2$, with Lewis bases. Importance of outer-sphere hydrogen bonding in adduct structures. Inorg Chem 47:11417–11422
172. Sano S, Sugiura Y, Meada Y, Ogawa S, Morishima I (1981) Electronic states of iron oxyporphyrin and verdohemochrome obtained by coupled oxidation of iron porphyrin. J Am Chem Soc 103:2888–2890
173. Morishima I, Fujii H, Shiro Y, Sano S (1986) NMR studies of metalloporphyrin radicals. Iron (II) oxophlorin radical formed from iron(III) *meso*-hydroxyoctaethylporphyrin. J Am Chem Soc 108:3858–3860
174. Morishima I, Fujii H, Shiro Y, Sano S (1995) Studies on the iron(II) *meso*-oxyporphyrin π-neutral radical as a reaction intermediate in heme catabolism. Inorg Chem 34:1528–1535
175. Balch AL, Koerner R, Noll BC (1996) A role for electron transfer in heme catabolism? Structure and redox behavior of an intermediate, (pyridine)$_2$Fe(octaethyloxophlorin). J Am Chem Soc 118:2760–2761
176. Szterenberg L, Latos-Grażyński L, Wojaczyński J (2002) Oxophlorin and metallooxophlorin radicals – DFT studies. Chemphyschem 3:575–583
177. Rath SP, Olmstead MM, Balch AL (2004) The effects of axial ligands on electron distribution and spin states in iron complexes of octaethyloxophlorin, intermediates in heme degradation. J Am Chem Soc 126:6379–6386
178. Rath SP, Olmstead MM, Balch AL (2006) Electron distribution in iron octaethyloxophlorin complexes. Importance of the Fe(III) oxophlorin trianion form in the bis-pyridine and bis-imidazole complexes. Inorg Chem 45:6083–6093
179. Gheidi M, Safari N, Zahedi M (2012) Effect of axial ligand on the electronic configuration, spin states, and reactivity of iron oxophlorin. Inorg Chem 51:7094–7102
180. Rath SP, Koerner R, Olmstead MM, Balch AL (2003) Reversible binding of nitric oxide and carbon–carbon bond formation in a *meso*-hydroxylated heme. J Am Chem Soc 125:11798–11799
181. Rath SP, Olmstead MM, Balch AL (2004) Reactions of *meso*-hydroxyhemes with carbon monoxide and reducing agents in search of the elusive species responsible for the $g = 2.006$ resonance of carbon monoxide-treated heme oxygenase. Isolation of diamagnetic iron(II) complexes of octaethyl-*meso*-hydroxyporphyrin. Inorg Chem 43:6357–6365
182. Rath SP, Olmstead MM, Latos-Grażyński L, Balch AL (2003) Formation and isolation of an iron-tripyrrole complex from heme degradation. J Am Chem Soc 125:12678–12679
183. Ortiz de Montellano PR, Auclair K (2003)) Heme oxygenase structure and mechanism. In: Kadish KM, Smith KM, Guilard R (eds) The porphyrin handbook, vol 12. Academic, San Diego, pp 183–210 (Chapter 75)

184. Saito S, Itano HA (1986) Cyclization of biliverdins to verdohaemochromes. J Chem Soc Perkin Trans 1 1–7
185. Saito S, Sumita S, Iwai K, Sano H (1988) Preparation of mesoverdohemochrome IXα dimethyl ester and Mössbauer spectra of related porphyrins. Bull Chem Soc Jpn 61:3539–3547
186. Balch AL, Noll BC, Safari N (1993) Structural characterization of low-spin iron(III) complexes of octaethyloxoporphyrin. Inorg Chem 32:2901–2905
187. Balch AL, Mazzanti M, Olmstead MM (1994) Preparation of a cobalt analogue of verdoheme by coupled oxidation of cobalt(II) octaethylporphyrin. J Chem Soc Chem Commun 269–270
188. Balch AL, Koerner R, Olmstead MM (1995) Crystallographic characterization of octaethylverdohaem. J Chem Soc Chem Commun 873–874
189. Lord PA, Latos-Grażyński L, Balch AL (2002) Reactivity of iron verdohemes with phenylmagnesium bromide. Formation of paramagnetic iron–phenyl complexes. Inorg Chem 41:1011–1014
190. Rath SP, Olmstead MM, Balch AL (2004) Oxidative verdoheme formation and stabilization by axial isocyanide ligation. Inorg Chem 43:7648–7655
191. Khorasani-Motlagh M, Safari N, Noroozifar M, Saffari J, Biabani M, Rebouças JS, Patrick BO (2005) New class of verdoheme analogues with weakly coordinating anions: the structure of (μ-oxo)bis[(octaethyloxoporphinato)iron(III)] hexafluorophosphate. Inorg Chem 44:7762–7769
192. Khorasani-Motlagh M, Safari N, Noroozifar M, Shahroosvand H, Parsaii Z, Patrick BO (2007) Formation and stabilization of five-coordinate iron(II) verdoheme analogues by axial weakly coordinating anion ligation. X-ray crystal structures of $[(OEOPFe)_2O](X)_2$ ($X = AsF_6$, SbF_6). Inorg Chim Acta 360:2331–2338
193. Balch AL, Bowles FL (2010)) Coordination chemistry of verdohemes and open-chain oligopyrrole systems involved in heme oxidation and porphyrin destruction. In: Kadish KM, Smith KM, Guilard R (eds) Handbook of porphyrin science with applications to chemistry, physics, materials science, engineering biology and medicine, vol 8. World Scientific, Singapore, pp 293–342 (Chapter 40)
194. Saito S, Itano HA (1982) Verdohemochrome IXα: preparation and oxidoreductive cleavage to biliverdin IXα. Proc Natl Acad Sci USA 79:1393–1397
195. Nguyen KT, Rath SP, Latos-Grażyński L, Olmstead MM, Balch AL (2004) Formation of a highly oxidized iron biliverdin complex upon treatment of a five-coordinate verdoheme with dioxygen. J Am Chem Soc 126:6210–6211
196. Fuhrhop J-H, Krüger P (1977) 1-Oder 19-methoxy-1-oder 19-amino-und 1-oder 19-thio-desoxybiliverdine. Liebigs Ann 360–370
197. Latos-Grażyński L, Johnson J, Attar S, Olmstead MM, Balch AL (1998) Reactivity of the verdoheme analogues, 5-oxaporphyrin complexes of cobalt(II) and zinc(II) with nucleophiles: opening of the planar macrocycle by alkoxide addition to form helical complexes. Inorg Chem 37:4493–4499
198. Koerner R, Latos-Grażyński L, Balch AL (1998) Models for verdoheme hydrolysis. Paramagnetic products from the ring opening of verdohemes, 5-oxaporphyrin complexes of iron(II) with methoxide ion. J Am Chem Soc 120:9246–9255
199. Johnson JA, Olmstead MM, Stolzenberg AM, Balch AL (2001) Ring-opening and *meso* substitution from the reaction of cyanide ion with zinc verdohemes. Inorg Chem 40:5585–5595
200. Latos-Grażyński L, Wojaczyński J, Koerner R, Johnson JJ, Balch AL (2001) Verdoheme reactivity. Remarkable paramagnetically shifted ^1H NMR spectra of intermediates from the addition of hydroxide or methoxide with Fe^{II} and Fe^{III} verdohemes. Inorg Chem 40:4971–4977
201. Johnson JA, Olmstead MM, Balch AL (1999) Reactivity of the verdoheme analogues. Opening of the planar macrocycle by amide and thiolate nucleophiles to form helical complexes. Inorg Chem 38:5379–5383

202. Saito S, Itano HA (1987) Autoxidation and solvolysis products of octaethylverdohaemochrome. J Chem Soc Perkin Trans 1 1183–1188
203. Davari MD, Bahrami H, Zahedi M, Safari N (2010) Effect of the axial ligands on the structure and reactivity of tin verdoheme in the ring opening process. Inorg Chim Acta 363:1577–1586
204. Jamaat PR, Safari N, Ghiasi M, Naghavi SS, Zahedi M (2008) Noninnocent effect of axial ligand on the heme degradation process: a theoretical approach to hydrolysis pathway of verdoheme to biliverdin. J Biol Inorg Chem 13:121–132
205. Davari MD, Bahrami H, Zahedi M, Safari N (2009) How tin metal prevents verdoheme ring opening? Comparative study of various nucleophiles. J Mol Struct (THEOCHEM) 908:1–11
206. Davari MD, Bahrami H, Zahedi M, Safari N (2009) Theoretical investigations on the hydrolysis pathway of tin verdoheme complexes: elucidation of tin's ring opening inhibition role. J Mol Model 15:1299–1315
207. Bonfiglio JV, Bonnett R, Hursthouse MB, Malik KMA (1977) Crystal structure of the nickel complex of 2,3,7,8,12,13,17,18-octaethyl-1,19(21*H*,24*H*)-bilindione (octaethylbilatriene-*abc*). J Chem Soc Chem Commun 83–84
208. Bonfiglio JV, Bonnett R, Buckley DG, Hamzetash D, Hursthouse MB, Malik KMA, McDonagh AF, Trotter J (1983) Linear tetrapyrroles as ligands: syntheses and X-ray analyses of boron and nickel complexes of octaethyl-21*H*,24*H*-bilin-1,19-dione. Tetrahedron 39:1865–1874
209. Balch AL, Mazzanti M, Noll BC, Olmstead MM (1993) Geometric and electronic structure and dioxygen sensitivity of the copper complex of octaethylbilindione, a biliverdin analog. J Am Chem Soc 15:12206–12207
210. Balch AL, Mazzanti M, Noll BC, Olmstead MM (1994) Coordination patterns for biliverdin-type ligands. Helical and linked helical units in four-coordinate cobalt and five-coordinate manganese(III) complexes of octaethylbilindione. J Am Chem Soc 116:9114–9122
211. Attar S, Balch AL, Van Calcar PM, Winkler K (1997) Electron transfer behavior and solid state structures of the helical cobalt complexes of the open-chain tetrapyrrole ligand, octaethylbilindione. J Am Chem Soc 119:3317–3323
212. Attar S, Ozarowski A, Van Calcar PM, Winkler K, Balch AL (1997) Axial ligation modulates the electron distribution in helical cobalt complexes derived from octaethylbilindione. Chem Commun 1115–1116
213. Koerner R, Olmstead MM, Van Calcar PM, Winkler K, Balch AL (1998) Carbon monoxide production during the oxygenation of cobalt complexes of linear tetrapyrroles. Formation and characterization of Co^{II}(tetraethylpropentdyopent anion)$_2$. Inorg Chem 37:982–988
214. Lord PA, Olmstead MM, Balch AL (2000) Redox characteristics of nickel and palladium complexes of the open-chain tetrapyrrole octaethylbilindione: a biliverdin model. Inorg Chem 39:1128–1134
215. Lord PA, Noll BC, Olmstead MM, Balch AL (2001) A remarkable skeletal rearrangement of a coordinated tetrapyrrole: chemical consequences of palladium π-coordination to a bilindione. J Am Chem Soc 123:10554–10559
216. Fuhrhop J-H, Krüger P, Sheldrick WS (1977) Darstellung Struktur und Eigenschaften der 5-aza-5-oxonia-und 5-thioniamesoporphin-dimethylester und -protoporphin-dimethylester. Liebigs Ann 339–359
217. Kakeya K, Nakagawa A, Mizutani T, Hitomi Y, Kodera M (2012) Synthesis, reactivity and spectroscopic properties of *meso*-triaryl-5-oxaporphyrins. J Org Chem 77:6510–6519
218. Zahedi M, Bahrami H, Shahbazian S, Safari N, Ng SW (2003) An ab initio/hybrid (ONIOM) investigation of biliverdin isomers and metal–biliverdin analogue complexes. J Mol Struct (THEOCHEM) 633:21–33
219. Wasbotten I, Ghosh A (2006) Biliverdine-based metalloradicals: sterically enhanced noninnocence. Inorg Chem 45:4914–4921
220. Szterenberg L, Latos-Grażyński L, Wojaczyński J (2003) Metallobiliverdin radicals – DFT studies. Chemphyschem 4:691–698

221. Zahedi M, Safari N, Haddadpur S (2000) Semiempirical molecular orbital calculations of biliverdin: study of dynamics and energetics of the self-association of a two-electron oxidation product. J Mol Struct (THEOCHEM) 531:79–88
222. Zahedi M, Kamalipour M, Safari N (2002) Theoretical studies of biliverdin: energetics of the reduction pathways to bilirubin. J Mol Model 8:113–118
223. Bonnett R, McDonagh AF (1970) Oxidative cleavage of the haem system: the four isomeric biliverdins of the IX series. Chem Commun 237–238
224. Bonnett R, McDonagh AF (1973) The *meso*-reactivity of porphyrins and related compounds. Part VI. Oxidative cleavage of the haem system. The four isomeric biliverdins of the IX series. J Chem Soc Perkin Trans 1 881–888
225. Crusats J, Suzuki A, Mizutani T, Ogoshi H (1998) Regioselective porphyrin bridge cleavage controlled by electronic effects. Coupled oxidation of 3-demethyl-3-(trifluoromethyl) mesohemin IX and identification of its four biliverdin derivatives. J Org Chem 63:602–607
226. Niemevz F, Alvarez DE, Buldain GY (2002) Chemical oxidation of synthetic iron(III)-complex of 15-phenyl protoporphyrin IX. Heterocycles 57:697–704
227. Niemevz F, Vazquez MS, Buldain G (2008) Synthesis of a series of mono-*meso*-arylmeso-porphyrins III of biological interest and their biliverdin derivatives. Synthesis 875–882
228. Wang J, Niemevz F, Lad L, Huang L, Alvarez DE, Buldain G, Poulos TL, Ortiz de Montellano PR (2004) Human heme oxygenase oxidation of 5- and 15-phenylhemes. J Biol Chem 279:42593–42604
229. Hudson MF, Smith KM (1975) Bile pigments. Chem Soc Rev 4:363–399
230. Frankenberg N, Lagarias JC (2003)) Biosynthesis and biological functions of bilins. In: Kadish KM, Smith KM, Guilard R (eds) The porphyrin handbook, vol 13. Academic, San Diego, pp 211–235 (Chapter 83)
231. Lagarias JC (1982) Bile pigment-protein interactions. Coupled oxidation of cytochrome c. Biochemistry 21:5962–5967
232. van der Horst MA, Hellingwerf KJ (2004) Photoreceptor proteins "Star actors of modern times": a review of the functional dynamics in the structure of representative members of six different photoreceptor families. Acc Chem Res 37:13–20
233. Tenhunen R, Marver HS, Schmid R (1968) The enzymatic conversion of heme to bilirubin by microsomal heme oxygenase. Proc Natl Acad Sci USA 61:748–755
234. Frydman RB, Frydman B (1987) Heme catabolism: a new look at substrates and enzymes. Acc Chem Res 20:250–256
235. Yoshida T, Migita CT (2000) Mechanism of heme degradation by heme oxygenase. J Inorg Biochem 82:33–41
236. Kikuchi G, Yoshida T, Noguchi M (2005) Heme oxygenase and heme degradation. Biochem Biophys Res Commun 338:558–567
237. Rivera M, Zeng Y (2005) Heme oxygenase steering dioxygen activation toward heme hydroxylation. J Inorg Biochem 99:337–354
238. Morse D, Choi AMK (2005) Heme oxygenase-1. From bench to bedside. Am J Respir Crit Care Med 172:660–670
239. Matsui T, Iwasaki M, Sugiyama R, Unno M, Ikeda-Saito M (2010) Dioxygen activation for the self-degradation of heme: reaction mechanism and regulation of heme oxygenase. Inorg Chem 49:3602–3609
240. Schuller DJ, Zhu W, Stojiljkovic I, Wilks A, Poulos TL (2001) Crystal structure of heme oxygenase from the Gram-negative pathogen *Neisseria meningitidis* and a comparison with mammalian heme oxygenase-1. Biochemistry 40:11552–11558
241. Friedman J, Lad L, Li H, Wilks A, Poulos TL (2004) Structural basis for novel δ-regioselective heme oxygenation in the opportunistic pathogen *Pseudomonas aeruginosa*. Biochemistry 43:5239–5245
242. Uchida T, Sekine Y, Matsui T, Ikeda-Saito M, Ishimori K (2012) A heme degradation enzyme, HutZ, from *Vibrio cholerae*. Chem Commun 48:6741–6743

243. Lad L, Schuller DJ, Shimizu H, Friedman J, Li H, Ortiz de Montellano PR, Poulos TL (2003) Comparison of the heme-free and -bound crystal structures of human heme oxygenase-1. J Biol Chem 278:7834–7843
244. Schuller DJ, Wilks A, Ortiz de Montellano PR, Poulos TL (1999) Crystal structure of human heme oxygenase-1. Nat Struct Mol Biol 6:860–867
245. Unno M, Matsui T, Ikeda-Saito M (2012) Crystallographic studies of heme oxygenase complexed with an unstable reaction intermediate, verdoheme. J Inorg Biochem 113:102–109
246. Lad L, Ortiz de Montellano PR, Poulos TL (2004) Crystal structures of ferrous and ferrous–NO forms of verdoheme in a complex with human heme oxygenase-1: catalytic implications for heme cleavage. J Inorg Biochem 98:1686–1695
247. Sugishima M, Sakamoto H, Higashimoto Y, Noguchi M, Fukuyama K (2003) Crystal structure of rat heme oxygenase-1 in complex with biliverdin-iron chelate. Conformational change of the distal helix during the heme cleavage reaction. J Biol Chem 278:32352–32358
248. Lad L, Friedman J, Li H, Bhaskar B, Ortiz de Montellano PR, Poulos TL (2004) Crystal structure of human heme oxygenase-1 in a complex with biliverdin. Biochemistry 43:3793–3801
249. Sharma PK, Kevorkiants R, de Visser SP, Kumar D, Shaik S (2004) Porphyrin traps its terminator! Concerted and stepwise porphyrin degradation mechanisms induced by heme-oxygenase and cytochrome P450. Angew Chem Int Ed 43:1129–1132
250. Kumar D, de Visser SP, Shaik S (2005) Theory favors a stepwise mechanism of porphyrin degradation by a ferric hydroperoxide model of the active species of heme oxygenase. J Am Chem Soc 127:8204–8213
251. Chen H, Moreau Y, Derat E, Shaik S (2008) Quantum mechanical/molecular mechanical study of mechanisms of heme degradation by the enzyme heme oxygenase: the strategic function of the water cluster. J Am Chem Soc 130:1953–1965
252. Matsui T, Nakajima A, Fujii H, Mansfield Matera K, Migita CT, Yoshida T, Ikeda-Saito M (2005) O_2- and H_2O_2-dependent verdoheme degradation by heme oxygenase. Reaction mechanisms and potential physiological roles of the dual pathway degradation. J Biol Chem 280:36833–36840
253. Lai W, Chen H, Matsui T, Omori K, Unno M, Ikeda-Saito M, Shaik S (2010) Enzymatic ring-opening mechanism of verdoheme by the heme oxygenase: a combined X-ray crystallography and QM/MM study. J Am Chem Soc 132:12960–12970
254. Damaso CO, Bunce RA, Barybin MV, Wilks A, Rivera M (2005) The ferrous verdoheme–heme oxygenase complex is six-coordinate and low-spin. J Am Chem Soc 127:17582–17583
255. Gohya T, Sato M, Zhang X, Migita CT (2008) Variation of the oxidation state of verdoheme in the heme oxygenase reaction. Biochem Biophys Res Commun 376:293–298
256. Zhou H, Migita CT, Sato M, Sun D, Zhang X, Ikeda-Saito M, Fujii H, Yoshida T (2000) Participation of carboxylate amino acid side chain in regiospecific oxidation of heme by heme oxygenase. J Am Chem Soc 122:8311–8312
257. Zeng Y, Deshmukh R, Caignan GA, Bunce RA, Rivera M, Wilks A (2004) Mixed regioselectivity in the Arg-177 mutants of *Corynebacterium diphtheriae* heme oxygenase as a consequence of in-plane heme disorder. Biochemistry 43:5222–5238
258. Lad L, Wang J, Li H, Friedman J, Bhaskar B, Ortiz de Montellano PR, Poulos TL (2003) Crystal structures of the ferric, ferrous and ferrous-NO forms of the Asp140Ala mutant of human heme oxygenase-1: catalytic implications. J Mol Biol 330:527–538
259. Ratliff M, Zhu W, Desmukh R, Wilks A, Stojiljkovic I (2001) Homologues of neisserial heme oxygenase in gram-negative bacteria: degradation of heme by the product of the *pigA* gene of *Pseudomonas aeruginosa*. J Bacteriol 183:6394–6403

260. Caignan GA, Deshmukh R, Wilks A, Zeng Y, Huang H, Moënne-Loccoz P, Bunce RA, Eastman MA, Rivera M (2002) Oxidation of heme to β- and δ-biliverdin by *Pseudomonas aeruginosa* heme oxygenase as a consequence of an unusual seating of the heme. J Am Chem Soc 124:14879–14892
261. Deshmukh R, Zeng Y, Furci LM, Huang H, Morgan BN, Sander S, Alontaga AY, Bunce RA, Moënne-Loccoz P, Rivera M, Wilks A (2005) Heme oxidation in a chimeric protein of the α-selective *Neisseriae meningitidis* heme oxygenase with the distal helix of the δ-selective *Pseudomonas aeruginosa*. Biochemistry 44:13713–13723
262. Peng D, Ma L-H, Smith KM, Zhang X, Sato M, La Mar GN (2012) Role of propionates in substrate binding to heme oxygenase from *Neisseria meningitidis*: a nuclear magnetic resonance study. Biochemistry 51:7054–7063
263. Zhang X, Fujii H, Mansfield Matera K, Migita CT, Sun D, Sato M, Ikeda-Saito M, Yoshida T (2003) Stereoselectivity of each of the three steps of the heme oxygenase reaction: hemin to meso-hydroxyhemin, meso-hydroxyhemin to verdoheme, and verdoheme to biliverdin. Biochemistry 42:7418–7426
264. Torpey J, Ortiz de Montellano PR (1996) Oxidation of the *meso*-methylmesoheme regioisomers by heme oxygenase. Electronic control of the reaction regiospecificity. J Biol Chem 271:26067–26073
265. Evans JP, Niemevz F, Buldain G, Ortiz de Montellano PR (2008) Isoporphyrin intermediate in heme oxygenase catalysis. Oxidation of α-meso-phenylheme. J Biol Chem 283:19530–19539
266. Torpey J, Ortiz de Montellano PR (1997) Oxidation of α-*meso*-formylmesoheme by heme oxygenase. Electronic control of the reaction regiospecificity. J Biol Chem 272:22008–22014
267. Lemberg R, Legge JW, Lockwood WH (1939) Coupled oxidation of ascorbic acid and haemoglobin. I. Biochem J 33:754–758
268. Avila L, Huang H, Damaso CO, Lu S, Moënne-Loccoz P, Rivera M (2003) Coupled oxidation vs heme oxygenation: insights from axial ligand mutants of mitochondrial cytochrome b5. J Am Chem Soc 125:4103–4110
269. O'Carra P (1975)) Heme cleavage: biological systems and chemical analogs. In: Smith KM (ed) Porphyrins and metalloporphyrins. Elsevier, Amsterdam, pp 123–153 (Chapter 4)
270. Brown SB, Docherty JC (1978) Haem degradation in abnormal haemoglobins. Biochem J 173:985–987
271. Murakami T, Morishima I, Matsui T, Ozaki S, Hara I, Yang H-J, Watanabe Y (1999) Effects of the arrangement of a distal catalytic residue on regioselectivity and reactivity in the coupled oxidation of sperm whale myoglobin mutants. J Am Chem Soc 121:2007–2011
272. Vernon DI, Brown SB (1984) Formation of bile pigments by coupled oxidation of cobalt-substituted haemoglobin and myoglobin. Biochem J 223:205–209
273. Rice JK, Fearnley IM, Barker PD (1999) Coupled oxidation of heme covalently attached to cytochrome b_{562} yields a novel biliprotein. Biochemistry 38:16847–16856
274. Avila L, Huang H, Rodríguez JC, Moënne-Loccoz P, Rivera M (2000) Oxygen activation by axial ligand mutants of mitochondrial cytochrome b5: oxidation of heme to verdoheme and biliverdin. J Am Chem Soc 122:7618–7619
275. Kräutler B (2003)) Chlorophyll breakdown and chlorophyll catabolites. In: Kadish KM, Smith KM, Guilard R (eds) The porphyrin handbook, vol 13. Academic, San Diego, pp 183–209 (Chapter 82)
276. Masuda T, Fujita Y (2008) Regulation and evolution of chlorophyll metabolism. Photochem Photobiol Sci 7:1131–1149
277. Moser S, Müller T, Oberhuber M, Kräutler B (2009) Chlorophyll catabolites – chemical and structural footprints of a fascinating biological phenomenon. Eur J Org Chem 21–31
278. Barry CS (2009) The stay-green revolution: recent progress in deciphering the mechanisms of chlorophyll degradation in higher plants. Plant Sci 176:325–333

279. Hörtensteiner S, Wüthrich KL, Matile P, Ongania K-H, Kräutler B (1998) The key step in chlorophyll breakdown in higher plants. Cleavage of pheophorbide a macrocycle by a monooxygenase. J Biol Chem 273:15335–15339
280. Pružinská A, Tanner G, Anders I, Roca M, Hörtensteiner S (2003) Chlorophyll breakdown: pheophorbide *a* oxygenase is a Rieske-type iron–sulfur protein encoded by the accelerated cell death 1 gene. Proc Natl Acad Sci USA 100:15259–15264
281. Troxler RF, Smith KM, Brown SB (1980) Mechanism of photo-oxidation of bacteriochlorophyll-*c* derivatives. Tetrahedron Lett 21:491–494
282. Iturraspe J, Gossauer A (1991) Dependence of the regioselectivity of photo-oxidative ring opening of the chlorophyll macrocycle on the complexed metal ion. Helv Chim Acta 74:1713–1717
283. Okamoto Y, Tamiaki H (2011) C3- and C13-substituent effects on electronic absorption spectra of linear tetrapyrroles produced by photooxidation of zinc chlorophyll derivatives. J Photochem Photobiol A Chem 219:250–254
284. Saito S, Furukawa K, Osuka A (2010) Fully π-conjugated helices from oxidative cleavage of *meso*-aryl-substituted expanded porphyrins. J Am Chem Soc 132:2128–2129
285. Berlicka A, Latos-Grażyński L, Szterenberg L, Pawlicki M (2010) Photooxidation of dithiaethyneporphyrin. Eur J Org Chem 5688–5695
286. Pawlicki M, Bykowski D, Szterenberg L, Latos-Grazyński L (2012) From 21,23-dioxaporphyrin to a 3-pyranone dioxacorrole skeleton: the Achmatowicz rearrangement in the porphyrin frame. Angew Chem Int Ed 51:2500–2504

Top Heterocycl Chem (2014) 33: 203–230
DOI: 10.1007/7081_2013_109

Published online: 8 May 2013

Synthetic Routes to Unsymmetrical Porphyrins

Sara Nardis

Abstract The richness of properties showed by porphyrins and related compounds has attracted the attention of researchers for many years, because the attempt to mimic the essential role that these macrocycles have in nature make them of potential interest as organic materials in a number of application fields. Many of these exploitations require the presence of diverse functionalities on the peripheral positions of the tetrapyrrolic macrocycles, to finely tune the porphyrin properties or to covalently attach them to a particular substrate. For this reason in the last decade a particular attention has been focused on the developments of rational or statistical synthetic protocols, aimed for the preparation of unsymmetrically substituted porphyrin derivatives.

Keywords Functionalization reactions · Synthetic protocols · Unsymmetrical porphyrins

Contents

S. Nardis (✉)
Department of Chemical Science and Technologies, University of Rome Tor Vergata, 00133 Rome, Italy
e-mail: nardis@scienze.uniroma2.it

1 Introduction

The wide range of chemical and spectroscopic properties of porphyrins and their related compounds has given a strong boost to their study and application in many fields, such as medicine (PDT, BNCT, radioimmunotherapy and imaging) [1, 2], photophysical applications for antenna devices or reaction center models [3, 4], catalysis [5, 6], or, even more, in the realization of chemical sensors [7]. As a result, complex nanostructures composed of porphyrin subunits, such as large multiporphyrin arrays [8–12], polymers [13–15], dendrimers [16, 17], mechanically linked systems [18, 19], and molecular machines [20, 21], have been reported. The results obtained in the preparation of such a huge number of functionalized porphyrin systems are mainly due to the development of reliable, efficient, and specific methods for the synthesis of these macrocycles. Based on the historical contribution of many chemists [22], theoretically almost any desired β-substituted porphyrin can be prepared, while some initial limitation has been found in the preparation of *meso*-substituted porphyrins. However *meso*-tetraarylporphyrins have been the most exploited macrocycles for practical applications, because of the facile preparation routes starting from commercially available precursors and to their higher robustness in different chemical environments than their corresponding β-alkyl-substituted counterparts. Due to their symmetry and to the ease in the preparation steps, 5,10,15,20-tetrasubstituted porphyrins (**1**) [22] or the *trans*-A_2disubstituted porphyrins (**3** with B=D=H) [23] (Scheme 1) have been the most applied macrocycles and for this reason the functionalization of these species has been mostly oriented to their β-positions, with a significant body of methods has been developed for such a modification.[24] However current applications indicate a considerable demand for unsymmetrically substituted porphyrins: examples for this are amphiphilic porphyrins in photodynamic cancer therapy (PDT) [25, 26], synthons for crystal engineering and supramolecular chemistry [27, 28], push–pull systems for biomimetic studies on electron transfer [29], and uses in optics, especially in nonlinear optics (NLO) [30]. Unsymmetrical substitution is used to either position donor/acceptor residues in a manner suitable for the desired interaction, or to yield systems with a strong intermolecular dipole moment to enhance optical effects, or to create an anchoring site for the interaction with the substrates in order to realize stable monolayers [31].

The most important contribution in the area of unsymmetrical porphyrins has been made by the groups of Lindsey and Senge, by using completely different synthetic approaches; nevertheless several other groups have contributed to the development of these preparative approaches by increasing the yields of the reactions, introducing different substituents in order to expand the porphyrin aromatic system, or by increasing selectivity of the reactions.

In principle the access to these less symmetrical porphyrins can be achieved in two different ways: by the complete synthesis of the macrocycle, which could be statistical (Retrosynthesis a, in Scheme 1) or rational (Retrosynthesis b, in Scheme 1), or by subsequent functionalization (Retrosynthesis c, in Scheme 1) of a simpler, pre-formed porphyrin framework using C–C coupling reactions.

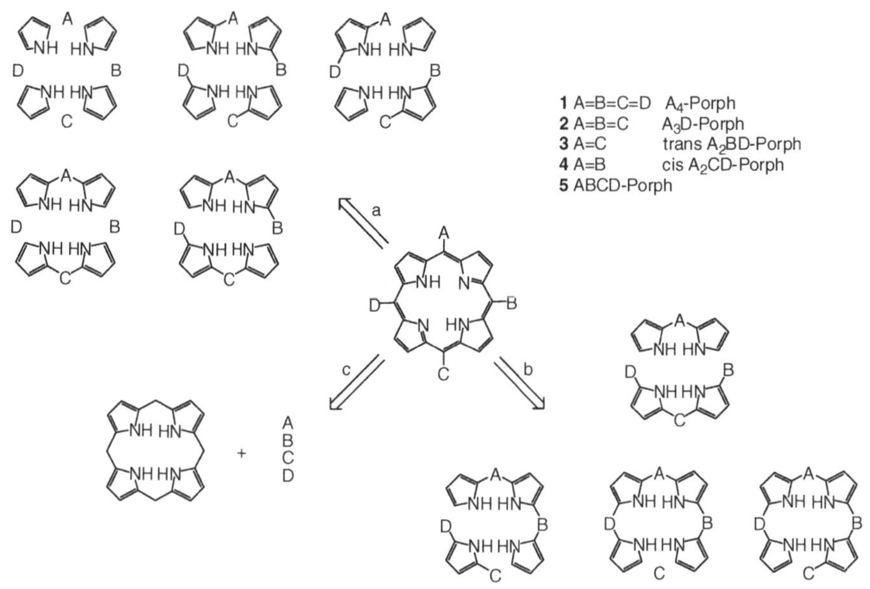

Scheme 1 Retro-synthetic pathways to ABCD-porphyrins

2 Synthetic Routes to Unsymmetrical Porphyrins

2.1 Statistical Approach

The complete synthesis of the macrocycle, depending on the desired pattern of substitution, can be theoretically obtained by mixed condensation of pyrrole, or dipyrromethane, with different aldehydes (Scheme 1, route a). Condensation of different aldehydes with pyrrole, when the reaction is performed under kinetic control, leads sometimes to the formation of the desired product, but in general results in a mixture of various isomers [32]. The use of 2-substituted pyrrole, bearing in the α-position a reactive species having the functionalities of interest [22, 33], can be an alternative for the achievement of the porphyrin macrocycle. Although this pathways is in some way reminiscent of the biological route starting from the porphobilinogen precursor, unfortunately also in this case, either the number of possible regioisomers is too high, requiring a cumbersome separation step, or the yields of the products are too low.

Dipyrromethanes and 1-acyldipyrromethanes are important building blocks as well, since simple procedures for their preparation are now available. Dipyrromethanes can be prepared [34–38], as shown in Scheme 2, by directed syntheses, employing an α-alkylthio group to direct the reaction, by one flask synthesis (on the left), or by a stepwise approach, starting from the preparation of an α-acylpyrrole: the procedures are quite simple, the only challenge lies in stopping

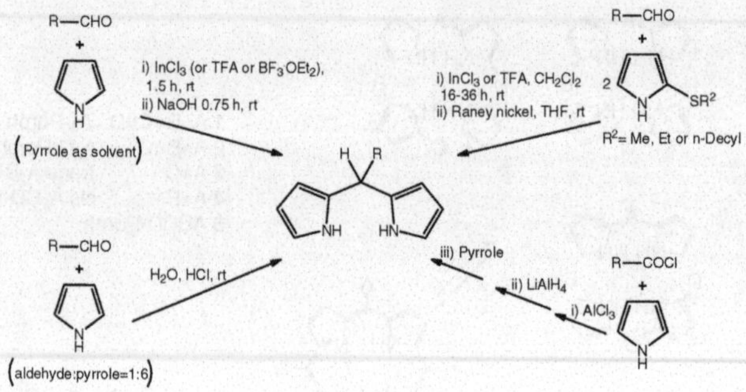

Scheme 2 Syntheses of dipyrromethanes

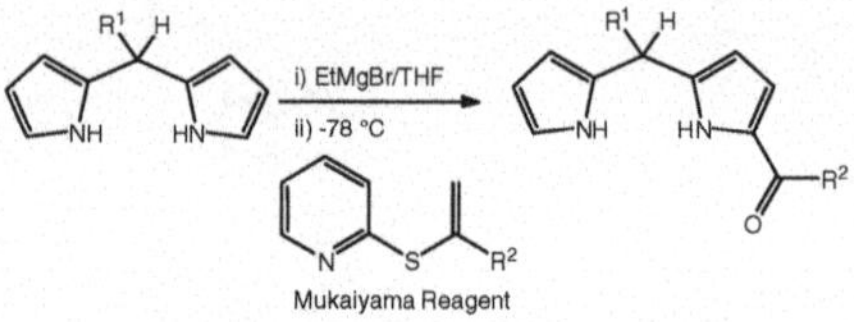

Scheme 3 1-Acylation of dipyrromethanes

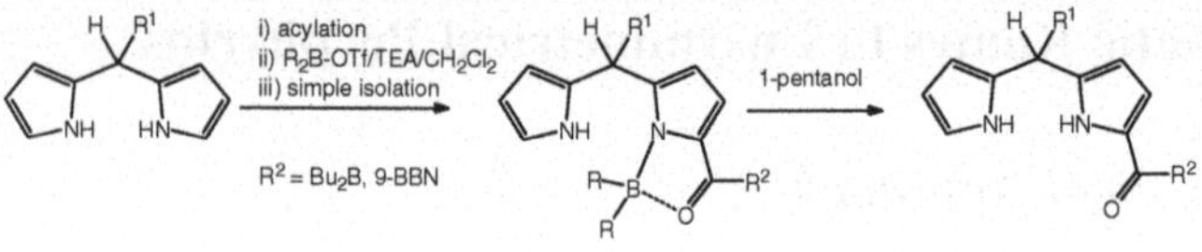

Scheme 4 Syntheses of *B*,*B*-dialkyl-*B*-(1-acyldipyrromethane)boron(III) complex

oligomerization at the dipyrromethane stage. 1-Acyldipyrromethanes, prepared by acylation of dipyrromethane carrying out the reaction of dipyrromethane-Grignard reagent at −78°C with a S-2-pyridyl thiolate (Mukaiyama reagent) or acid chloride (Scheme 3) [39, 40], are instead difficult to purify.

A solution to this problem has been achieved by reacting the 1-acyldipyrromethane with a dialkylboron triflate (e.g., Bu_2B-OTf or 9-BBN-OTf) to give the corresponding *B*,*B*-dialkyl-*B*-(1-acyldipyrromethane)boron(III) complex (Scheme 4). The 1-acyldipyrromethane–boron complexes, which are hydrophobic, stable to routine handling, and soluble in common organic solvents, present the advantage to crystallize readily and chromatograph without streaking: after purification, the 1-acyldipyrromethane can be liberated in high yield from the boron complex upon treatment with 1-pentanol [41].

As an example, the synthesis of various *meso*-substituted porphyrins can be achieved by condensation of two nonidentical 1-acyldipyrromethanes (Scheme 5):

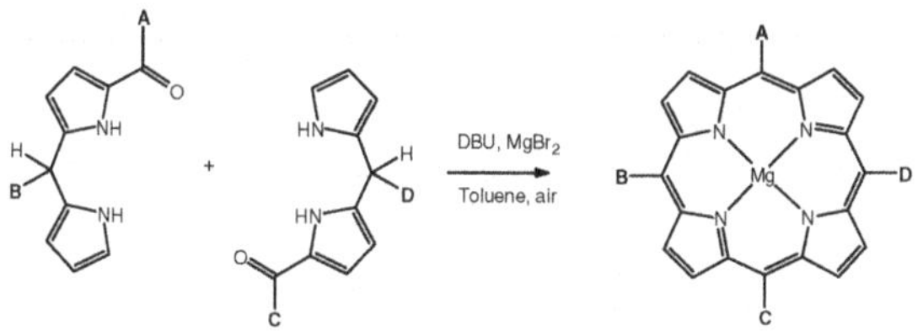

Scheme 5 Condensation of two different dipyrromethanes

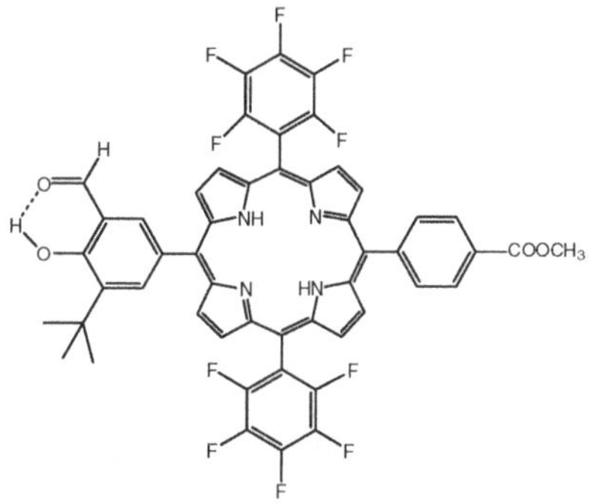

Fig. 1 Intramolecular hydrogen bond to drive the selectivity of the reaction

this statistic route has the advantage to afford the *cis*-A_2CD-Porph **4** (where A=B=R or H) (Scheme 1) without the *trans* isomer. Unfortunately the self-condensation of the involved dipyrromethanes results in the formation of other two porphyrins (A_2B_2-Porph and C_2D_2-Porph) leading to yields that are usually lower than 30% [42].

Moreover, the acidic conditions normally involved in the precursor condensation to obtain the porphyrin macrocycle, even if mild, are often not compatible with some functional groups such as phenol, formyl, terminal ethynyl, or carboxylic acid.

In order to introduce this kind of functionalities, protecting groups or further synthetic modifications are required, making the synthesis of multifunctionalized porphyrins even more challenging. An effort to overcome this problem has been done by Megiatto et al. [43], developing a procedure based on the formation of intramolecular hydrogen bonds to impart selectivity in the preparation of a *trans* A_2BD-Porph, but the yields are still low (11%) (Fig. 1).

Furthermore, the acidic conditions, besides the effects on labile substituents, often results in a significant scrambling (fragmentation and undesired recombination of fragments) (Scheme 6) of the intermediates and consequently in an increased number of possible products [22].

For these reasons, based on the methodologies developed for alkyl-substituted porphyrins, several methods have been proposed to provide a rational access to porphyrins bearing different types of *meso* substituents in order to avoid, or reduce, scrambling or polymerization phenomena.

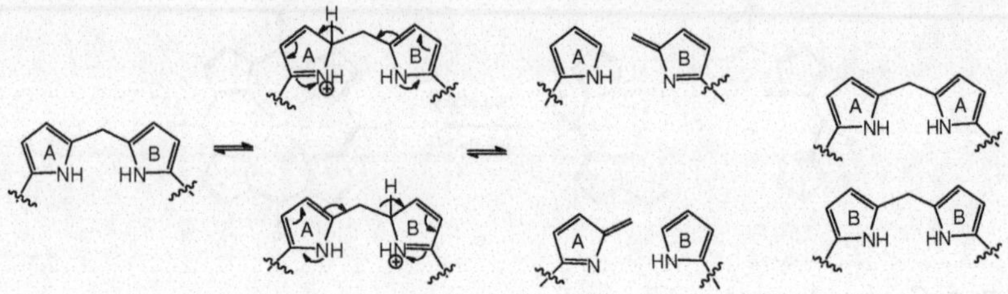

Scheme 6 Scrambling of the pyrrolic units

2.2 *[2+2] Routes*

This method employs the reaction between a dipyrromethane-1,9-dicarbinol and a dipyrromethane in a two step, one-flask process: the acid-catalyzed condensation leads to the formation of the porphyrinogen, and the following oxidation by DDQ, at room temperature, is necessary to get the corresponding free base porphyrin (Scheme 7).

The success of this procedure, useful for the preparation of porphyrins **2**, **3**, **4**, and **5** (Scheme 1), relies on a series of advances made in the development of efficient syntheses for the preparation of dipyrromethanes [34–38], 1-acyldipyrromethanes [39–41], 1,9-diacyldipyrromethanes [44, 45], and dipyrromethane-dicarbinols [46]. Further progress, having as a result the increasing of the process yields, has been done identifying the best reaction conditions for the condensation (i.e., type and amount of the acid catalyst), thus avoiding the acidolysis and the successive scrambling, responsible for the formation of porphyrins mixtures [40]. A huge number of different substituents has been used for the preparation of the pyrromethane species, but a tactical consideration concerning the order of introduction of the diverse groups should be done: the substituent on the dipyrromethane only encounters conditions for mild acid catalysis (dipyrromethane and porphyrin formation), while each of the substituents on the peripheral positions of the acyldipyrromethane also encounters the conditions for acylation and keto reduction. Thus a very broad range of substituents can be incorporated at the D site of the dipyrromethane, whereas a more limited number of functional groups can be inserted in positions A, B, and C, on the acyldipyrromethane.

Complementary routes can be used to *trans*-substituted porphyrins **3** (Scheme 1), and in particular to prepare sparsely porphyrins, that is, those with fewer than four *meso* substituents such as A_2BD-Porph with A=H (Scheme 8).

"Route 1" entails formylation of a dipyrromethane, imination with propylammine, and reaction with a second dipyrromethane, in refluxing ethanol containing zinc acetate [47]. The "Route 2," instead, requires treatment of a dipyrromethane with Eschenmoser's reagent (*N,N*-dimethylmethyleneiminium iodide), at room temperature, to give the 1,9-bis(*N,N*-dimethylaminomethyl)dipyrromethane, whose condensation with a second dipyrromethane and subsequent oxidation afford the desired porphyrin complex [48]. In both the cases the reaction is compatible with different

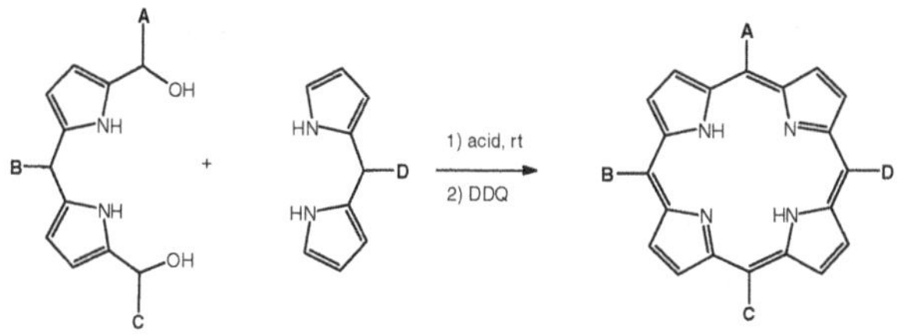

Scheme 7 [2+2]-Condensation to *meso*-tetrasubstitutedporphyrins

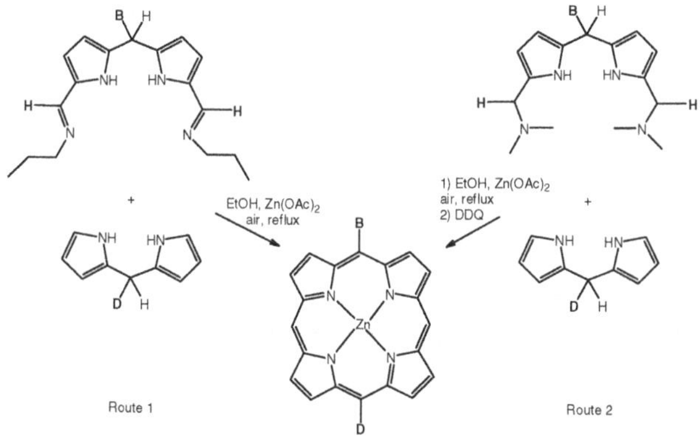

Scheme 8 Complementary routes to sparsely porphyrins

combination of aryl and alkyl substituents and, among the others, can afford porphyrin **2** (A_3D-Porph, with A=H) with a unique substituent on the *meso*-positions. Because of their compact size and ability to incorporate hydrophilic or amphipathic groups, such molecules are ideal for biological applications, as shown hereafter.

2.3 [3+1] Route

Introduced by Johnson [49] as a "3+1" variation on the MacDonald condensation, this methodology, after a period of disinterest due in part to the difficulties involved in the preparation of the required intermediate, has been successfully applied to the synthesis of alkylporphyrins [50] and carbaporphyrins [51]. The route is based on the acid-catalyzed cyclization of a tripyrrane, in general obtained by condensation of pyrrole and a 2,5-bis(hydroxymethyl)pyrrolc, with another unit of diol (Scheme 9).

Such a method has been successfully used for the preparation of the symmetric *cis*-A_2Porph (A=B and C=D=H) [52]: the insertion of both alkyl and aryl groups

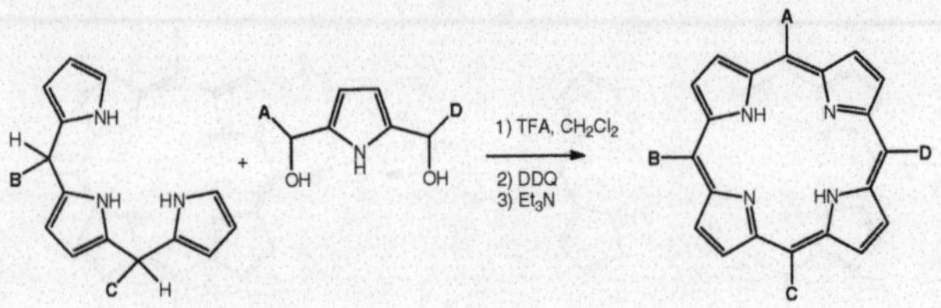

Scheme 9 [3+1] Protocol for the preparation of ABCD-Porphyrins

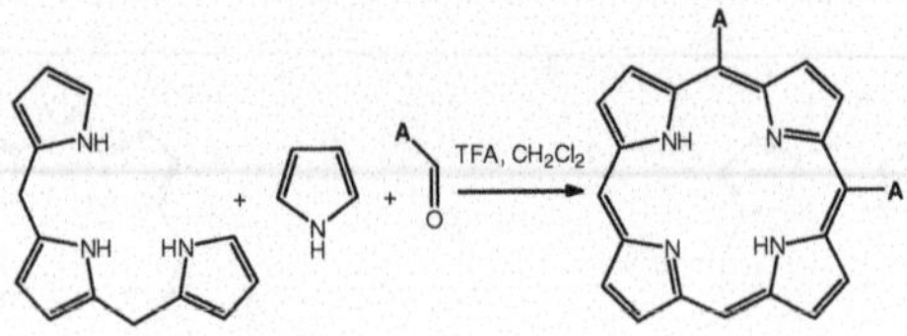

Scheme 10 Synthesis of a A_2-Porphyrin by the [3+1] route

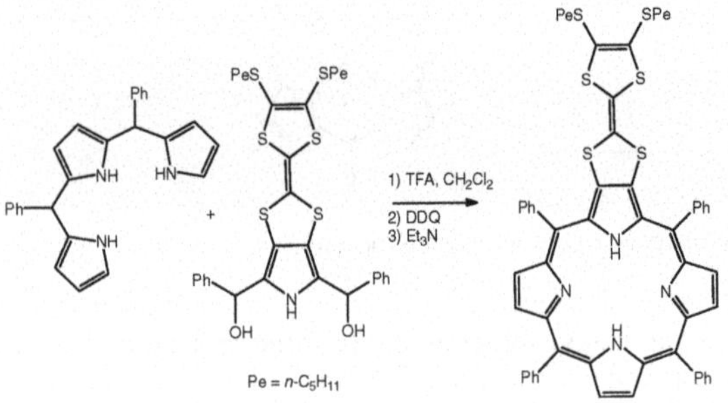

Scheme 11 Example of [3+1] approach to porphyrins

has been obtained condensing an unsubstituted tripyrrane (B=C=H) with pyrrole and the desired aliphatic or aromatic aldehyde in a 1:1:2 molar ratio (Scheme 10). The yields are generally lower than 11%.

The synthesis of porphyrins with substituents solely at the 5,10-positions, previously prepared by elaborate procedures [53], is particularly important since these macrocycles can be used as starting materials for the preparation of more complex and eventually unsymmetric systems like push–pull structures to be applied in nonlinear optic applications [54].

Another example of the [3+1] route to porphyrins is the reaction of the 5,10-diphenyltripyrrane [55] with a tetrathiafulvalene functionalized 2,5-bis (hydroxymethylphenyl)-pyrrole [56]: in this case the asymmetry is due to the introduction of a β-functionalized diol unit during the cyclization leading to the *meso*-5,10,15,20-tetrasubstitutedporphyrin (Scheme 11).

Scheme 12 Intramolecular cyclization of bilane

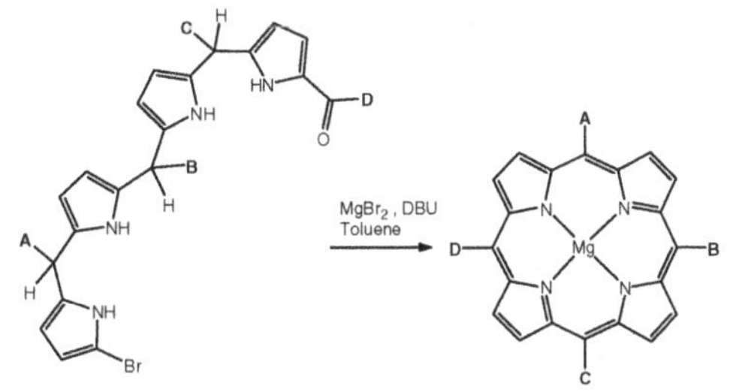

2.4 *Via Bilanes*

A bilane route relies on the acid-catalyzed reaction of a 1-acyldipyrromethane, bearing the A and B substituents, and a 9-X-dipyrromethane-1-carbinol (X = thiocyanato, bromo, and ethylthio) functionalized with the C and D groups, to give the corresponding 1-protected-19-acylbilane: among the examined protecting groups, bromo results the most effective protecting group [57]. For this step, the order for the introduction of the substituents on the peripheral positions of both the acyldipyrromethane and the dipyrromethanecarbinol follows the considerations made for the above-mentioned [2+2] condensation. The second key reaction (Scheme 12) entails a one-flask intramolecular cyclization of the protected bilane: the presence of a metal-template and the use of non-nucleophilic base in toluene afford the corresponding metalloporphyrin.

Such an approach, particularly useful for the preparation of porphyrins bearing four different substituents, has several advantages with respect to the conventional [2+2] and [3+1] methods: the chance to perform large-scale synthesis and the chance to avoid the use of acid catalyst and quinone oxidant are maybe the more interesting ones, while the necessity of high temperature (115°C) to achieve the macrocycle formation seems to be the most important limitation [58].

3 Functionalization of the Macrocycle

Beside the total synthesis of unsymmetrically substituted porphyrins, where the yields are often not satisfying, in the last 10 years a combination of well-developed condensation methods, with subsequent functionalization by organolithium compounds or transition-metal-catalyzed coupling protocols, have been studied.

Such an approach is based on the concept that previously described methods require highly involved pyrrole and dipyrrin chemistry, followed by the appropriate [2+1] or [3+1] protocol, while the use of easily available reagents, such as the symmetric porphine, 5,10,15,20-tetraarylporphyrin, or 2,3,7,8,12,13,17,18-octaethylporphyrin, and their functionalization, should lead to more appreciable results.

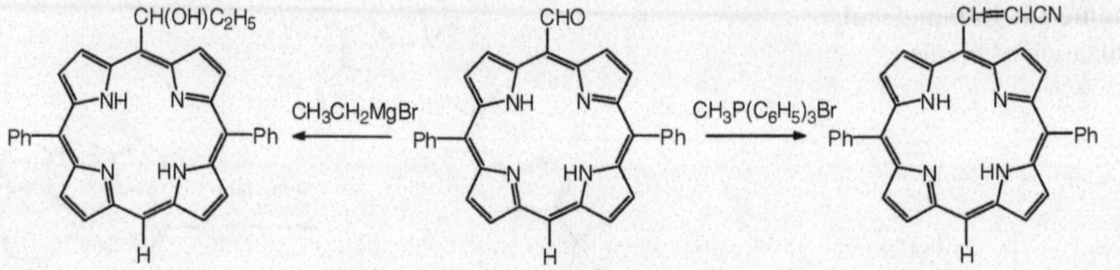

Scheme 13 Grignard and Witting reactions carried out on the *meso*-formyl group

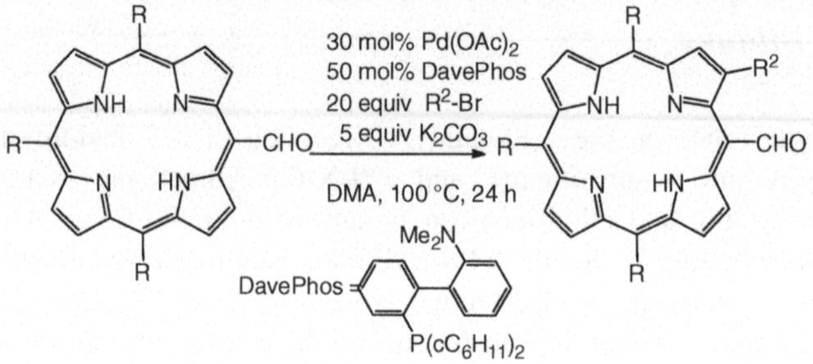

Scheme 14 Arylation of *meso*-formyl porphyrins

Unfortunately, porphyrins have proven to be very resilient towards many reactions that on paper should proceed quite easily, for example, Friedel–Crafts or Grignard reactions [59], but over the years several electrophilic substitution reactions have been described ([60] and reference therein) that can be used as entry reactions into many other porphyrin derivatives. Among these, one of the most widely used reactions is the formylation, which can be obtained by Vilsmeier reaction or by the insertion of the 1,3-dithian-2-yl residue as a precursor for the –CHO group (in this case the formylation is under nucleophilic conditions) and used as starting material for reaction with organomagnesium or organophosphorous compounds (Scheme 13) [61] or in palladium-catalyzed intermolecular direct arylation (Scheme 14) [62, 63].

Other important reactions are the nitration or halogenation both in *meso*- and β-positions: the latter in particular has been used in recent years as entry point for modern C–C coupling reactions [64]. Among these, the Heck reaction has been used by Senge, for example, to prepare *meso*-vinylacceptor-substituted porphyrin reacting a Zn(II)dibromo-ditolylporphyrin with butylacrylate (Fig. 2) [54, 65, 66]; the Suzuki reaction [66] on bromo derivatives has been used for the preparation of unsymmetrically substituted porphyrins both with arylboronic acids [67] and with organotrifluoroborates [68] which have the advantage to present a higher air stability with respect to the conventional Suzuki–Miyaura reagents. Moreover, large and well-ordered assemblies have been obtained by multiple coordination bonds between Zn-porphyrins and triazole moieties: for this purpose a *meso*-ethynylated porphyrin has been obtained by a Sonogashira reaction and then, by the so-called click chemistry, using a Cu(I)-catalyzed 1,3-dipolar cycloaddition reaction of benzyl azide, the porphyrin has been functionalized (Fig. 3) [69–71].

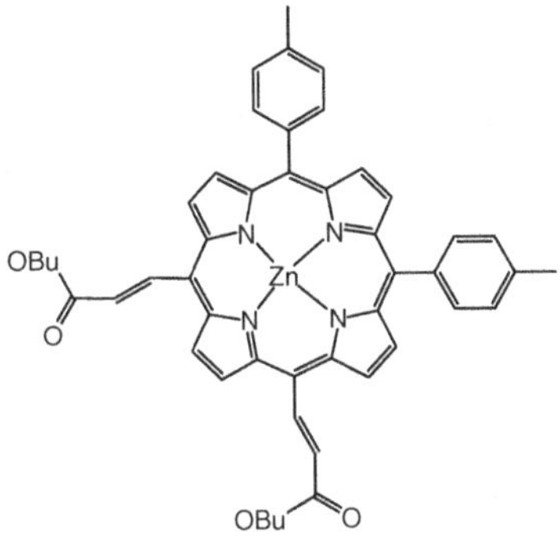

Fig. 2 {(all-E)-5,10-bis(4-(2-butoxycarbonylethenyl)-15,20-di(*p*-tolylporphyrinato} zinc(II)

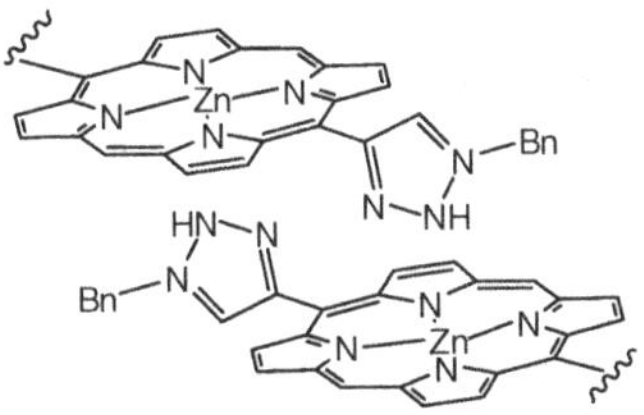

Fig. 3 Self-assembly of *meso*-triazole Zn(II)-porphyrins

On the other hand, only few examples for nucleophilic addition or substitution reactions have been described [72, 73] and generally require the activation of the macrocycle via electron-withdrawing groups, the insertion of appropriate central metals or steric effects. Takanami et al. reported the direct conversion of 5,15 disubstituted porphyrins into *meso*-formylporphyrins via a sequential S_NAr reaction with (2-pyridyldimethylsyilyl)methyllithium ($PyMe_2SiCH_2Li$) followed by oxidation with DDQ [74]. The same authors have recently reported a facile protocol for the preparation of silylmethyl-substituted porphyrins via Pd(0)-catalyzed Kumada cross-coupling of bromoporphyrins with silylmethyl magnesium reagents. These reactions have been accomplished by a novel catalytic system consisting of a $Pd_2(dba)_3$ and phosphine oxide ligand, $Ph_2P(O)H$, affording a multipurpose synthone for CHO, CH_2OH, CH_2OMe, and CH_2F functionalities as well as the fluoride ion-mediated desilylative introduction of C–C single and double bonds (Scheme 15) [75].

Good results have also been obtained with organozinc and organotin reagents [76].

In the last 10 years [77] the element of novelty is represented by the development, by Senge and coworkers, of a general method for the direct *meso* substitution of inactivated alkyl- and arylporphyrins with aryl or alkyl residues: in particular the use of organolithium reagents can be successfully applied to the functionalization of the porphyrin ring to introduce almost any desired residue, often quantitatively.

The addition-oxidation sequence has been initially studied on different metal complexes of 2,3,7,8,12,13,17,18-octaethylporphyrin using different organolithium compounds: the best results, in terms of compromise between yields and solubility,

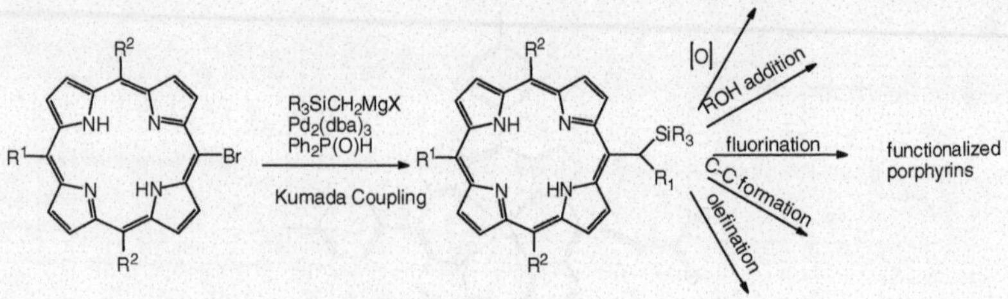

Scheme 15 Functionalization via Kumada coupling

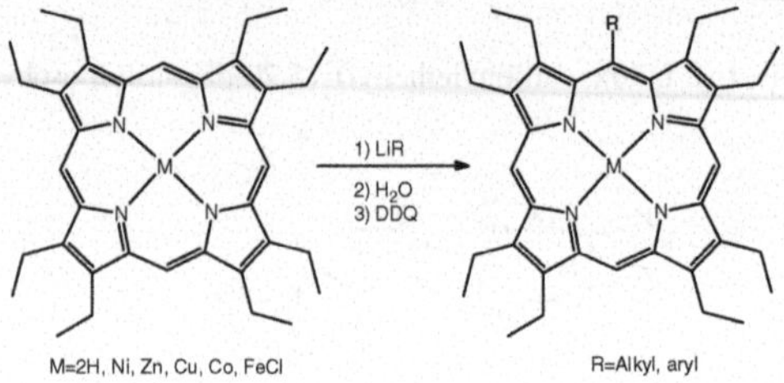

Scheme 16 *Meso*-functionalization by organolithium compounds

have been obtained with Ni complexes, while any lithium compound tested gave satisfactory results (Scheme 16). The results have been then extended by testing the reaction on both tetraarylporphyrins and tetraalkylporphyrins: *meso* and beta positions demonstrated to be reactive, with a preference for the *meso* substitution. The most important result of this method is the introduction of groups amenable for further modification and, furthermore, that the reaction, after the insertion of the first group, can be repeated again on the isolated porphyrin in order to afford the introduction of 2-, 3-, and 4-*meso* substituents.

A full analysis of the synthesis of all members of the ABCD-type porphyrin family has been done by comparing the results obtained either with the Pd-catalyzed coupling reactions or by the nucleophilic reaction with organolithium reagents [78]. Based on the reported results, the synthesis of AB-Porph can be achieved, starting from A-Porph, by Heck C–C coupling on the mono bromo derivative or by reaction with RLi: in both the cases the yields are low due, for the former reaction, to the problematic monobromination and, for the latter, to the formation of regioisomers and products of poly-substitution. For this reason the best method to afford the AB-Porph is still the complete synthesis, but the resulting porphyrin is suitable starting materials to obtain ABC-type systems through nucleophilic substitution, in quite good yields [79]. The ABC-systems can be converted into the ABCD-Porph, but the yields of the reaction with organolithium compounds are unsatisfactory, even when using the one-pot method (RLi/RI) on AB-Porph (Scheme 17) [80, 81]: in this case the anionic intermediate, coming from the

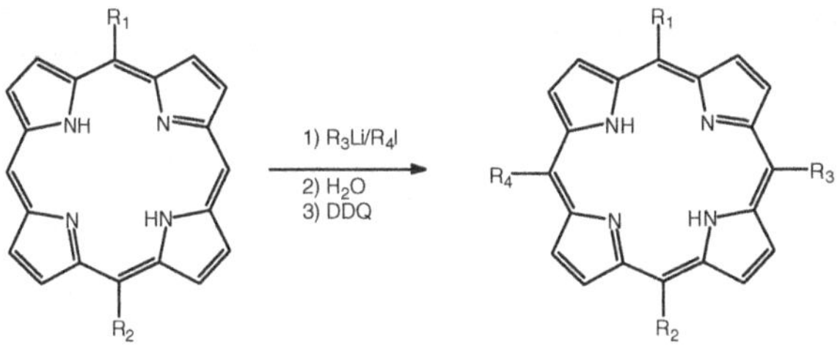

Scheme 17 One-pot RLi/RI method

reaction with RLi in THF, is trapped with electrophiles (for example, alkyl iodide such as *n*-propyl iodide or *n*-pentyl iodide) leading to the formation of the *meso*-tetrasubstituted porphyrin. For ABCD-porphyrins, beside the total synthesis, the best results are obtained with Pd-catalyzed reactions, even if the RLi method gives easier access to alkyl-substituted porphyrins. A combination of the two methods is necessary to obtain, by the transformation sequence A → AB → ABC → ABCD-Porph, almost any desired poly-functionalized porphyrin.

4 Other Methods

Other strategies can be used to functionalize a symmetric porphyrin in order to get an asymmetry in the system.

Coutsolelos reported the functionalization of the α,α,α,α atropisomer of the 5,10,15,20-tetrakis(2-aminophenyl)porphyrin by reaction with N^1-benzyl-N^2, N^2-bis[2-(benzylamino)-ethyl]ethane-1,2-diamine [82] and 2-methoxyphenylamine (Scheme 18) in order to prepare and characterize a mimetic model for Cytochrome *c* oxidase: such an enzyme is a member of the heme-Cu terminal oxidase super-family involved in the respiratory chains of mitochondria and aerobic bacteria.

The same porphyrin, α4 atropisomer, has been used by Hosseini and coworkers [83] for the synthesis of polynucleating ligands to be used in NIR imaging applications. The porphyrin, used as the antenna, has been combined with 8-hydroxyquinolinyl chelating groups, for the binding of lanthanide ions, by amide junctions: among the symmetric products, the asymmetric trisubstituted porphyrin has been isolated in a 50% yield (Fig. 4).

A facile modification of the porphyrin macrocycle to produce π-extended systems has been developed: the method is based on radical cyclization reactions through the installation of alkynes at the β-pyrrole periphery of a symmetric *meso*-tetraarylporphyrin. After the introduction of triple bonds in β-positions, subsequent thermal or photochemical, $PdCl_2$-mediated, activation of the alkynes leads, by benzannulation process, to the picenoporphyrins extended structures (Scheme 19).

These properties enable some systems to exhibit large two-photon absorption cross sections, which can be relevant for applications like in vivo imaging [84].

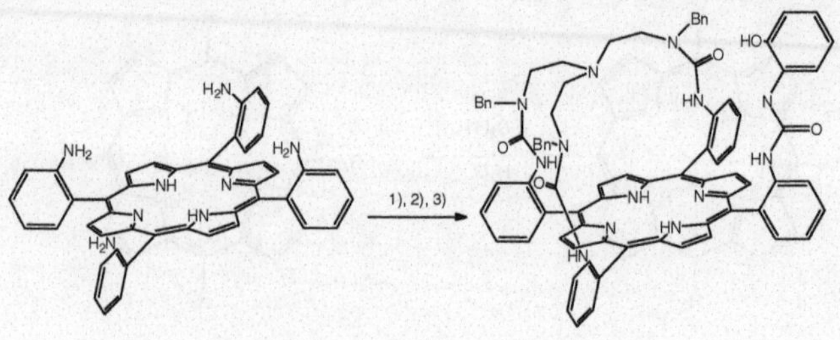

Scheme 18 Functionalization of an α,α,α,α atropisomer. Reagents: (1) N^1-benzyl-N^2, N^2-bis [2-(benzylamino)-ethyl]ethane-1,2-diamine, triphosgene, Et_3N, CH_2Cl_2, r.t.; (2) 2-methoxy-phenylamine, triphosgene, Et_3N, CH_2Cl_2, r.t.; (3) BBr_3, CH_2Cl_2, −78°C to 0°C

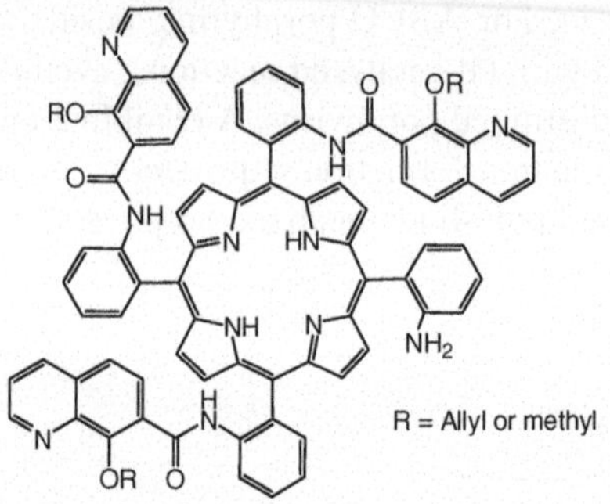

Fig. 4 Porphyrin-based polynucleating ligands

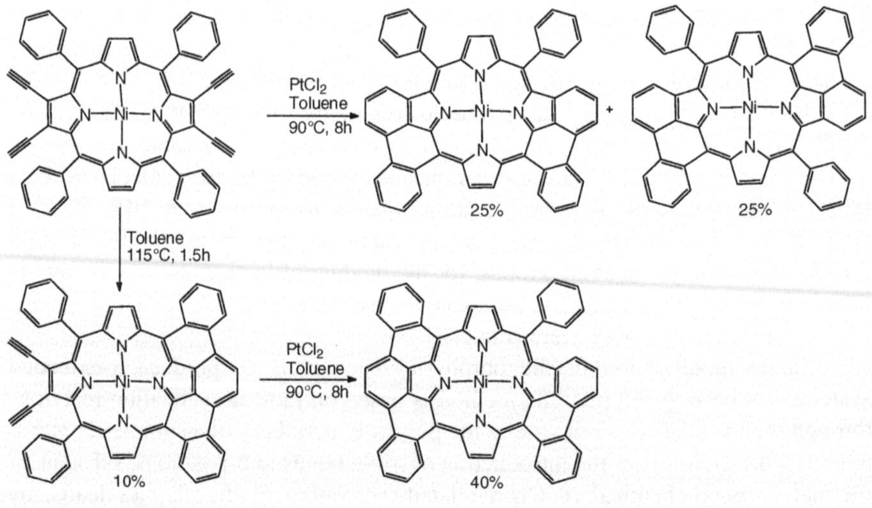

Scheme 19 Synthesis of extended π structures

5 Applications

The availability of these unsymmetrical compounds allows for the preparation of further functionalized compounds or even different related macrocycles, often in higher yields with respect to the established procedures. For example, a series of *meso*-5,10,20-triaryl-2,3-chlorins [85] or carbohydrate-porphyrin conjugates [86] for PDT applications has been studied: glycoconjugated chlorins or carbohydrate-porphyrin oligomers could be prepared, respectively, by reduction [87] or by functionalization (Scheme 20) of the corresponding asymmetric porphyrins, obtained by addition of 4-hydroxyphenyllithium to a 5,15-diarylporphyrin.

For the latter compound, deprotection of the trimethylsilyl group with tetrabutylammonium fluoride and immediate reaction, by Gaser–Hay oxidative coupling, Heck cross-coupling, or in Zemplèn conditions, led the authors to the preparation of a series of oligomers useful as photosensitizers, in particular, for two-photon excited PDT.

Mono glycoporphyrins and 5,10-bis-modified heterogeneous glycoporphyrins (Fig. 5) [88] have also been accessed by microwave-mediated “click” and stepwise “double click” reactions [89].

The use of asymmetrical porphyrin can be also useful for the preparation of ditopic receptor for the development of catalytic systems [90]: mimicking the natural occurring systems, by the introduction of a combination of electrostatic and hydrophobic interactions with a possible anchoring through coordination to a metal center, the receptor has been prepared by coupling the large size and structural flexibility of the calixarene cavity with the peculiar properties of the porphyrin ring (Scheme 21).

In this case, a triphenyliodoporphyrin has been transformed into the corresponding alkyne-functionalized zinc porphyrin (Scheme 22): the porphyrin has been obtained by a Suzuki–Miyaura cross-coupling reaction with a silyl-protected ethynylphenylboronic pinacolic ester followed by cleavage of the silyl group and metalation. The resulting tetrapyrrolic macrocycle has been then tethered with a functionalized calixarene to give the final receptor.

In the field of molecular motors and machines [91–93], Hosseini and coworkers reported porphyrin-based switchable molecular turnstiles (Fig. 6) obtained by using a A_3B-Porph, a 5-(4-pyridyl)-10,15,20-triphenylporphyrin, as the starting material: in particular, the dihydroxy tin complex of the mentioned porphyrin has been condensed with handles of different length composed of two resorcinol units connected to a 2,6-pyridinecarboxamide tridentate coordinating site by oligoethyleneglycol spacers [94].

In an apolar solvents (such as CD_2Cl_2) the turnstile is closed (Fig. 6, on the right) due to the formation of hydrogen bond between the donor tridentate unit on the handle and the acceptor site on the pyridyl moiety of the porphyrin: in presence of a hydrogen-bond disrupting solvent (DMSO, for example), the free rotation of the handle around the porphyrin leads to the turnstile opening (Fig. 6, on the left).

Another rotor–stator combination has been reported as a model for biological self-locking systems, such as Ca^{2+}/calmodulin-dependent protein kinase II.

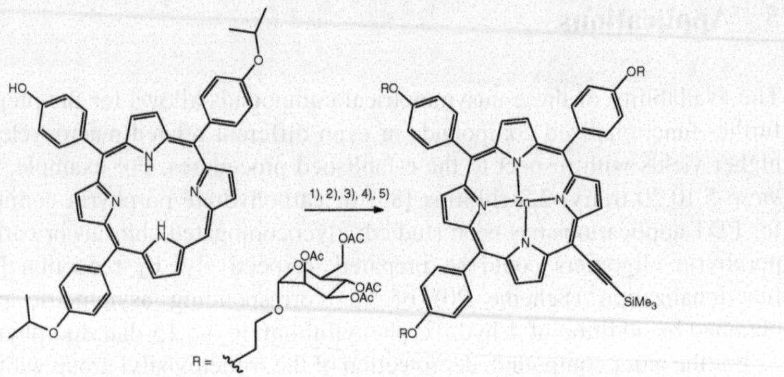

Scheme 20 Reagents: (1) BBr_3, CH_2Cl_2; (2) 2-bromoethoxy-*o*-2′,3′,4′,6′-tetraacetyl-α-D-mannose, Cs_2CO_3, DMF; (3) NBS, Pyridine; (4) $Zn(OAc)_2$, $CHCl_3/MeOH$; (5) trimethylsilylacetylene, CuI, $Pd(PPh_3)_2Cl_2$, THF/triethylammine

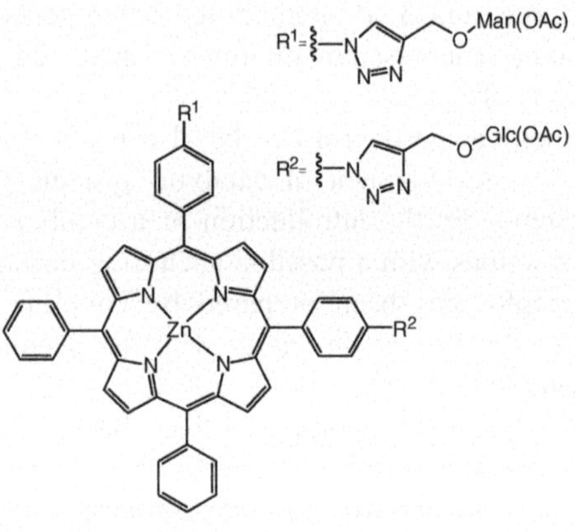

Fig. 5 5,10-bis-modified glycoporphyrin

An essential feature of these enzymes is that they are normally kept dormant and locked, but they can be unlocked with proper keys whenever necessary: the binding of Ca^{2+} ions promotes a conformational change in the calmodulin, leading to the dissociation of the autoinhibitory segment and to the activation of the enzyme. In the reported model, the rotary module is composed of an enantiomer of a ferrocene functionalized with a zinc complex of the asymmetric 5-(4,4,5,5-tetramethyl-1,3,2-dioxaborolan-2-yl)-10,15,20-tris(4-methylphenyl)porphyrin and an aniline units at each cyclopentadienyl ring (Scheme 23).

Photochromic 1,2-bispyridylethylene (Scheme 23, on the right) acts as an external key: UV irradiation allows for its *trans*-to-*cis* isomerization, while visible light, in the presence of triplet sensitizers such as Zn-porphyrins, allows for the backward isomerization. The system is internally locked in apolar solvents by a double intramolecular Zn–N coordination between the Zn-porphyrins and the aniline units (Scheme 23), but the external key, bearing two pyridine units, when

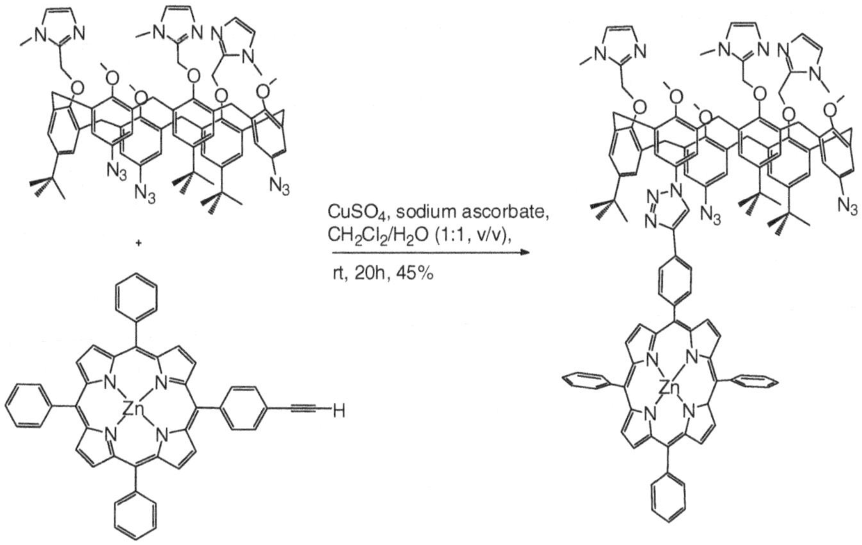

Scheme 21 Ditopic receptor

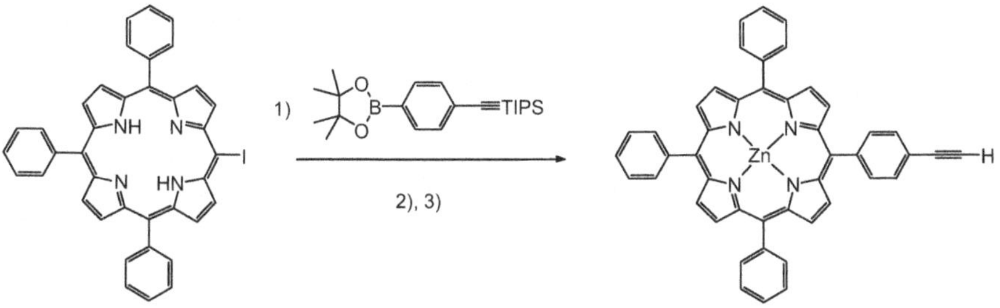

Scheme 22 (1) DMF/toluene (1:1, v/v), $[Pd_2dba_3]\cdot CHCl_3$, PPh_3, Cs_2CO_3, 60 h, 90°C, 55%; (2) NBu_4F (1 M in THF), CH_2Cl_2, 20 h, room temperature, 100%; (3) $Zn(OAc)_2$, CH_2Cl_2/MeOH (1:1, v/v), room temperature, 1 h, 92%

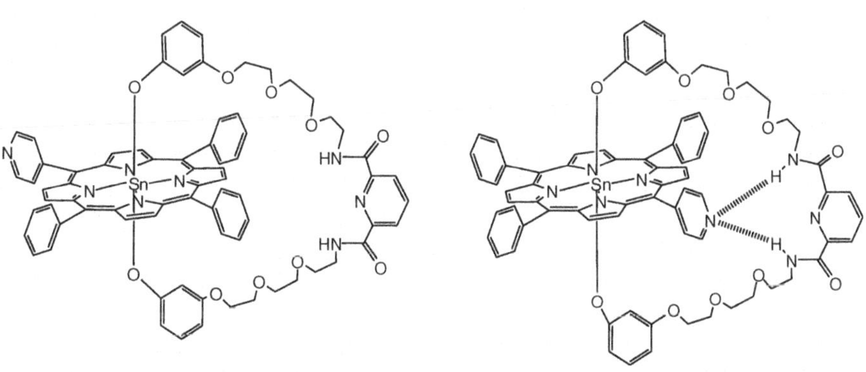

Fig. 6 Porphyrin turnstile open (*left*) and closed (*right*)

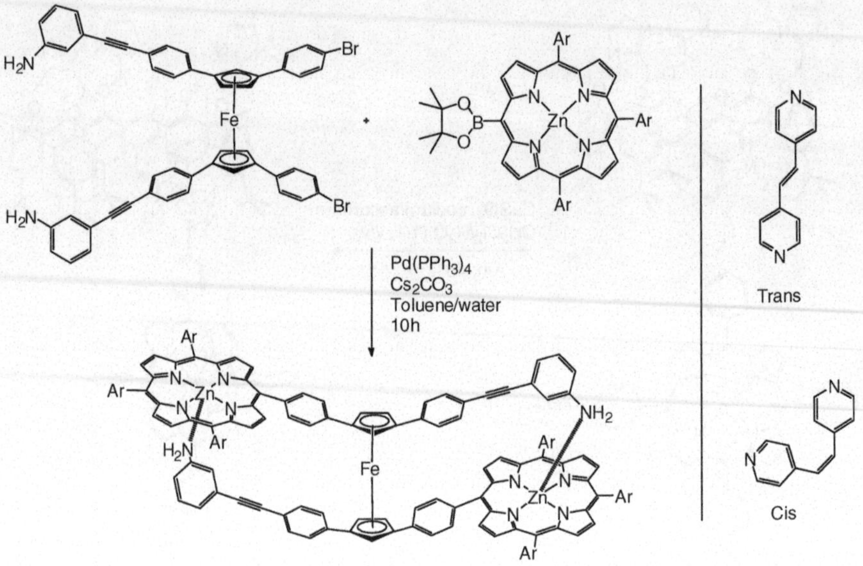

Scheme 23 Self-locking ferrocene–porphyrin system

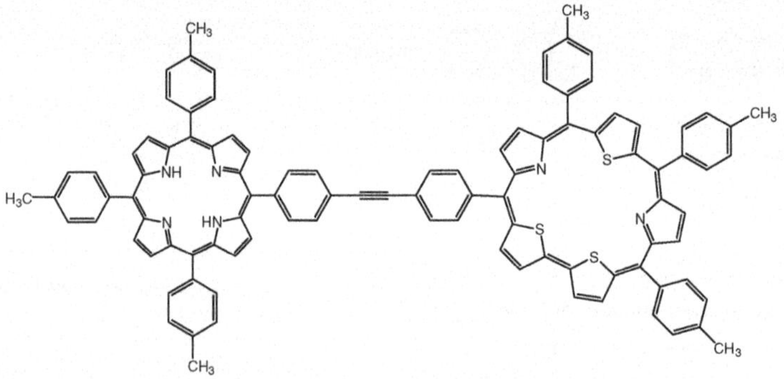

Fig. 7 Porphyrin–smaragdyrin dyad

in its *cis*-conformation, is capable to coordinate the zinc porphyrin units inducing a competition between *intra*- and *inter*molecular interaction and leading to an externally locked state. When the *cis*-1,2-bispyridylethylene is isomerized into its *trans* form, the key detaches from the rotary system and the external lock is disabled [95].

The unsymmetric A_3B-Porph, 20-(4-ethynylphenyl)-5,10,15-tri(4-methylphenyl) porphyrin, prepared in the paper according to the mixed condensation of different aldehyde with a dipyrromethane [96] has been used as starting material for the preparation of covalently linked porphyrin-expanded heteroporphyrin dyads (Fig. 7): in particular, a porphyrin–sapphyrin dyad and a porphyrin–smaragdyrin dyad have been synthesized as potential fluorescent anion sensors [97].

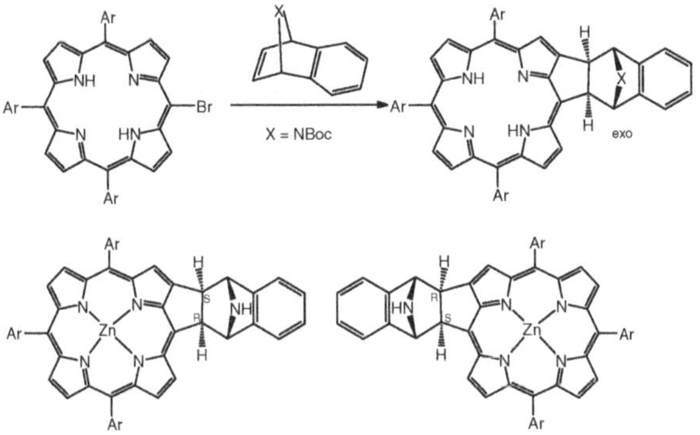

Scheme 24 Pd-catalyzed annulation of bromoporphyrin with norbornene derivatives and racemic mixture of benzoazanorbornene-fused porphyrin

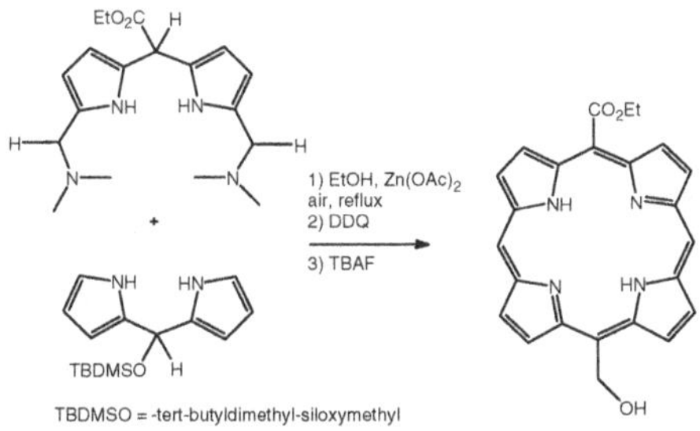

Scheme 25 Synthesis of compact architecture

The A_3-Porph 5,10,15-triphenylporphyrin and the corresponding Ni complex have been brominated and then used in a Pd-catalyzed [3+2] annulation with norbornene derivatives to afford chiral fused structures: chiral porphyrins have in fact received much attention as scaffold for precise molecular recognition, construction of supramolecular structures, and asymmetric catalysis [98].

In particular, the authors examined the self-assembly of a racemic mixture of Zn benzoazanorbornene-fused porphyrin (Scheme 24) obtained after the annulation by removal of the *tert*-butoxycarbonyl (Boc) group. The following insertion of Zn led to the formation of a stable heterochiral dimer resulting from complementary coordination both in the solid and solution states.

As previously mentioned, for some applications ranging from medicine, wherein small molecules should passively cross the blood–brain barrier, to molecular information storage applications, where the small size of the molecule permits a high charge density, the use of porphyrins equipped with compact and unsymmetrical substituents is desirable. In this area Lindsey et al. reported the preparation of porphyrins bearing one carbon oxygenic substituent (hydromethyl, formyl, or ester), directly attached to the macrocycle, to afford a series of symmetrical and asymmetrical porphyrins such as the hydroxymethyl-porphyrin reported in Scheme 25, prepared according to a [2+2] method [99].

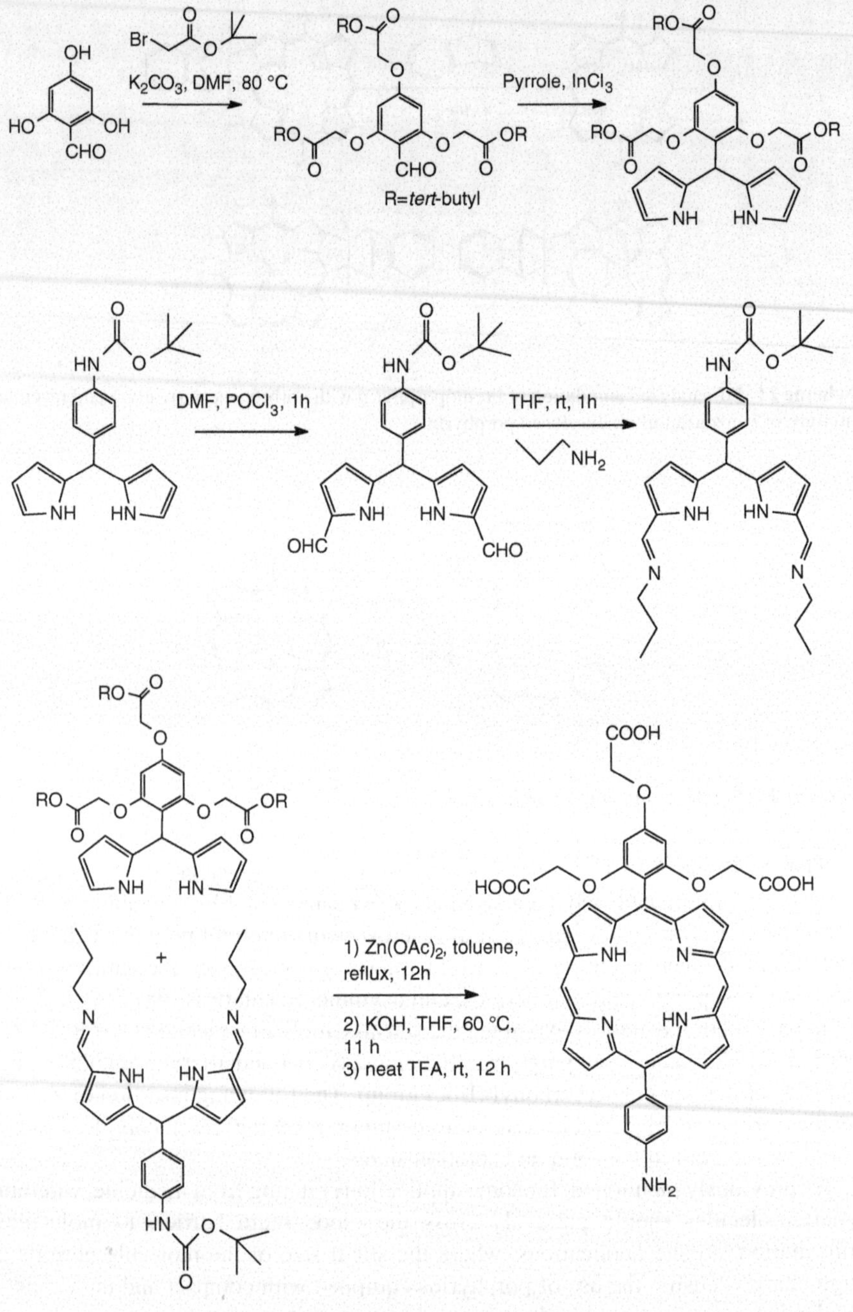

Scheme 26 Water-soluble bioconjugatable porphyrins

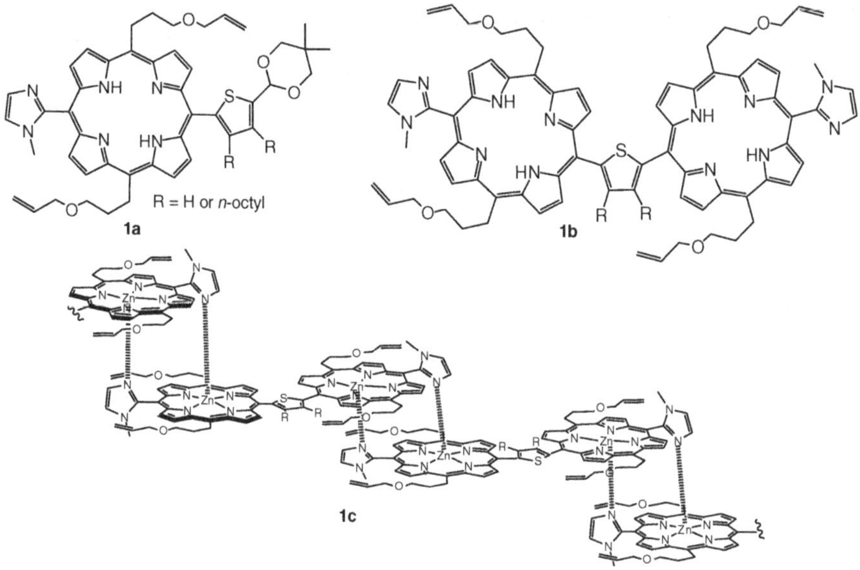

Fig. 8 Self-assembled giant porphyrin macroring

Using "Route 1" (Scheme 4) [100], instead, *trans*-AB-porphyrin architectures tailored with a water-solubilizing 2,4,6-tri(carboxymethoxy)phenyl substituent and an aldehyde (or amine) group, suitable for reductive amination in aqueous media, have been prepared as bioconjugatable systems (Scheme 26). To perform bioconjugation in aqueous solution, and avoid solubility problems arising with mixed aqueous-organic media, the carboxy moieties must be deprotected at the end using TFA.

Another important field of application for the porphyrin macrocycle is the construction of models for the bacterial photosynthetic light-harvesting antennae, in view of their importance in biological energy transformation events.

Among the various covalently and non-covalently linked multiporphyrin systems, a five-membered aromatic thiophene has been explored as the linker between bis[imidazolylporphyrinatozinc(II)] molecules **1b** for making larger self-assembled macrorings with enlarged internal angles. Starting from the asymmetric acetal-protected thiophenylporphyrin **1a** (Fig. 8), the deprotection of the acetal group followed by the condensation with another dipyrromethane unit and mixed aldehyde affords the dyad **1b** that was metalated to lead, by strong complementary coordination of imidazolyl to zinc, to the giant porphyrin macroring **1c** [101].

Asymmetric indolo[3,2-*b*]carbazole-porphyrin and corrole conjugates, endowed with a different number of indolocarbazole units (Fig. 9), have been prepared in order to evaluate the effect of the progressive substitution on the photophysical properties of the materials, for a potential application in optoelectronic devices [102].

It is worth mentioning that, since the coordination chemistry of metalloporphyrins is not necessarily restricted to the tetrapyrrolic core, but can also occur at its periphery,

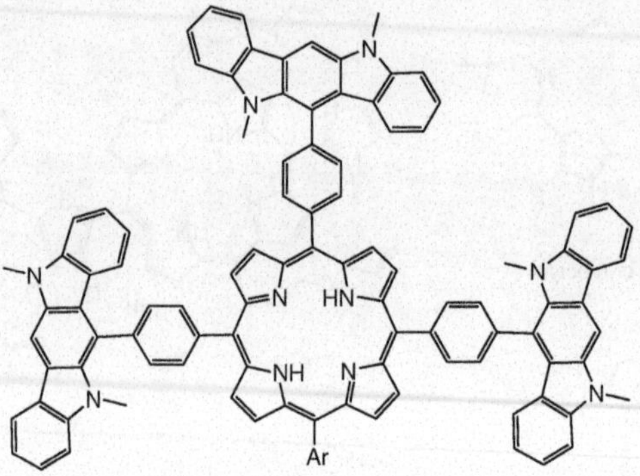

Fig. 9 *meso*-Indolo[3,2-*b*]carbazolyl-substituted porphyrin

numerous ligand sites have been attached to, or incorporated into, the porphyrin systems, symmetric and asymmetric, in order, for example, to develop libraries of supramolecular catalyst or to construct multiporphyrin assemblies [103].

6 Conclusion

The enormous variety of application fields, where porphyrins play a leading role, has driven the research, beside the improvement of the consolidated synthetic strategies, towards the development of new approaches for the preparation of specific, focused tetrapyrrolic systems. Based on traditional syntheses of symmetrical structures, new methods have been studied in the last decade in order to obtain asymmetrical poly-functionalized porphyrins: together with the total synthesis of the macrocycle, via [2+2], [3+2], or bilane routes, a particular attention has been devoted to the organometallic cross-coupling reactions as new methodologies to transform symmetrical porphyrins into ABCD-systems.

References

1. Sharman WM, van Lier JE (1999) Photodynamic therapeutics: basic principles and clinical applications. Drug Discov Today 4:507–517
2. (2010) Phototherapy, Immunotherapy and Imaging. In: Kadish KM, Smith KM, Guilard R (eds) Handbook of porphyrin science, vol 4. Academic, Singapore
3. Moore TA, Moore AL, Gust D (2009) Fuels via artificial photosynthesis. Acc Chem Res 42:1890–1898
4. Collman JP, Fu L (2001) Mimicking photosynthetic solar energy transduction. Acc Chem Res 34:40–48
5. Meunier B (1992) Metalloporphyrins as versatile catalysts for oxidation reactions and oxidative DNA cleavage. Chem Rev 92:1411–1456

6. Brulè E, de Miguel YR (2006) Supported metalloporphyrin catalysts for alkene epoxidation. Org Biomol Chem 4:599–609
7. Paolesse R, Monti D, Nardis S, Di Natale C (2011) Porphyrin-based chemical sensors. In: Kadish KM, Smith KM, Guilard R (eds) The porphyrin handbook, vol 12. Academic, Singapore
8. Maeda C, Kamada T, Aratani N, Osuka A (2007) Chiral self-discriminative self-assembling of meso–meso linked diporphyrins. Coord Chem Rev 251:2743–2752
9. Yoon ZS, Yoon MC, Kim D (2005) Excitonic coupling in covalently linked multiporphyrin systems by matrix diagonalization. J Photochem Photobiol C 6:249–263
10. Wagner RW, Johnson TE, Lindsey JS (1996) Soluble synthetic multiporphyrin arrays. 1. Modular design and synthesis. J Am Chem Soc 118:11166–11180
11. Kodis G, Terazono Y, Bhushan K, Zaks J, Madden C, Moore AL, Moore TA, Fleming GR, Gust D (2011) Mimicking the role of the antenna in photosynthetic photoprotection. J Am Chem Soc 133:2916–2922
12. Hyakutake T, Okura I, Asai K, Nishide H (2008) Dual-mode oxygen-sensing based on oxygen-adduct formation at cobaltporphyrin–polymer and luminescence quenching of pyrene: an optical oxygen sensor for a practical atmospheric pressure. J Mater Chem 18:917–922
13. Liddell PA, Gervaldo M, Bridgewater JW, Keirstead AE, Lin S, Moore TA, Moore AL, Gust D (2008) Porphyrin-based hole conducting electropolymer. Chem Mater 20:135–142
14. Brennan BJ, Kenney MJ, Liddell PA, Cherry BR, Li J, Moore AL, Moore TA, Gust D (2011) Oxidative coupling of porphyrins using copper(II) salts. Chem Commun 47:10034–10036
15. Li WS, Aida T (2009) Dendrimer porphyrins and phthalocyanines. Chem Rev 109:6047–6076
16. Fukuzumi S, Saito K, Ohkubo K, Khoury T, Kashiwagi Y, Absalom MA, Gadde S, D'Souza F, Araki Y, Ito O, Crossley MJ (2011) Multiple photosynthetic reaction centres composed of supramolecular assemblies of zinc porphyrin dendrimers with a fullerene acceptor. Chem Commun 47:7980–7982
17. Megiatto JD Jr, Spencer R, Schuster DI (2009) Efficient one-pot synthesis of rotaxanes bearing electron donors and [60]fullerene. Org Lett 11:4152–4155
18. Faiz JA, Heiz V, Sauvage JP (2009) Design and synthesis of porphyrin-containing catenanes and rotaxanes. Chem Soc Rev 38:422–442
19. Megiatto JD Jr, Abwandner S, de Miguel G, Guldi DM (2010) [2]Catenanes decorated with porphyrin and [60]fullerene groups: design, convergent synthesis, and photoinduced processes. J Am Chem Soc 132:3847–3861
20. Thordarson P, Bijsterveld EJA, Rowan AE, Nolte RJM (2003) Epoxidation of polybutadiene by a topologically linked catalyst. Nature 424:915–918
21. Sourav Saha AHF, Stoddart JF, Impellizzeri S, Silvi S, Venturi M, Credi A (2007) A redox-driven multicomponent molecular shuttle. J Am Chem Soc 129:12159–12171
22. Smith KM (2000) Strategies for the synthesis of octaalkylporphyrin systems. In: Kadish KM, Smith KM, Guilard R (eds) The porphyrin handbook, vol 1. Academic, San Diego
23. Brückner C, Posakony JJ, Johnson CK, Boyle RW, Dolphin D (1998) Novel and improved syntheses of 5,15-diphenylporphyrin and its dipyrrolic precursors. J Porphyr Phthalocyanines 2:455–465
24. Jaquinod L (2000) Functionalization of 5,10,15,20-tetrasubstituted porphyrins. In: Kadish KM, Smith KM, Guilard R (eds) The porphyrin handbook, vol 1. Academic, San Diego
25. Wiehe A, Shaker YM, Brandt JC, Mebs S, Senge MO (2005) Lead structures for applications in photodynamic therapy. Part 1: synthesis and variation of m-THPC (Temoporfin) related amphiphilic A_2BC-type porphyrins. Tetrahedron 61:5535–5564
26. Wiehe A, Simonenko EJ, Senge MO, Röder B (2001) Hydrophilicity vs hydrophobicity – varying the amphiphilic structure of porphyrins related to the photosensitizer m-THPC. J Porphyr Phthalocyanines 5:758–761

27. Drain CM, Goldberg I, Sylvain I, Falber A (2005) Synthesis and applications of supramolecular porphyrinic materials. Top Curr Chem 245:55–88
28. Drain CM, Varotto A, Radivojevic I (2009) Self-organized porphyrinic materials. Chem Rev 109:1630–1658
29. Holten D, Bocian DF, Lindsey JS (2002) Probing electronic communication in covalently linked multiporphyrin arrays. A guide to the rational design of molecular photonic devices. Acc Chem Res 35:57–69
30. Senge MO, Fazekas M, Notaras EGA, Blau WJ, Zawadzka M, Locos OB, Ni Mhuircheartaigh EM (2007) Nonlinear optical properties of porphyrins. Adv Mater 19:2737–2774
31. Sariola-Leikas E, Hietala M, Veselov A, Okhotnikov O, Semjonov SL, Tkachenko NV, Lemmetyinen H, Efimov A (2012) Synthesis of porphyrinoids with silane anchors and their covalent self-assembling and metalation on solid surface. J Colloid Interface Sci 369:58–70
32. Lindsey S (2000) Synthesis of meso-substituted porphyrins. In: Kadish KM, Smith KM, Guilard R (eds) The porphyrin handbook, vol 1. Academic, San Diego
33. Geier GR III, Lindsey JS (2001) Investigation of porphyrin-forming reactions. Part 2 examination of the reaction course in two steps, one flask syntheses of meso substituted porphyrins. J Chem Soc Perkin Trans 2:687–700
34. Balakumar A, Muthukumaran K, Lindsey JS (2004) A new route to *meso*-formyl porphyrins. J Org Chem 69:5112–5115
35. Laha JK, Dhanalekshmi S, Tamiguchi M, Ambroise A, Lindsey JS (2003) A scalable synthesis of meso-substituted dipyrromethanes. Org Process Res Dev 7:799–812
36. Wickramasinghe A, Jaquinod L, Nurco DJ, Smith KM (2001) Investigations on the directive effects of a single meso-substituent via nitration of 5,12,13,17,18-pentasubstituted porphyrins: syntheses of conjugated β-nitroporphyrins. Tetrahedron 57:4261–4269
37. Trova MP, Gauuan PJF, Pechulis AD, Bubb SM, Bocckino SB, Crapo JD, Day B (2003) Superoxide dismutase mimetics. Part 2: synthesis and structure–activity relationship of glyoxylate- and glyoxamide-derived metalloporphyrins. J Bioorg Med Chem 11:2695–2707
38. Kral V, Vasek P, Dolensky B (2004) Green chemistry for preparation of oligopyrrole macrocycle precursors: novel methodology for dipyrromethanes and tripyrromethanes synthesis in water. Collect Czech Chem Commun 69:1126–1136
39. Rao PD, Littler BJ, Geier GR III, Lindsey JS (2000) Efficient synthesis of monoacyl dipyrromethanes and their use in the preparation of sterically unhindered trans-porphyrins. J Org Chem 65:1084–1092
40. Lindsey JS (2010) Synthetic routes to meso-patterned porphyrins. Acc Chem Res 43:300–311
41. Muthukumaran K, Ptaszek M, Noll B, Scheidt WR, Lindsey JS (2004) Boron-complexation strategy for use with 1-acyldipyrromethanes. J Org Chem 69:5354–5364
42. Dogutan DK, Ptaszek M, Lindsey JS (2008) Rational or statistical routes from 1-acyldipyrrimethanes to meso-substituted porphyrins. Distinct patterns, multiple pyridyl substituents, and amphipathic architectures. J Org Chem 73:6187–6201
43. Melgiatto JD, Patterson D, Sherman BD, Moore TA, Gust D, Moore A (2012) Intramolecular hydrogen bonding as a synthetic tool to induce chemical selectivity in acid catalyzed porphyrin synthesis. Chem Commun 48:4558–4560
44. Tamaru SI, Yu L, Youngblood WJ, Muthukumaran K, Taniguchi M, Lindsey JS (2004) A tin-complexation strategy for use with diverse acylation methods in the preparation of 1,9-diacyldipyrromethanes. J Org Chem 69:765–777
45. Zaidi SHH, Muthukumaran K, Tamaru SI, Lindsey JS (2004) 9-Acylation of 1-acyldipyrromethane containing a dialkylboron mask for the α-acyldipyrrole motif. J Org Chem 69:8356–8365
46. Rao PD, Dhanalekshmi S, Littler BJ, Lindsey JS (2000) Rational syntheses of porphyrins bearing up to four different meso substituents. J Org Chem 65:7323–7344

47. Taniguchi M, Balakumar A, Fan D, McDowell BE, Lindsey JS (2005) Imine-substituted dipyrromethanes in the synthesis of porphyrins bearing one or two *meso* substituents. J Porphyr Phthalocyanines 9:554–574
48. Fan D, Taniguchi M, Yao Z, Dhanalekshmi S, Lindsey JS (2005) 1,9-bis(N, N-dimethylaminomethyl)dipyrromethanes in the synthesis of porphyrins bearing one or two meso-substituents. Tetrahedron 61:10291–10302
49. Broadhurst MJ, Grigg R, Johnson AW (1971) Synthesis of porphin analogues containing furan and/or thiophen rings. J Chem Soc C 3681–3690
50. Nguyen LT, Senge MO, Smith KM (1996) Simple methodology for syntheses of porphyrins possessing multiple peripheral substituents with an element of symmetry. J Org Chem 61:998–1003
51. Lash TD (1996) Porphyrin synthesis by the "3+1" approach: new applications for an old methodology. Chem Eur J 2:1197–1200
52. Wiehe A, Ryppa C, Senge MO (2002) A practical synthesis of meso-monosubstituted, β-unsubstituted porphyrins. Org Lett 4:3807–3809
53. Briñas RP, Brückner C (2002) Synthesis of 5,10-diphenylporphyrin. Tetrahedron 58:4375–4381
54. Senge MO, Ryppa C, Fazekas M, Zawadzka M, Dahms K (2011) 5,10-A_2B_2-type meso-substituted porphyrins-A unique class of porphyrins with a realigned dipole moment. Chem Eur J 17:13562–13573
55. Brückner C, Sternberg ED, Boyle RW, Dolphin D (1997) 5,10-Diphenyltripyrrane, a useful building block for the synthesis of meso-phenyl substituted expanded macrocycles. Chem Commun 1689–1690
56. Nielsen KA, Levillain E, Lynch VM, Sessler JL, Jeppesen JO (2009) Tetrathiafulvalene porphyrins. Chem Eur J 15:506–516
57. Dogutan DK, Zaidi SHH, Thamyongkit P, Lindsey JS (2007) New route to ABCD-porphyrins via bilanes. J Org Chem 72:7701–7714
58. Dogutan DK, Lindsey JS (2008) Investigation of the scope of a new route to ABCD-bilanes and ABCD-porphyrins. J Org Chem 73:6728–6742
59. Fuhrhop J-H (1978) Irreversible reactions on the porphyrin periphery (excluding oxidations, reductions, and photochemical reactions). In: Dolphin D (ed) The porphyrins, vol II. Academic, New York
60. Senge MO, Kalisch WW, Bischoff I (2000) The reaction of porphyrins with organolithium reagents. Chem Eur J 6:15
61. Dahms K, Senge MO, Bakar MB (2007) Exploration of meso-substituted formylporphyrins and their Grignard and Wittig reactions. Eur J Org Chem 3833–3848
62. Kawamata Y, Tokuji S, Yorimitsu H, Osuka A (2011) Palladium-catalyzed β-selective direct arylation of porphyrins. Angew Chem Int Ed 50:8867–8870
63. Tokuji S, Awane H, Yorimitsu H, Osuka A (2013) Direct arylation of meso-formyl porphyrin. Chem Eur J 19:64–68
64. Sharman WM, Van Lier JE (2000) Use of palladium catalysis in the synthesis of novel porphyrins and phthalocyanines. J Porphyr Phthalocyanines 4:441–453
65. Locos OB, Arnold DP (2006) The Heck reaction for porphyrin functionalization: synthesis of meso-alkenyl monoporphyrins and palladium-catalyzed formation of unprecedented meso-β ethene-linked diporphyins. Org Biomol Chem 4:902–916
66. Notaras EGA, Fazekas M, Doyle JJ, Blau WJ, Senge MO (2007) A_2B_2-type push-pull porphyrins as reverse saturable and saturable absorbers. Chem Commun 2166–2168
67. Shi B, Boyle RW (2002) Synthesis of unsymmetrically substituted meso-phenylporphyrins by Suzuki cross coupling reactions. J Chem Soc Perkin Trans 1:1397–1400
68. Horn S, Cundell B, Senge MO (2009) Exploration of the reaction of potassium organotrifluoroborates with porphyrins. Tetrahedron Lett 50:2562–2565
69. Tomizaki K, Lysenko AB, Taniguchi M, Lindsey JS (2004) Synthesis of phenylethyne-linked porphyrin dyads. Tetrahedron 60:2011–2023

70. Chinchilla R, Nagera C (2007) The Sonogashira reaction: a booming methodology in synthetic organic chemistry. Chem Rev 107:874–922
71. Maeda C, Kim P, Cho S, Park JK, Lim JM, Kim D, Vura-Weis J, Wasielewski MR, Shinokubo H, Osuka A (2010) Large Porphyrin Squares from the self-assembly of meso-triazole-appended L-shaped meso-meso linked Zn^{II}-triporphyrins: synthesis and efficient energy transfer. Chem Eur J 16:5052–5061
72. Shea KM, Jaquinod L., Khoury R.G., Smith KM (1998) Dodecasubstituted metallochlorins (metallodihydroporphyrins). Chem Commun 759
73. Callot HJ (1972) Stereochimie de l'addition de carbenes sur la meso-tetraphenylporphine. Tetrahedron Lett 13:1011
74. Takanami T, Wakita A, Sawaizumi A, Iso K, Onodera H, Suda K (2008) One-pot synthesis of *meso*-formylporphyrins by S_NAr reaction of 5,15-disubstituted porphyrins with (2-pyridyldimethylsilyl)methyllithium. Org Lett 10:685–687
75. Sugita N, Hayashi S, Hino F, Takanami T (2012) Palladium-catalyzed Kumada coupling reaction of bromoporphyrins with silylmethyl Grignard reagents: preparation of silylmethyl-substituted porphyrins as a multipurpose synthon for fabrication of porphyrin systems. J Org Chem 77:10488–10497
76. Sergeeva NN, Scala A, Bakar MA, O'Riordan G, O'Brien J, Grassi G, Senge MO (2009) J Org Chem 74:7140–7147
77. Senge MO (2011) Stirring the porphyrin alphabet soup-functionalization reactions for porphyrins. Chem Commun 47:1943–1960
78. Senge MO, Shaker YM, Pintea M, Ryppa C, Hatscher SS, Ryan A, Sergeeva Y (2010) Synthesis of meso-substituted ABCD-type porphyrins by functionalization reactions. Eur J Org Chem 237–258
79. Senge MO, Feng X (2000) Regioselective reaction of 5,15-disubstituted porphyrins with organolithium reagents-synthetic access to 5,10,15-trisubstituted porphyrins and directly meso-meso linked bisporphyrins. J Chem Soc Perkin Trans 1:3615–3621
80. Feng X, Senge MO (2000) One-pot synthesis of functionalized asymmetric 5,10,15,20-substituted porphyrins from 5,15-diaryl- or dialkyl-porphyrins. Tetrahedron 56:587–590
81. Feng X, Bischoff I, Senge MO (2001) Mechanistic studies on the nucleophilic reaction of porphyrins with organolithium reagents. J Org Chem 66:8693–8700
82. Charalambidis G, Ladomenou K, Boitrel B, Coutsolelos AG (2009) Synthesis and studies of a super-structured porphyrin derivative- a potential building block for CcO mimic models. Eur J Org Chem 1263–1268
83. Eckes F, Deiters E, Metivet A, Bulach V, Hosseini MW (2011) Synthesis and structural analysis of porphyrin-based polynucleating ligands bearing 8-methoxy- and 8-(allyloxy) quinoline units. Eur J Org Chem 2531–254
84. Boerner LJK, Nath M, Pink M, Zaleski JM (2011) Synthesis of unique extended π structures by Pt-mediated benzannulation of nickel tetraalkynylporphyrins. Chem Eur J 17:9311–9315
85. Varamo M, Loock B, Maillards P, Grierson DS (2007) Developments of strategies for the regiocontrolled synthesis of meso-5,10,15,20-triaryl-2,3-chlorins. Org Lett 23:4689–4692
86. Achell S, Couleaud P, Baldeck P, Teulade-Fichou MP, Maillard P (2011) Carbohydrate-porphyrin conjugates with two-photon absorption properties as potential photosensitizing agents for photodynamic therapy. Eur J Org Chem 1271–1279
87. Whitlock HW, Hanauer R, Oster MY, Bower BK (1969) Diimide reduction of porphyrins. J Am Chem Soc 91:7485–7489
88. Locos OB, Heindl CC, Corral A, Senge MO, Scanlan EM (2010) Efficient synthesis of glycoporphyrins by microwave-mediated "click" reactions. Eur J Org Chem 1026–1028
89. Huisgen R (1984) In: Padwa A (ed) 1,3-dipolar cycloaddition chemistry. New York, Wiley
90. Monnereau C, Rebilly J-N, Reinaud O (2011) Synthesis and first studies of the host-guest and substrate recognition properties of a porphyrin-tethered calyx[6]arene ditopic ligand. Eur J Org Chem 166–175
91. Stoddart JF (2001) Molecular machines. Acc Chem Res 34:410–411

92. Flood AH, Ramirez RJA, Deng WQ, Muller RP, Goddard WA III, Stoddart JF (2004) Meccano on the nanoscale—a blueprint for making some of the world's tiniest machines. Aust J Chem 57:301–322
93. Sauvage JP (2001) A light-driven linear motor at the molecular level. Science 291:2105–2106
94. Guenet A, Graf E, Kyritsakas N, Hosseini MW (2011) Porphyrin-based switchable molecular turnstiles. Chem Eur J 17:6443–6452
95. Muraoka T, Kinbara K, Aida T (2006) A self-locking molecule operative with a photoresponsive key. J Am Chem Soc 128:11600–11605
96. Lindsey JS, Prathapan S, Johnson JS, Wagner RW (1994) Porphyrin building blocks for modular construction of bioorganic model systems. Tetrahedron 50:8941–8968
97. Rajeswara Rao M, Ravikanth M (2011) Synthesis and studied of covalently linked porphyrin-expanded heteroporphyrin dyads. Eur J Org Chem 1335–1345
98. Mizumura M, Shinokubo H, Osuka A (2008) Synthesis of chiral porphyrins through Pd-catalyzed [3+2] annulation and heterochiral self-assembly. Angew Chem Int Ed 47:5378–5381
99. Yao Z, Bhaumik J, Dhanalekshmi S, Ptaszek M, Rodriguez P, Lindsey JS (2007) Synthesis of porphyrins bearing 1–4 hydroxymethyl groups and other one-carbon oxygenic substituents in distinct patterns. Tetrahedron 63:10657–10670
100. Muresan AZ, Lindsey JS (2008) Design and synthesis of water-soluble bioconjugatable trans-AB-porphyrins. Tetrahedron 64:11440–11448
101. Fujisawa K, Satake A, Hirota S, Kobuke Y (2008) Construction of giant porphyrin macrorings self-assembled from thiophenylene-linked bisporphyrins for light-harvesting antennae. Chem Eur J 14:10735–10744
102. Maes W, Ngo TH, Starukhin AS, Kruk MM, Dehaen W (2010) *meso*-Indolo[3,2-*b*] carbazolyl-substituted porphyrinoids: synthesis, characterization and effect of the number of indolocarbazole moieties on the photophysical properties. Eur J Org Chem 2576–2586
103. Suijkerbuijk BMJM, Gebbink JMK (2008) Merging porphyrins with organometallics: synthesis and applications. Angew Chem Int Ed 47:7396–7421

Top Heterocycl Chem (2014) 33: 231–292
DOI: 10.1007/7081_2013_110

Published online: 17 May 2013

Recent Advancements in Chiral Porphyrin Self-Assembly

Donato Monti

Abstract The supramolecular chirogenesis in porphyrin self-assembled architectures is a field of high importance, with wide application in technology, material chemistry and medicine. In this chapter the more recent aspect of this issue will be covered, with emphasis on the experimental protocols, on the properties of the suprastructures obtained, and on their applications.

Keywords Porphyrinoids · Supramolecular chemistry · Porphyrin aggregates · Chirality

Contents

D. Monti (✉)
Department of Sciences and Chemical Technologies, University of Tor Vergata, Rome 00133, Italy
e-mail: monti@stc.uniroma2.it

1 Introduction

The supramolecular chemistry of porphyrin derivatives is an important issue of these tetrapyrrolic macrocycles. The concept of Supramolecular Chemistry, i.e. "*The Chemistry Beyond the Molecules*", was developed by Jean-Marie Lehn, as the chemistry of molecular assemblies and the intermolecular bonds [1]. The architectures are held by non-covalent interactions, and the magnitude of the enthalpies involved lies in the range of about 0.3–130 kJ mol^{-1}, enabling the systems to self-correction and self-repairing [2–9]. Within this field, Supramolecular Chirality deals with asymmetric tools, at building block or at final complex systems level. It is based on the transfer of asymmetry information through non-covalent interactions, from a specific internal (chiral substituent) or external vector (template, solvents or physical forces) [10]. This is also known as "*Supramolecular Chirogenesis*" and it has broad application in multidisciplinary fields such as materials technology and nanotechnology, catalysis, medicine, and quite recently, in data storage and processing. Its paramount importance manifests itself in the highest level of life evolution, which is constituted by the DNA double helix.

Porphyrin derivatives play a fundamental part in this scenario, owing to their unique physicochemical and structural features. They are known in fact as the pigments of life, playing key roles in systems of fundamental importance as photosynthetic apparatus [11], heme proteins [12, 13], and Cytochrome family [14]. The possibility to undergo straightforward skeleton transformation and to coordinate in the core a huge variety of ions [15], make them versatile building blocks for the construction of systems of wide and interesting applications, from sensing [16], bio-mimetic catalysis [17], photovoltaic [18] and medicine [19].

This chapter will deal with the different protocols pursued for obtaining chiral porphyrin suprastructures, with some emphasis on the kinetic and thermodynamic aspects that control their formation.

2 The Origin of Chirality

Chirality (from the Greek term *kheir*, meaning "hand") is a concept that has been known in chemistry since the second half of nineteenth century. The term "Chiral" and "Chirality" was coined by Sir William Thompson (later Lord Kelvin), and appeared for the first time in a lecture delivered to the Oxford University Junior Scientific Club in 1893 [20] (for a critical report on stereochemical terminology see [21]) in which is stated ... "I call any geometrical figure, a group of points chiral, and say it has chirality, if its image in a plane mirror, ideally realised, cannot be brought to coincide with itself". The concept of "dissymmetry" (lately turned into "asymmetry") was formerly developed by Louis Pasteur in his works devoted to the resolution of racemic mixtures of crystals of tartaric acid salts. Pasteur recognised that the two stereoisomers (enantiomers) were "optically active", polarising light

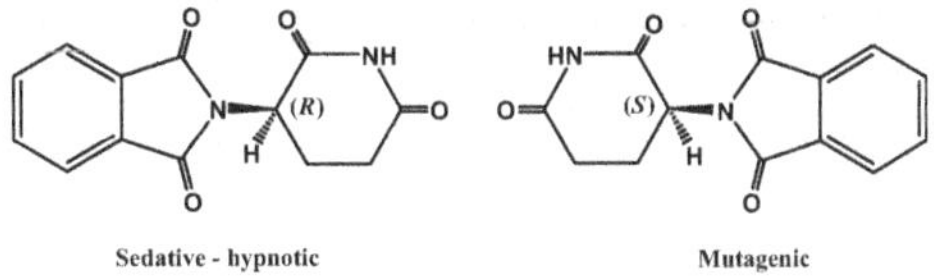

Fig. 1 Molecular structure of thalidomide enantiomers

differently (left vs. right), as a consequence of the asymmetric spatial grouping of the atoms [22].

First of all, we would not start a review on supramolecular chirality overlooking the striking importance of the role of the asymmetry in the Life. Just by citing Louis Pasteur sentence "Optical activity is the signature of Life", we can recognise the importance of this issue. Flavours and fragrances serve as typical examples of the physiological recognition of molecular asymmetry [23]. *R*(−)-carvone, for example, shows the odour of caraway, whereas its enantiomer *S*(+)-carvone is responsible of the odour of spearmint. *R*- and *S*-limonene (orange vs. lemon odour), *R*- and L-menthol, or *R*- and L-asparagine are just other examples of differences in organoleptic properties. The fact that the two enantiomers are perceived as smelling differently is proof that olfactory receptors must contain chiral groups, allowing them to respond differently (i.e. more strongly) to one enantiomer than to the other. However, it must be said that not all enantiomers have distinguishable odours by human nose.

Chirality represents the major issue, for example, in the development of pharmaceutical compounds. More than 50% of the drugs currently in use are in fact chiral compounds, although most of them are consumed as racemate. The effect of the presumed "not active" enantiomer is not always known, and this assumption may lead to deleterious effects.

A well-known example is constituted by thalidomide drug. Thalidomide is an anti-nausea and sedative drug that was introduced in the late 1950s to be used as a sleeping pill, and was quickly discovered to help pregnant women with the effects of morning sickness. It was sold from 1957 until 1962, when it was withdrawn after being found to be a teratogen, which caused many different forms of birth defects. Later studies showed that the noxious effects were due to the "in vivo" epimerisation of the *R* form into its *S* isomer (Fig. 1).

The negative effects of thalidomide led to the development of more structured drug regulations and controls over drug uses and developments [24].

The "emergence of chirality", or better, of the "biomolecular homochirality", has been taken as the "fingerprint of life" and its evolutionary origin. The basic building blocks of life, natural aminoacids and sugars, that are the monomers of the most important biopolymers (proteins and DNA or RNA), are (almost) exclusively found in their L or D configurations, respectively. As stated in the book of Klabunovsky: " … the use of handedness is an example of evolutionary deliberate simplification" [25]. By taking into consideration, for example, a racemic mixture of aminoacids (AA) used to build polypeptides in a "racemic world", the starting AA(1) has (almost) equal chances of being D- or L-isomer; after the addition of a second AA, the probability to have the homochiral sequence L-AA(1)-L-AA(2) are

- Theory
  - Chance
    - Mirror-Symmetry Breaking
    - ***Chiral Amplification***
    - Chiral Transmission
  - Deterministic
    - ***Mirror-Symmetry Breaking***
    - Chiral Amplification
    - Chiral Transmission

Fig. 2 Comparison between Chance (Stochastic) and Deterministic (fostered by an effector) theories of homochirality. The crucial steps for each scenario are reported in *bold*

$1/2^2$. The same holds for the D-sequence. For a sequence of n AAs, the probability of the homochiral chain formation is $1/2^n$. It is evident that, in a scenario of Life based on the molecular recognition and self-replication, this would rapidly turn out in the impossibility of life propagation and preservation.

Two theories on the generation of the chirality have been given, based on whether there has been, or not, a specific bias which triggered the mirror-symmetry breaking observed in the biomolecules. If the answer to the question is negative, we are facing theories based on "chance" mechanisms. Conversely, if the case of a positive answer, we are dealing with a cause–effect relationship, as a consequence of an earlier chiral influence, i.e. "deterministic" mechanisms and theories. This concept is visually summarised in Fig. 2.

In both theories, the initial Mirror-Symmetry Breaking is followed by two other consecutive steps, i.e. Chiral Amplification and Chiral Transmission. In the Deterministic scenario the crucial step is the first event, due to the effect of a "cause", whereas the hypotheses of random generation considers the amplification stage to have crucial importance.

A simple example would explain the differences of the two events. The irreversible generation of a racemic mixture (of strictly equal amounts of both of the enantiomers) from an achiral (i.e. pro-chiral) precursor, can be regarded as a binomial bernoullian distribution, in which one event do not influence the subsequent trial. This is closely related to the act of tossing a coin, with the "head" or "tail" results being of equal probability ($p = 1/2$). For N trials, with N sufficiently large, we can apply a statistical Gaussian probability distribution, which gives expectedly the most probable results of $N/2$. This indicates that the formation of a racemic mixture is the most probable state. But we also have to remind that the dispersion of the results, i.e. the "standard deviation" (σ) is a function of the number of trials ($\sigma = N^{1/2}/2$). For a "mol" of trials ($N = N_A$) we end up with $\sigma \approx 4 \times 10^{11}$. This means that the stochastic achievement of an exact racemic mixture become a rather unlike event as the number of trials N increase. Of course a one hundred of billion of molecule over an Avogadro number, as outlined in the example, represents a tiny imbalance, but it could be of high significance if a subsequent step of amplification is successfully on set.

Other examples are constituted by the spontaneous symmetry breaking in crystallisation, which is determined by a stochastic chiral nucleation effect [26–29]. The so formed enantiomorphic enriched crystals would have act as a template in the biochemical evolution of chiral molecules.

The Deterministic Theory is based on the effect of the influence of a "chirality bias" (chirality source) that imposed the parity violation on the evolving pre-biotic molecules.

The most important effect is the inherent dissymmetry of the Nature due to the asymmetry of the Weak Forces mediated by W^+, W^- and Z° bosons. This is known as The Parity Violation Principle [30], and it has been shown in the β-Decay of ^{60}Co, in which a daughter nucleus ^{60}Ni, with increased atomic number but with unchanged atomic mass, is formed. The process is mediated by W^- boson and is accompanied by emission of an electron and an antineutrino [31]. Neutrinos and antineutrinos are the simplest (archetypal) example of chirality, being particles featuring left-handed (−1) and right-handed (+1) helicity, respectively. Later the "weak" forces involved in the β-Decay have been unified with the electromagnetic one in the "electroweak theory" of the Standard Model [32]. This predicted the presence of a "weak neutral current" mediated by neutral Z° boson, which also generates the parity violating interactions. This can be extended to systems as atoms and molecules, suggesting that parity violation could be the origin of the molecular chirality [33]. In terms of exact energy content a true enantiomer of a molecule, i.e. the one with topological inverted configuration, should be constituted by antiparticles, therefore belonging to the realm of antimatter. Parity violation, in terms of optical activity, has been found in vapours of heavy metals, as the strength of the weak neutral current increases with the third power of the atomic number (Z^3) [34].

Quantitative ab initio calculations on aminoacids have been developed [35, 36] giving a very small energy preference of L- over D-aminoacids and D- over L-sugars. The energy difference lies in the range of 10^{-17} to 10^{-15} kJ mol^{-1}, corresponding to an enantiomeric excess ee ≈ 10^{-16} to 10^{-14} (i.e. 10^6–10^8 molecules mol^{-1}). The calculated effects are larger for heavier molecules such as (*R*)- and (*S*)-bromochlorofluoromethane (ca. 10^{-14} kJ mol^{-1}) [37] and complexes of osmium and rhenium (ca. 5×10^{-11} kJ mol^{-1} and 2×10^{-10} kJ mol^{-1}, respectively) [38, 39].

Another recognised example of chirality sources is the Circularly Polarised Light (CPL), which is the specific "physical effector" in the asymmetric photolysis (photodegradation) of a racemic mixture, and in asymmetric photosynthesis from an achiral starting material. The interaction of CPL with a mixture of enantiomers produces significant and reproducible ee of an enantiomer, with respect to its counterpart. A recent example of asymmetric photodegradation of a D,L-Leucine racemic mixture confirmed this hypothesis. When the racemate is photolysed by r-CPL ($\lambda = 182$ nm), an excess of *R*-Leucine resulted in the final reaction mixture, whereas by irradiating with l-CPL, the isomer with opposite handedness is found [40].

On the Earth, a small degree of circular polarisation is produced as a consequence of the atmospheric scattering. An opposite polarisation at dawn and at dusk is caused by the Earth's rotation. A kinetic effect on the photodegradation, due to the different temperature of the sea surfaces at day and at night, would have been caused a stereochemical imbalance of the chiral products [41, 42]. Indeed, this mechanism is thought to be operative also in the extra terrestrial domain, in which the source of CPL is the scattering of the stellar light by dust grains [43, 44]. Moreover, the cold and dry condition of the interstellar space prevents the racemisation of the chiral material.

The Murchison meteorite (Murchison, Australia, 1969) is one of the most famous and studied collectors of chiral molecules from the outer space. This is a carbonaceous chondrite, dating from around 4,500 million years ago, and it is constituted by 10–20% of organic compounds, such as proteinogenic aminoacids, nitrogenated heterocycles (purines, pyrimidines, adenine and uracil), sugar derivatives and other significant substances [45–47]. Analyses showed a 7–9% prevalence of L-form for some aminoacids. The unnatural ^{15}N enrichment in the aminoacids confirms the extraterrestrial source of the species. L-Enantiomeric enrichment has been lately confirmed for other carbonaceous chondrites [48–50].

Another effective chiral input that is thought to be involved in the selection of the homochirality is the vortex flow in a liquid, as the ones generated in primordial sea by Coriolis forces. Of course, the total effects would cancel out if we consider the opposite chirality induced in the boreal and austral hemispheres, but this controversial point can be resolved by assuming that prebiotic molecules were formed in local spots in one hemisphere, due to the distribution of the known earliest supercontinent Vaalbara, between 3,300 and 3,600 million years ago [51].

This effect was studied at laboratory level by Ribò and coworkers, in the case of chiral homoassociation of porphyrin derivatives. The direction of the stirring of the solution (i.e. counter clockwise vs. clockwise motion) selectively drove the symmetry breaking in the formation of chiral porphyrin aggregates, in a predictable manner, as shown by CD spectroscopy [52].

Several theoretical and experimental models have been proposed for the amplification of a starting small asymmetric imbalance. Autocatalytic models, in which the formation of an enantiomer catalyses the production of itself, inhibiting the formation of the opposite enantiomer have been developed initially by Frank [53], and lately modified by Kondepudi and Prigogine, taking into account the "Bifurcation Theory" [54].

In the Frank model, an open flux reactor is fed with achiral reagents (A, B) that react to give chiral products L and D. These products catalyse their own formation (i.e. autocatalytic process) with equal preference and equal rate, due to the symmetry of the system. At this point the two enantiomers react with each counterpart, and are converted into an inactive product P, which is in turn pumped out from the reactor, maintaining a stationary state. This final step is necessary for the onset of chiral amplification. If the concentration of the reactants remains below a critical point, the system is purely racemic. Conversely, if the input A and B is increased over a critical point, the racemic state is unstable and spontaneously evolves into one of the possible chiral states (L or D enrichment) breaking the racemic symmetry.

The first experimental proof of amplification of initial imbalance was given by Kenso Soai in its seminal work [55]. This is an autocatalytic process of amplification by a tiny initial chiral imbalance of the product (i.e. a pyrimidyl alkanol) in the reaction of a pyrimidine aldehyde with diisopropylzinc (Fig. 3).

In this system, the chiral initiator in low enantiomeric excess (*R* or *S*) determines the final absolute configuration of the product with overwhelming enantiomeric excess (*R* or *S*), without the need of a chiral auxiliary. The study was further extended to a variety of pyrimidyl alkanols obtaining ees close to enantiomeric purity [56]. Studies devoted to the elucidation of the Soai's mechanism have been recently carried out, based on the Frank model and its further developments [57–59].

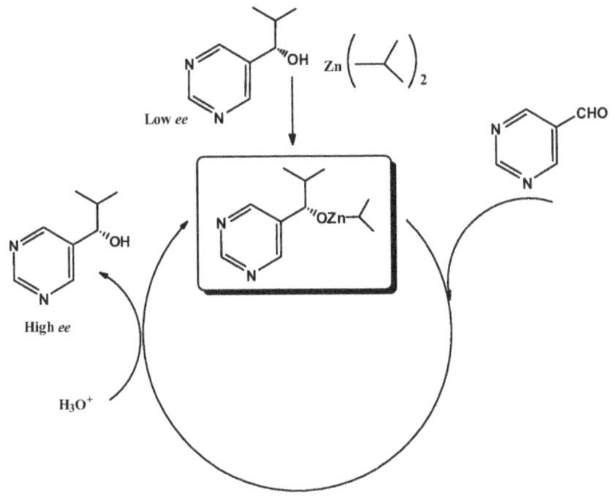

Fig. 3 Reaction scheme of the proposed autocatalytic path: a small enantiomeric excess of the product (with *S*-configuration, in this case) induces an enhancement of the ee of the overall reaction (adapted from [55])

Further studies, on the effect of different chiral initiators, either organic or inorganic systems, have been carried out, showing very interesting results. Chiral inductors such as solid sodium chlorates, *d*-(+) or *l*-(−)-quartz, or even CPL, gave stereoselectivity of the products, above 95% ee [60, 61].

3 Formation of Chiral Porphyrin Assemblies

The achievement of chiral porphyrin suprastructures, either chiral or achiral, can be achieved by different means, all of them relying on the combination of non-covalent interactions. This is usually accomplished by a number of well-established procedures such as (1) template aggregation of porphyrin building blocks on chiral polymeric matrices, either natural or non-natural; (2) metal coordination by chiral ligands; (3) spontaneous symmetry breaking induced by stirring or (4) by Langmuir-Blodgett or Langmuir-Schäfer film compression; (5) by interaction with chiral surfactants; (6) chiral aggregation steered by the presence of a chiral functionality present in the frame of the tetrapyrrolic macrocycle. A brief survey on the effect of a chiral media is also presented. The aim of this review is to give a detailed critical report on the recent achievements in the different aspects of this issue.

3.1 Interaction of Porphyrin Derivatives with Chiral Surfactants

The transfer of chirality form a molecular to a higher complexity level (i.e. supramolecular) is a fundamental issue for the understanding of life molecular evolution. The hierarchical transfer of (chiral) information stored at molecular

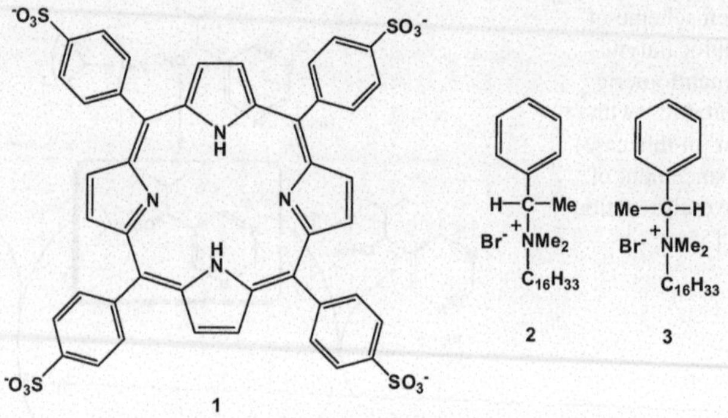

Fig. 4 Molecular structure of porphyrin **1** and chiral surfactants **2** and **3**

level is one of the most active areas of researches in porphyrin (supramolecular) chemistry. Besides this, the studies of the interaction of porphyrin derivatives with biological membrane mimics, such as micelle, liposome bilayers or vesicles, are of importance for their potential applications in Photodynamic Therapy (PDT) of tumours and related diseases. Reports on this topic have been offered by the groups of Ribò and Mancini, who jointly exploited the interaction between an anionic, water soluble porphyrin derivative **1**, with chiral cationic surfactants **2** and **3** (Fig. 4), with the aim to give insights on the mechanism of transfer of molecular information from molecular level (stored in the surfactant units) to larger nanoscale of polymolecular porphyrin assemblies [62].

Porphyrin derivative **1**, the sodium salt of 5,10,15,20-tetrakis(4-sulfonatophenyl) porphyrin, is a well-known building block largely used in the studies of chirally induced homoassociation [63]. This work pointed out other interesting features of the systems, which are dependent on several factors, such as the concentration of the units, their molar ratios and the mixing order followed for the preparation of the solutions. Noteworthy, above the critical micellar concentration of the surfactants (cmc), the macrocycle is included in the surfactant aggregates in monomeric form, as revealed by UV–visible and Fluorescence spectroscopies. However, Circular Dichroism spectra of the formed systems were silent, indicating the absence of transfer of chirality in this condition. Experiments carried out below the cmc, at [Surf]/[**1**] = 4:1 and 6:1 ratios ([**1**] = 10 μM), resulted in an induced CD effect (Fig. 5).

Broadening and red shift of the UV–visible features indicate the formation of "pre-micellar" porphyrin heteroaggregates. This finding is confirmed by both fluorescence and Resonance Light Scattering (RLS) spectroscopies, which showed a decrease of the typical emission intensity, and the appearance of a strong scattering signal on the porphyrin absorption region, respectively. The heteroaggregates, constituted by an extended array of macrocycles in a long-range electronic coupling, are held by a combination of electrostatic and hydrophobic interactions.

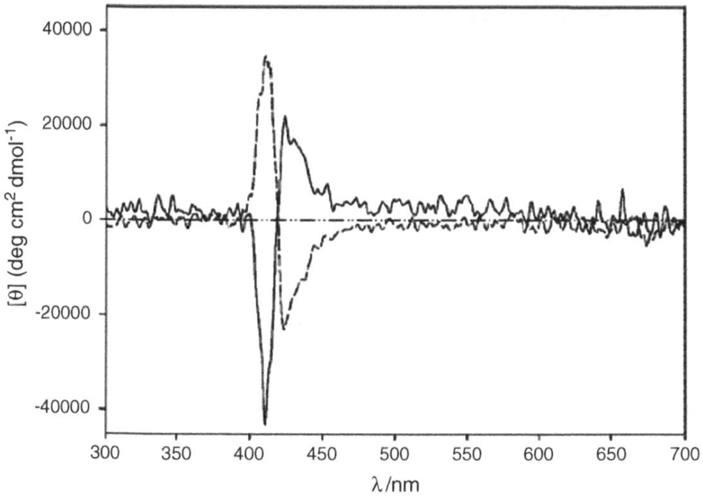

Fig. 5 CD spectrum of **1** (12.5 μM) at [surfactant]/[**1**] ratio of 4:1 in the presence of surfactant **2** (*solid line*) and **3** (*dashed line*) (reproduced with permission from [62])

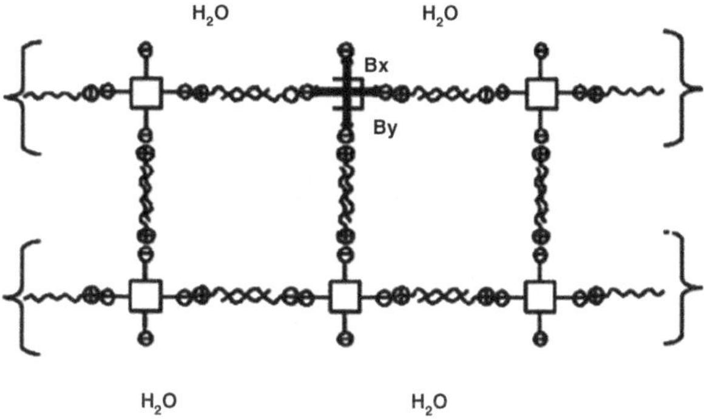

Fig. 6 A possible structure of the hetero-assemblies **1**-chiral surfactants (reproduced with permission from [62])

Very importantly, the sign of the spectral features are opposed for the two enantiomeric surfactants **2** and **3**, indicating a specific induced symmetry breaking, exerted by the chiral micelle head groups. It has been proposed by the authors that the surfactants strongly interact with the oppositely charged anionic groups present on the periphery of the tetrapyrrolic macrocycles, favouring an asymmetric out of plane tilting of the *meso*-phenyl groups. A Job-plot experiment reveals the formation of a heteroaggregates with a surfactant/porphyrin stoichiometry between 3:1 and 4:1, with a 2D and 3D orthogonal structure, respectively. The 2D arrangement is represented in Fig. 6.

Fig. 7 Molecular structure of porphyrin **1-SA-Phe**

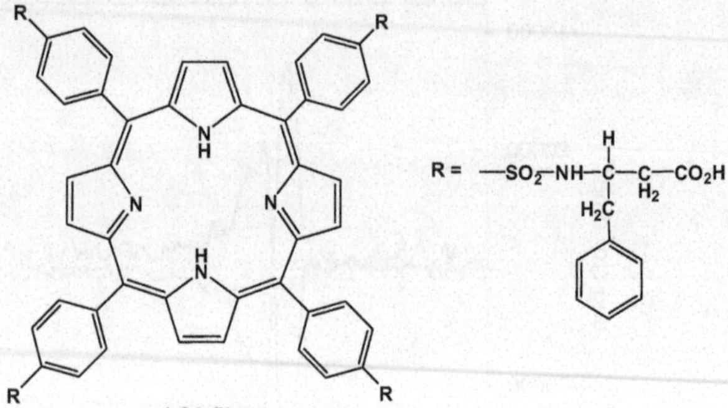

The complex network would explain the remarkable stability of the proposed assemblies toward inclusion in micellar phase, as revealed by further addressed experiments carried out in the presence of an excess of surfactants. The studies with related surfactant **3** showed similar results, in terms of the formation of the heteroaggregates species and chiral induced effect. However, the formation of the supramolecular structures occurs more easily, and this implies the occurrence of an effect of the intimate nature of the surfactant in the organisation of the heteroaggregates.

The diacid form of the former porphyrin **1** has been employed by the same authors as a "chirality probe" for the detection of traces of biological chiral contaminant of ultrapure water [64]. Trace levels of contaminants present in the bulk water used in the preparation of the solutions have been thought to be responsible, in many cases, for the spontaneous symmetry breaking observed in the case of formation of water-soluble achiral porphyrin homoassociates. These chiral polarisers are likely constituted charged phospholipids or lipid-protein fractions that act as chiral nucleation centres, as revealed by Atomic Force Microscopy (AFM). Competitive experiments by means of chiral surfactants (**2** and **3**, former work) below their cmc, revealed the soundness of this hypothesis. In the case of an excess of surfactant **2** (*R*-isomer) the CD spectra showed same "positive" spectral pattern of that obtained in "pure" water, with increased intensities. In the case of the addition of **3** (*S*-isomer) the appearance on the CD features with opposite negative signs occurred only at a concentration above a certain threshold. The threshold values depend on the quality of the water samples and on their ageing, determining the level of chiral biological debris. These results, beyond their high analytical importance, would give insights into the role of chiral effectors in the emergence of life.

Analogous works have been carried out by Andrade and coworkers, who studied the aggregation behaviour of a sulfonamide-derivative of **1**, bearing appended phenylalanine residues on the four *meso*-positions of the molecular frame (**1-SA-Phe**), either in the absence or in the presence of 1,2-dimirystoyl-*sn*-glycero-3-phosphocholine (**DMPC**) vesicles (Fig. 7) [65]. The studies have been carried out in water at different pH, by means of several spectroscopic techniques. The results

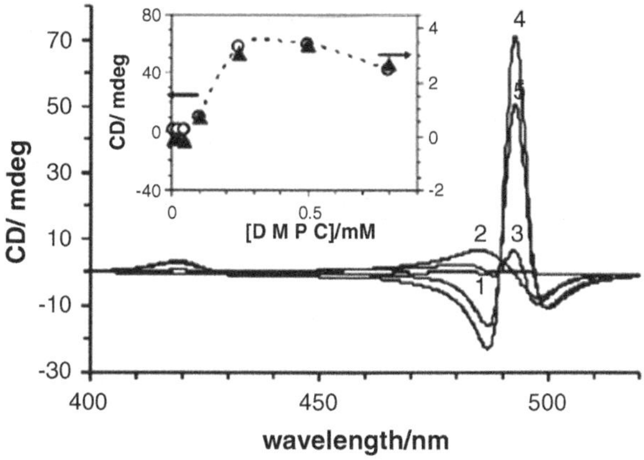

Fig. 8 Induced CD spectra of **1** (pH = 2.5) in the presence of increasing [**DMPC**], from 0 to 0.8 mM (1–4 traces). In the *inset* is reported the intensity of the peaks at 420 nm (*solid triangle*) and 490 nm (*open circle*) as a function of **DMPC** concentration (adapted from [65])

have been compared to those obtained in the case of the known porphyrin derivative **1**, to investigate the role of the different molecular structure on both the aggregation and inclusion features of the macrocycles in the biomembrane model.

The bulk pH of the solution strongly affects the aggregation state. At low pH, the derivative **1-SA-Phe** is mainly in aggregated form. The hypsochromic shift of the Soret band, by ca. 10 nm, indicates H-type face to face stacking. At intermediate pH ranges (2–4) scarce solubility of the porphyrin prevents accurate studies. The aggregation is disfavoured on increasing the temperature. Steady state and time-resolved fluorescence studies indicate the coexistence of aggregated species of different type. In the alkaline end (i.e. pH from 5 to 9), deprotonation of the carboxylic peripheral moieties occurs, increasing the solubility. Also in this case, the presence of different aggregates has been detected. Also in this pH range the increase of the temperature disfavours the aggregate state, with the formation of H-type porphyrin dimers.

The addition of **DMPC** vesicles promotes the inclusion in prevalent monomeric forms, by destabilisation of the aggregate species by compelling hydrophobic effect. This process has been found to be quite independent of the pH of the solution (from 3 to 7), with corresponding thermodynamic binding constant values being very similar, of ca. $3 \times 10^4\ M^{-1}$.

A completely different frame is found for the derivative **1**, which does not show any tendency to incorporate in **DMPC** vesicles in a wide range of pH values. The results indicate in this latter case the prevalence of aggregated J-type specie, presumably located at the vesicle surface, as indicated by the rise of the intense band at 490 nm. Interestingly, the interaction of **1** with **DMPC** surface causes the formation of structures featuring supramolecular chirality, as indicated by CD spectroscopy. The spectroscopic pattern (Fig. 8) indicates the formation of fibre-like structures.

This finding is different from that obtained in the case of interaction of the same substrate **1** with reverse micelles of surfactant sodium bis(2-ethylhexyl) sulfosuccinate (AOT-RM) studied in their previous work [66].

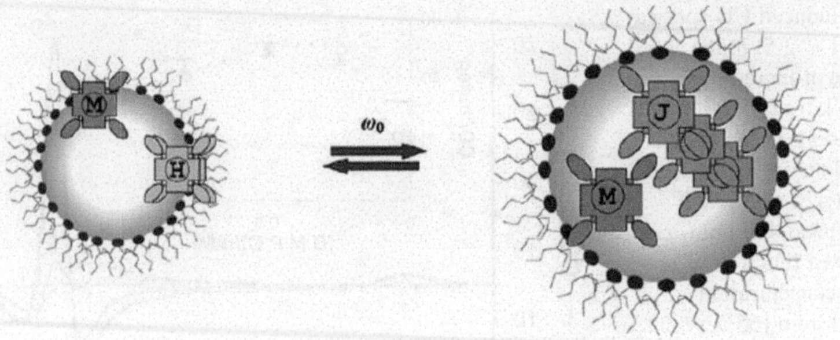

Fig. 9 Schematic drawing of the effect of the water uptake and aggregation state of porphyrin derivative **1** inside AOT RM in the absence of protein (reproduced with permission from [66])

The study has been carried out in ternary water/AOT/organic solvent mixture, at different water/AOT ratios (ω_0) and different pH. The effect of the presence of protein "drug-carriers" as human serum albumin (HSA) and β-lactoglobulin (βLG) has been investigated.

In the absence of protein the derivative **1** is in the form of H-type aggregate at low ω_0 values, at neutral or acidic pH. At higher ω_0 ratio (i.e. on increasing RM radii), the prevalent form of the macrocycle at acidic pH is the diacid monomeric and the J-aggregate forms, that are included in the inner water pool. The aggregation number of the supramolecular species depends on the size of RM, i.e. proportional to the amount of water (see Fig. 9). Calculations by spectroscopic means indicate an aggregation number of about 7–17 units. The addition of proteins to a porphyrin–AOT system was studied at different pH. At either pH $= 2$ or 7, the presence of protein (HSA or βLG) promotes the formation of J-aggregate at the expenses of H-counterparts for different ω_0 values. The increase of protein concentration up to 0.05 μM leads to deaggregation due to the quite strong protein–porphyrin binding ($K_B \approx 10^6$ M^{-1}).

The inclusion of **1** in AOT-RM resulted in an induced CD in acidic conditions, at $\omega_0 > 8$, that increase by rising the concentration of water. The presence of two positive coupled bands at 490 and 420 nm indicates a clockwise porphyrin arrangement, with the low energy transition being five times more intense than that in the blue spectral region. The presence of protein ($\omega_0 = 30$) resulted in a change in the CD spectra, with the same spectral pattern but with opposite sign, indicating a left-handed arrangement of the macrocycles. The effect depends also on the nature of the added drug-carrier, being somewhat higher for βLG. On increasing the protein concentration, a decrease of the CD intensities has been observed (Fig. 10). Comparative studies carried out with the water soluble, positively charged, *meso*-tetrakis (*N*-methyl-pyridinium-yl)porphyrin **4** showed the effect of the structure of the macrocycles on their interaction properties. This latter derivative does not show any tendencies to aggregate in the presence of proteins or to encapsulate in AOT-RM, probably due to repulsive coulombic interaction. Ion pair formation of cationic macrocycles with AOT could be detected only at premicellar surfactant concentration, i.e. below the cmc.

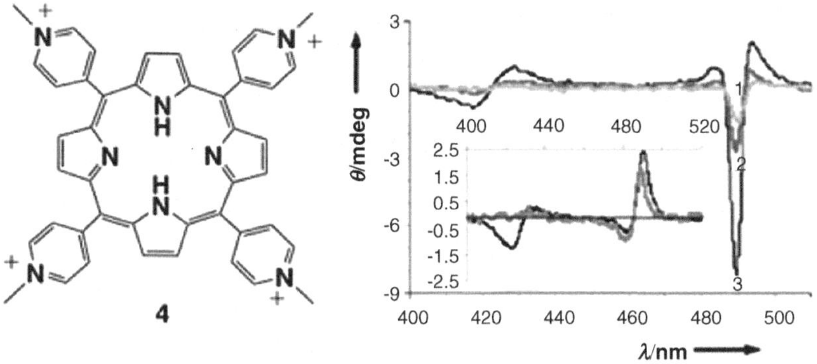

Fig. 10 Molecular structure of porphyrin **4**, and induced CD spectra of **1** in the absence of protein (1), and in the presence of HSA (2), and βLG (3) 0.10 μM, in AOT RM (pH = 2; ω_0 = 20) (adapted from [66])

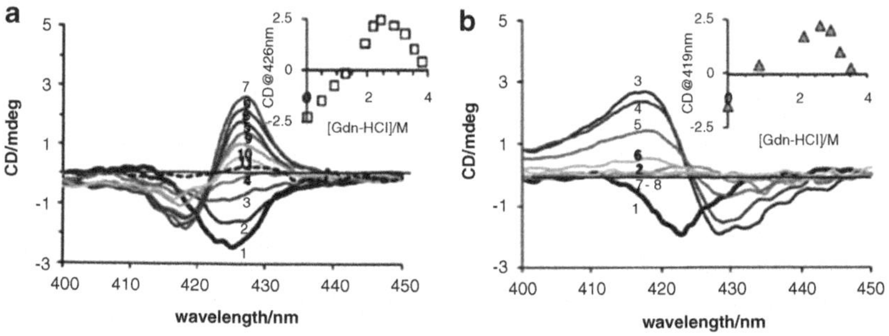

Fig. 11 Induced CD spectra of **1**/HAS (1:1 ratio) at pH = 7 (**a**) and pH = 2.5 (**b**) in the presence of different Gdn-HCl concentrations. *Insets*: corresponding dependence of ICD intensities vs. [Gdn-HCl] (reproduced with permission from [67])

Porphyrin **1** has been recently employed by same authors as a reporter for the HSA conformational changes upon denaturation promoted by guanidine hydrochloride (Gdm-HCl), at acidic pH [67]. The stability of HSA toward the denaturation effect is increased by interaction with **1**.

The interaction results in an induced CD, whose features depend on the pH of the solution and on the concentration of the denaturant species, as reported in Fig. 11. At neutral pH a blue shifted dichroic band appears (H-aggregates) which, upon addition of Gdn-HCl, evolve into bisignated positive spectra. This has been interpreted on the basis of the unfolding of the HSA. The same holds for the denaturation process at lower pH (2.5 units). In this case, opposite signs of the CD bands are observed. This has been tentatively explained by different local chirality of the binding site of the protein, although the effect of different mutual orientation of the macrocycles should also be taken into consideration. It is interesting to note the different pattern observed, with respect to that obtained in the case of AOT system [66]. In this present case the conservative bisignated dichroic bands would imply the formation of regularly arranged porphyrin subunits.

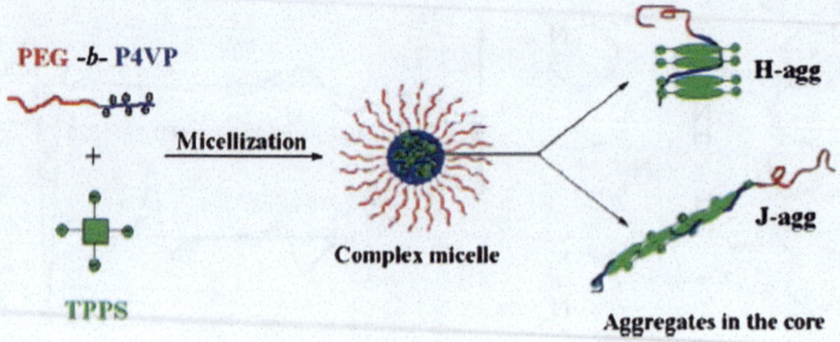

Fig. 12 Schematic drawing of **1**-induced micellisation of **PEG114-*b*-P4VP61**, with the formation of H- and J-aggregates (reproduced with permission from [68])

Another interesting series of work concerning the interaction of **1** and its diacid form **4** with surfactant systems have been published by Shi and coworkers [68]. The formation of aggregates of **1** was investigated in the presence of poly(ethylene glycol)-*block*-poly(4-vinylpyridine) (**PEG114-*b*-P4VP61**), in acidic condition. The resultant micelles have been characterised by spectroscopic means, as well as dynamic and static light scattering. The protonated derivative **1** is electrostatically embedded in the cationic **P4VP** core, which is protonated in the bulk aqueous media (pK_a ca. 4.5). The PEG units constitute the soluble shell of the micelle. The pH of the solution strongly influences the aggregation fashion of the included macrocycles, being of H-type at pH 3.0–4.0, and J-type in the range 1.5–2.5 (Fig. 12). These morphologies can be interchanged by tuning the pH of the solution, as a consequence of the protonation state of the inner nitrogen atoms of the porphyrin. The kinetic of the interconversion states depends on the fashion by which the change of pH is operated.

Noteworthy, the micellised J-aggregates show supramolecular chirality, as indicated by the presence of distinctive CD spectral features. Indicative results have been obtained at lower pH. The spectra show a peculiar bisignated positive pattern with weak bands at 490 and 420 nm. This finding has been interpreted by the authors in terms of the formation of both H- and J-chiral aggregates, although it is stated that the H-aggregates are CD silent. More probably, the spectral pattern is due to the formation of J-structures with complex overall morphology, as rod-like or related forms. The formation of chiral species would indicate the induction by some effectors, such as a small enantiomeric excess of the surfactants, or more likely, by the presence of some contaminants in the water used for the solutions [64]. However, quite interesting issue, the overall dissymmetry of the system can be effectively tuned by selecting the direction of the stirring of the solution during the formation of the aggregates. In particular, a clockwise stirring results in a positive Cotton effect, whereas the anticlockwise direction steers the appearance of negative CD features (Fig. 13). Notably, this important issue for the formation of chiral porphyrin aggregates will be described in deeper details in a subsequent section.

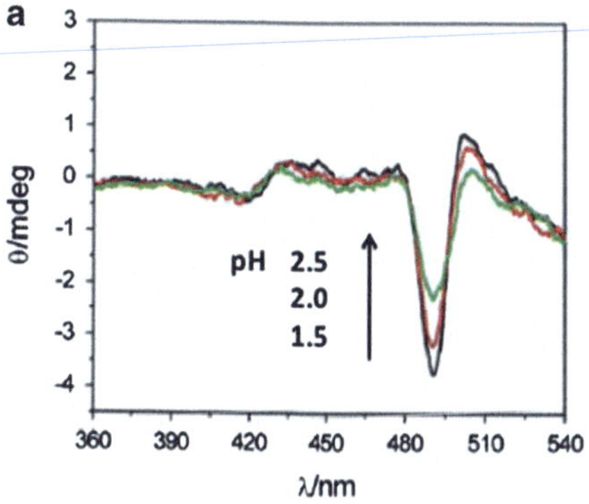

Fig. 13 CD spectra for **PEG_{114}-*b*-$P4VP_{61}$/1** micelles at different pH of the bulk solution (**a**), and a pH = 1.5 (**b**) under CW or ACW stirring (adapted from [68])

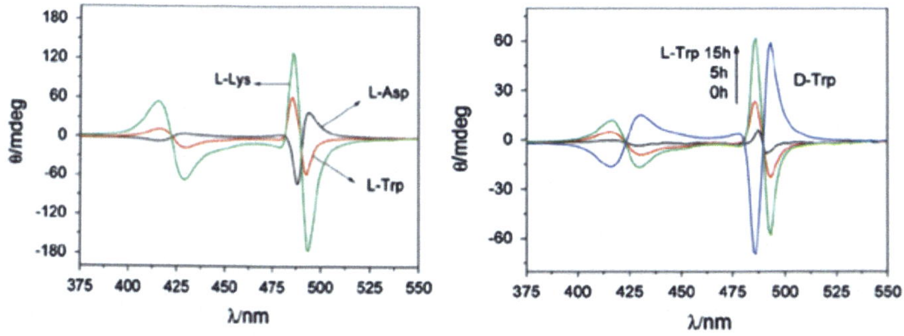

Fig. 14 (*Left*) CD spectra of **PEG_{114}-*b*-$P4VP_{61}$/1** system in the presence of various aminoacids. (*Right*) CD spectra of **PEG_{114}-*b*-$P4VP_{61}$/1** system in the presence of L- or D-Trp, at different times (adapted from [69])

This study has been extended to biomembrane models made by the former **PEG114-*b*-P4VP61** block polymer, in acidic conditions (pH = 2) with aminoacids such as aspartic acid (**Asp**), tryptophan (**Trp**) and lysine (**Lys**) that act as chiral inductors [69]. These species influence the aggregation of **1**. In particular, the protonated forms of the more basic aminoacids (**Trp** and **Lys**) promote the aggregation, whereas Asp acts as an inhibitor, due to the presence of residual negatively charged carboxylates. The extent of aggregation depends on the time of storage of the solutions (0–15 h), indicating the occurrence of kinetic effects. This finding is confirmed by light scattering studies, which show that the presence of the former two aminoacids increases the sizes of the J-type species, whereas in the case of **Asp** only aggregates of lower sizes are detected. The further addition of block-copolymer promotes and further stabilises the supramolecular architectures. An important issue is that the properties of the aminoacids determine the supramolecular chirality of the micelle embedded porphyrin J-aggregates (Fig. 14). The signs of the CD features depend on the type of the aminoacids, being negative in the case of

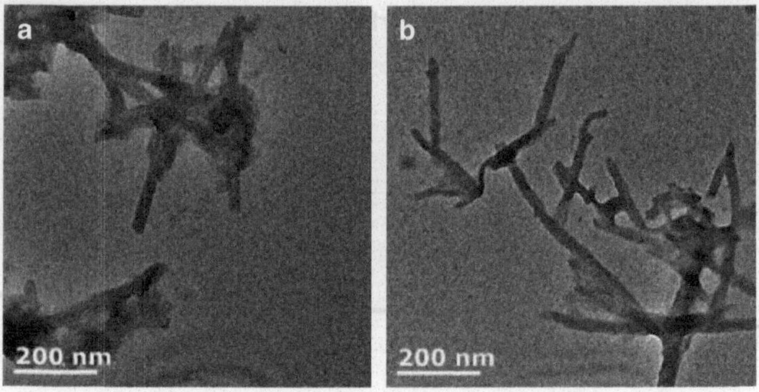

Fig. 15 TEM images of the micellar systems at DMA/**1** = 1, in the presence of **Trp** prepared by (**a**) protocol I; (**b**) protocol II (reproduced with permission from [70])

the more basic ones and positive for **Asp**. The concentration of aminoacids, and that of **1**, influences the overall chirality of the supramolecular species. Noteworthy, as shown in the case of L- or D-**Trp**, the configuration of the stereogenic centre determines the sign of the CD bands. Moreover, in line with the effect of the aggregate growth, the intensities of the bands increase with the ageing of the solutions, before being quenched by block-copolymer indicating the occurrence of a "scaling effect" of the asymmetry of the system.

The effect of the nature of the polymer was studied in a subsequent paper [70]. In this work micellar systems have been assembled with **1**, **Trp** and poly(ethylene glycol)-block-poly(2-(dimethylamino)ethyl methylacrylate (**PEG-*b*-PDMAEMA**), in aqueous solution at pH = 1.8. It has been proposed that the softer dimethylaminoside moieties would offer a more flexible backbone, beneficial for the chiral template effect exerted by **Trp**.

The protocol of preparation strongly influences the overall morphology of the supra-assembled systems; the addition of **1** to **PEG-*b*-PDMAEMA/Trp** solution (sequence I: porphyrin last) caused the formation of both small globular and rod structures (40–80 nm), whereas the addition of **PEG-*b*-PDMAEMA** to a solution of **1** and **Trp** (sequence II: polymer last) prevalently generates rod-like entities of larger size (90–150 nm), as showed by Transmission Electron Microscopy (TEM, Fig. 15).

However, the final extent of aggregation does not depend on the mixing procedures. Interestingly, the aggregates obtained in the presence of **Trp** show quite intense coupled CD signals with the typical bands centred at ca. 420 and 490 nm, indicating the formation of chiral structures (Fig. 16, left). Importantly, the sign of the low energy features and the intensities depends on either the stereochemistry of the chiral inductor or the mixing protocol followed. For example, for sequence I (porphyrin last), the sign of the J-type band is positive in the case of L-**Trp**, and negative in the case of D-**Trp** (Chirality I).

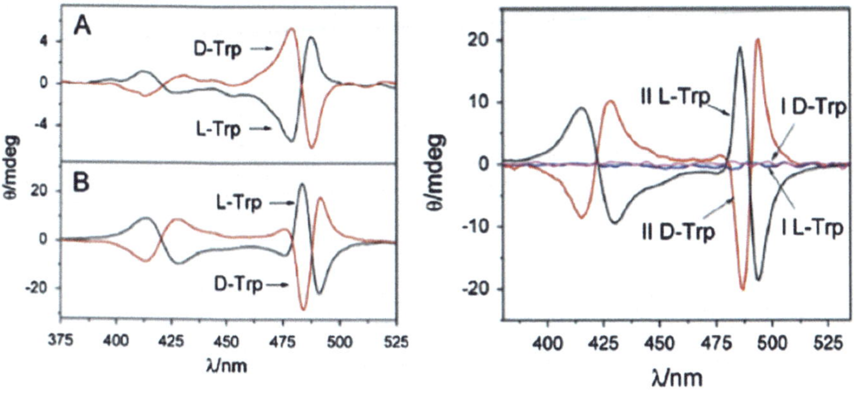

Fig. 16 (*Left plot*) CD spectra for **PEG-*b*-PDMAEMA/1** complex micelles prepared by protocol I (**a**) and II (**b**), in the presence of different stereoisomers of **Trp**. (*Right plot*) CD spectra for **PEG-*b*-P4VP/1** in the presence of **Trp** stereoisomers prepared by protocols I or II (adapted from [70])

Reversed signs, with higher (about fivefold) intensities, are found in the case of mixing procedure II (polymer last; Chirality II). The aggregates are thermodynamically stable, being the CD features unchanged by rising the temperature up to 70°C.

Noteworthy, the former examined polymer **PEG-*b*-P4VP**, shows chirality effects only by protocol II, as a likely consequence of the higher rigidity of the backbone. The differences in asymmetry of the aggregates have been interpreted on the basis of different hierarchical mode of interaction of anionic **1** with the charged cationic moieties of **PDMAEMA** chains. This hierarchical effect would be implied in the different size and morphology of the resulting species.

The signs of the CD features (i.e. the overall dissymmetry of the systems) are related also to the polymer/porphyrin ratios. A decrease of **DMA/1** (e.g. from 10 to 1) causes a reversal of the Chirality I in favour of Chirality II type. However, reversal of Chirality II, obtained by protocol II, could not be obtained by simply raising the **DMA/1** ratio. This "erase and rewrite" process could be effectively achieved by a concomitant change in the pH of the solution, from 1.8 to 5.5 and then back to 1.8. The increase of pH causes the deprotonation of **1**, and its disassembly into monomeric or oligomeric H-type achiral structures.

Finally, the chiral aggregated species shows, as stated, a "chirality memory" effect, upon addition of L- or D-**Trp** to the systems formed in the presence of D- or L-**Trp** counterparts, respectively. This would be called in a more proper way, a "chiral inertness" effect, as the reversal of chirality can be achieved in the presence of a large excess of specific **Trp**. However, this effect depends once again, on the mode of preparation, being achieved in the case of Chirality I only.

The chirality memory effect was further demonstrated by removing the **Trp** template by dialysis method. After the removal of the amino acid, the initial chirality can be switched off and on by the above mentioned pH cycles (1.8 to 5.5 to 1.8), with good reversibility. This effect has been explained by the persistence of chiral seeds at pH 5.5. However, prolonged standing at this condition for 1.5 h caused the complete destruction of these nuclei, causing the extinguishment of the effect.

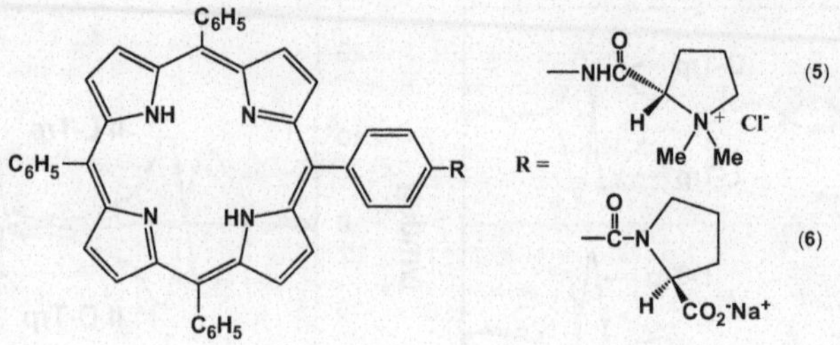

Fig. 17 Molecular structures of chirally functionalised porphyrins **5** and **6**

3.2 *Supramolecular Chirality Driven by Covalently Linked Chiral Substituents*

The effect of the presence of chiral functional groups on the molecular frame of porphyrin macrocycles on the overall dissymmetry of their aggregated species is an important field of studies. The understanding of these effects is undoubtedly of importance for the construction of functional self-assembled material at nano- and mesoscopic scale.

In the papers of Monti and coworkers the effect of the presence of charged proline derivatives as chiral molecular information, linked on the macrocycle frame has been examined (Fig. 17). These groups infer also amphiphilic properties to whole structure allowing for the study of the aggregation in aqueous solvent mixtures. The change of the polarity of the medium, controlled by the composition of the solvent (good vs. poor solvent ratio), drives the aggregation of the porphyrin platforms [71]. The red shift of the Soret band indicates the formation of J-like supramolecular architectures. Spectroscopic and structural (AFM) studies showed that the intimate nature of the appended chiral group, i.e. cationic (**5**) vs. anionic charge (**6**), has strong effect on the mechanism of the aggregation and on the morphology of the aggregates [72]. The self-aggregation of the cationic derivative **5** in EtOH/H_2O (25/75 v:v), for example, occurs by a autocatalytic kinetic [73] to give highly chiral assemblies with negative coupled CD bands. AFM topographies (on Highly Oriented Pyrolytic Graphite (HOPG)) showed the formation of fibril structures of tens of micrometres in length (Fig. 18). Conversely, the negatively charged counterpart shows faster diffusion-limited aggregation kinetics, with the formation of tighter globular structures.

Interestingly, the template aggregation of **5** onto preformed aggregates of **6** results in a strong amplification of chirality, with respect to the homoaggregation, likely arising from a strong electronic coupling among the macrocycles due to electrostatic effect exerted by the oppositely charged chiral functionalities [74].

Along this line, the authors extended the studies to the effect exerted by other chiral functional groups, such as sugars and steroids. Interesting results were reported in the case of the solvent-driven aggregation (EtOH/H_2O) of two derivatives bearing in 5,15-*meso* position two sugar residues, namely protected D-galacto (**7**) and D-gluco (**8**) pyranosides (Fig. 19) [75, 76].

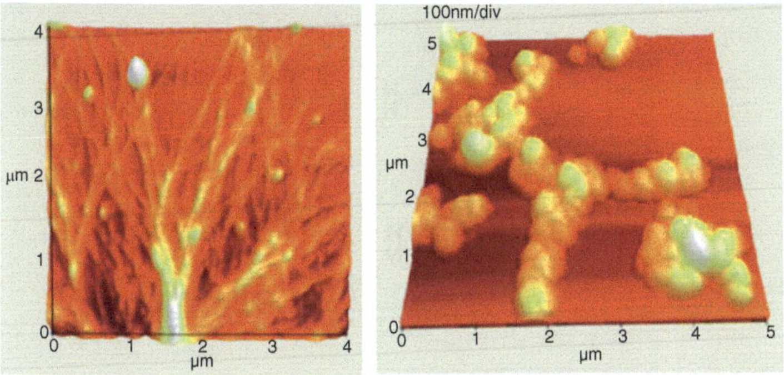

Fig. 18 AFM images for aggregates of **5** (*left*) and **6** (*right*) onto HOPG

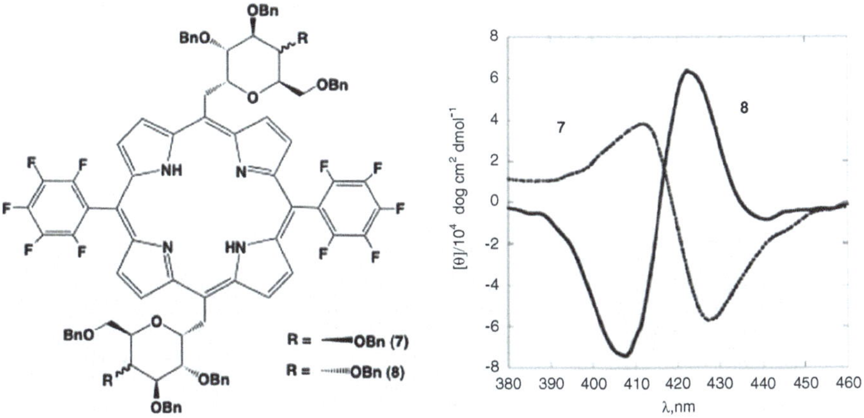

Fig. 19 (*Left*) Molecular structures of bis-D-galactoside derivative **7**, and bis-D-glucoside derivative **8**. (*Right*) CD spectra of the aggregate solution of **7** and **8** in EtOH/H_2O

Both of the species aggregate in mixed aqueous solvents by diffusion-limited kinetics, to give J-type assemblies featuring mirrored CD coupled spectra, due to the effect of stereochemical induction by the appended epimers.

The implementation of steroid moieties into the chiral side groups was studied (**9**, Fig. 20) [77].

The aggregation studies of **9**, carried out in dimethylacetammide (DMA)/water showed also in this case the formation of chiral J-type suprastructures. The concentration of the porphyrin influences the mechanism of aggregation, from autocatalytic to diffusion-limited (DLA) kinetics. This results in a modulation of the morphology of the final assemblies. AFM studies indicate that in the case of low concentration regime (0.8 μM), by autocatalytic path, long fibril structures of tens of micrometres in length are formed, whereas at higher regime (2.4 μM), by DLA decay, shorter rods of coalesced small unspecific globular shapes resulted (Fig. 21).

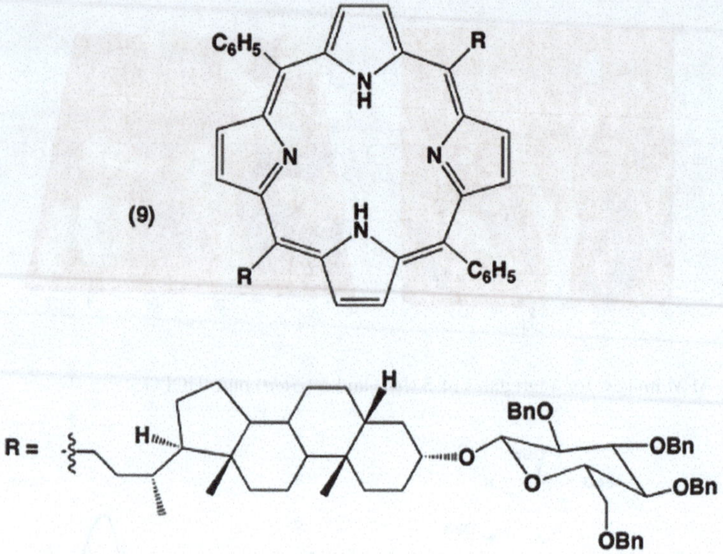

Fig. 20 Molecular structure of steroid appended porphyrin **9**

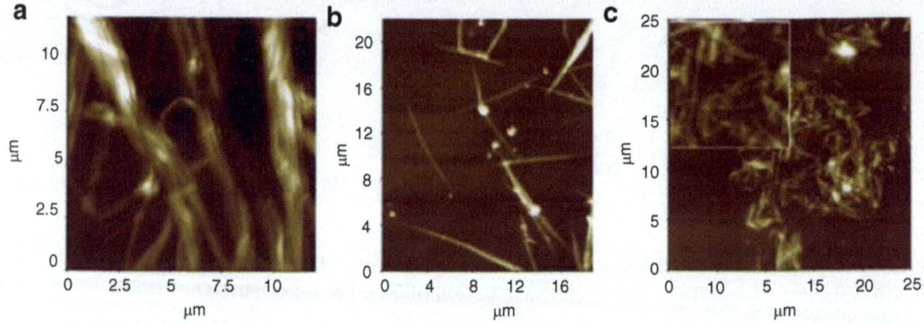

Fig. 21 AFM topographies on HOPG of aggregates of derivative **9** obtained at different concentrations in DMA/H_2O. (**a**) 0.8 μM; (**b**) 1.6 μM; (**c**) 2.4 μM

Furthermore, CD spectra revealed that the chirality of the final architectures are strictly related with the morphology of the structures, being higher in the case of the species obtained at low concentration. In all cases, negative coupled bands are featured, indicating an ACW disposition of the platforms in a left-handed helicity. Semiempirical calculations indicated that the driving force for the assembly is mainly due to Van der Waals interaction among the steroid substituents, with a preference for the left-handed arrangements of about 37 kJ mol^{-1}. The corresponding tetra-substituted porphyrin derivative, despite the larger number of chiral centres, showed reduced supramolecular chirality, as a probable consequence of increased steric hindrance of the molecular frame. These derivatives, and that of related structure, have shown interesting features in the selective interaction with bio-membrane models [78], and for saccharide sensing in protic media [79].

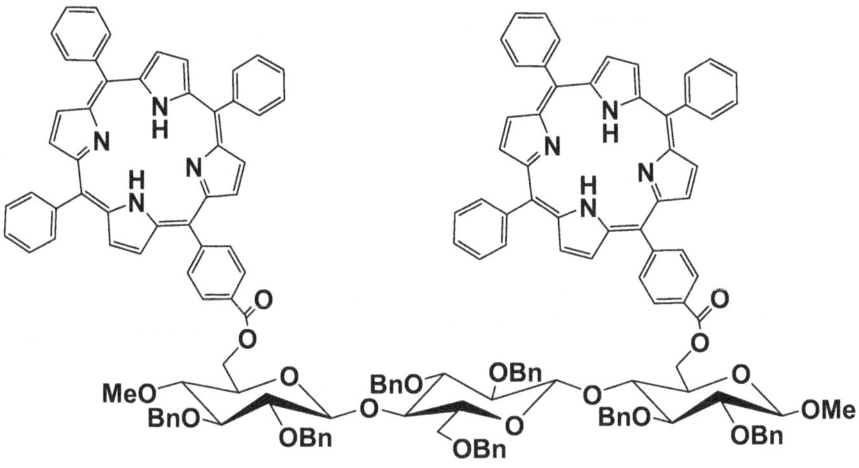

10

Fig. 22 Chemical structure of the cellotriose derivative **10**

The synthesis and the properties of a new cellotriose-functionalised bisporphyrin at O-6 and O″-6 positions (**10**) have been reported by Sakakibara and coworkers [80]. In their work the chiroptical properties of the bis-porphyrin derivative **10** have been explored, as well as their ability to include a fullerene moiety in the molecular tweezers (Fig. 22). The CD spectra feature a positive coupled Cotton effect, arising from a right-handed orientation of the porphyrin chromophores, although the backbone glycosidic structure presents a left-handed helix. A bis-glucose derivative, with high emission ability in Near-Infrared region (NIR), was reported by different authors to show good in vivo affinity for cellular membranes, opening interesting perspectives for the application of these systems as optical probe for cancer diagnosis and PDT [81].

Recently, Amabilino and coworkers reported on the effect of the number of stereogenic centres on the amplification of chirality [82]. Several derivatives have been studied, varying on the number and the position on the periphery of amide groups, prone to interact by hydrogen bond in hydrocarbon solvents (Fig. 23).

Furthermore, the presence of long hydrocarbon chains allows for the onset of Van der Waals interaction with large polarisable π-electron surfaces, and solubility in the solvents employed. The chiral self-assembly properties of the amidoporphyrin derivatives was studied on liquid (1-octanol)–solid interface on HOPG. Topographic images showed the presence of 2D-chiral domains. The assemblies appear as ordered rows with a variable angle of about 10° with respect to the main HOPG axis taken as reference (preliminary studies on one of the derivatives has been previously reported: [83]). The rows are composed by porphyrin units held together by Van der Waals interactions between the extended alkyl chains. The angles depend on the number of stereogenic centres (i.e. amido substituents) and decrease upon diminishing the number of the chiral centres. Results are reported in Fig. 24.

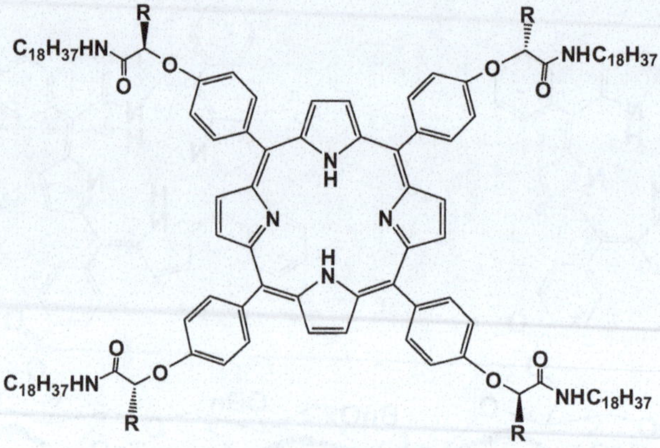

11) 5,10,15,20-R = CH_3 (*R,R,R,R*)
12) 5,10,15-R = CH_3; 20-R = H (*R,R,R*)
13) 5,15-R = CH_3; 10,20-R = H (*R,R*)
14) 5,10-R = CH_3; 15,20-R = H (*R,R*)
15) 5-R = CH_3; 10,15,20-R = H (*R*)
16) 5,10,15,20-R = H

Fig. 23 Molecular structures of derivatives **11–16**

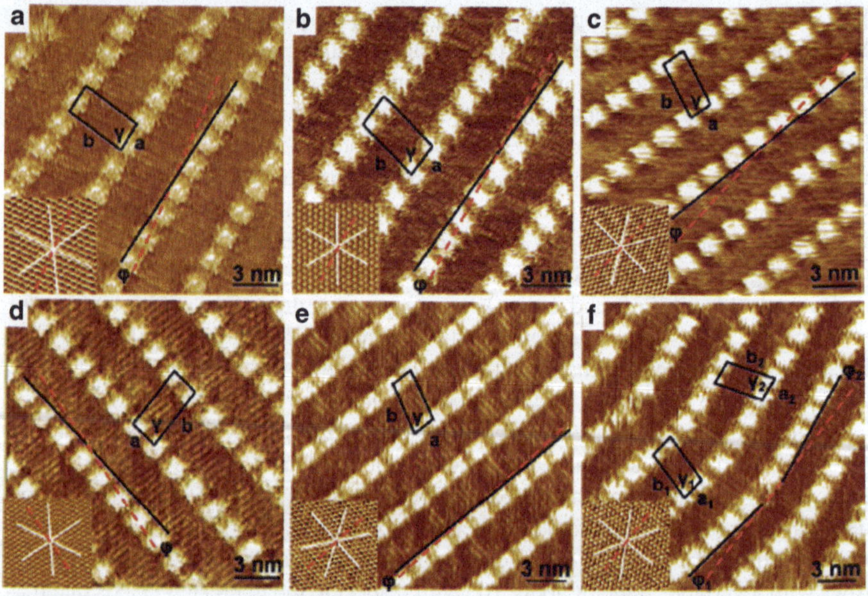

Fig. 24 STM images of porphyrins **12–16** at the HOPG-heptane interface: (**a**) **12**; (**b**) **13**; (**c**) **14**; (**d**) **14** (*S,S*-isomer); (**e**) **15**; (**f**) **16**. *Insets* show the related underneath HOPG orientation, with the direction of the main symmetry axes as *white lines* (reproduced with permission from [82])

However, regioisomers **14** and **15**, differing by 5,15- vs. 5,10-*meso* chiral substitution, gave same monolayers with same elements of symmetry.

Importantly, assemblies formed by opposite enantiomers show opposite angles, confirming the effect of the stereogenic centres on the overall chirality of the systems. Finally the achiral derivative does not show any average deviation from the HOPG main axis. MD calculations were in excellent agreement with the experimental findings. The formation of aggregates in solution has also been demonstrated in cyclohexane at 10 μM concentration. The formation of H-type architectures is clearly evident for the most substituted substrates, as indicated by a blue shift of the Soret band at ca. 400 nm. Also in solution most of the macrocycles show chiral structures, as revealed by addressed CD studies. Expectedly, the achiral compound does not show any bands in the spectroscopic range of the Soret B transition. Evidences for coupled, negative, CD features arose for the solutions of the others counterparts, with the spectral intensities increasing by decreasing the temperature down to $-10°C$, as an effect of a concomitant higher extent of aggregation. The negative sign of the CD bands indicates that the suprastructures are arranged in an anticlockwise fashion. Substantial hypsochromic effect is found for the derivative **15**, along with a non-conservative shape of the CD features. This has been explained on the basis of the formation of aggregates possessing different morphology with the interacting chromophores featuring different angles and/or distance.

A very interesting feature of these derivatives is the propensity to act as gelators in the solvent studied. The critical gel concentration increases with the number of stereogenic units, reflecting the solubility properties of the compounds. Also in this phase, a clear blue shift of the Soret bands is found, suggesting the formation of H-type aggregates. The material also showed some chirality, as indicated by Vibrational Circular Dichroism (VCD) Spectroscopy in the IR region of the amide absorption ($\nu_{C=O}$ ca. 1650 cm^{-1}). This technique, being of particular value for studying long-range interacting chiral supramolecular entities, has been widely employed in the assessment of the ternary and quaternary structure of proteins [84].

Also in this case, reflecting the behaviour in solution phase, the compound without chiral centres did not show any features, whereas in the case of the other counterparts, bands are indeed found in the amide region of the spectra. Derivative **12** shows a bisignated positive spectrum, as well as the derivative **14**, although with somewhat reduced entity, indicating for that compounds a twisted secondary morphology of the suprastructures. The other compounds feature only weaker negative bands. This has been ascribed to different modes of H-bond interaction, as also interpreted by small shifts on the band absorption maxima (Fig. 25).

The morphology of the dried gels (xerogels) has been investigated by SEM and AFM, indicating the occurrence of the effect of the monomer structures on the morphologies of the final nanoscopic phases, as composed by fibres of different length and thickness. Molecular modelling studies on all the compounds taken into consideration nicely support the overall experimental picture. It has been found that the stabilisation energy, in the range of 240 kJ mol^{-1}, depends on both the number of the chiral centres and on the sense of molecular screwing, i.e. clockwise (CW) vs. anticlockwise (ACW) mode. As far as this second issue is concerned, for

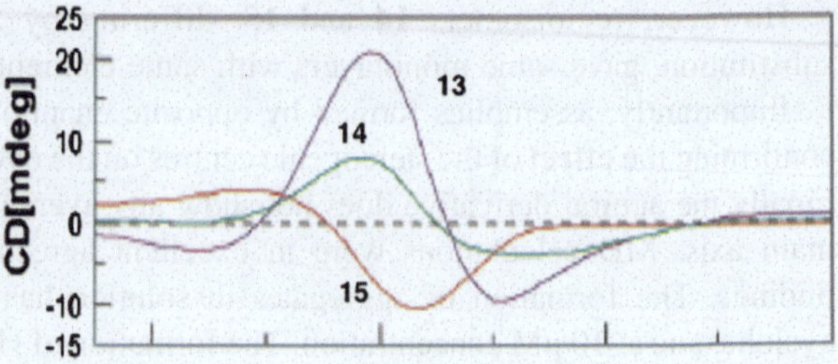

Fig. 25 CD spectra of compounds **13–15** in cyclohexane at 0°C (adapted from [82])

compound **11**, the one with the highest number of chiral substituents, a higher stabilisation energy of about 40 kJ mol^{-1} is found for the ACW twist, with respect to the CW, reflecting the negative features observed in the CD spectra. This is due to the steric hindrance exerted by the methyl group of the stereogenic centre in the unfavourable configuration, hampering the on set of stabilising hydrogen bond. On the other hand, the increasing of substituents causes only a slight increase of the stabilisation energies for the ACW configuration, within 10 kJ mol^{-1}, probably due to the increase of solubility of the species.

The chriroptical properties of the aggregates of compound **12**, bearing three chiral functionalities with *R* configuration, have been further studied by VCD spectroscopy [85]. The studies have been carried out in gel phase of methylcyclohexane at 500 μM concentration. Both IR and VCD experiments corroborate the formation of large stacked structures in helical organisation, held by hydrogen bonds between the amide side chains. The CD features indicate also a CCW mutual disposition of the macrocycles. Molecular mechanics simulations confirm the experimental findings.

Solid-state studies on similar tetra (*R*)-substituted derivatives, differing on the regiochemistry at 3- or 4-position on the *meso* phenyl rings, have been reported (Fig. 26). In this report the effect of the structural variation on the mesoscopic appearance of the layered morphologies on HOPG has been studied by AFM [86].

It has been shown that the overall morphology of the porphyrin layers depends on the regiochemistry of substitution, i.e. the 3- or 4-position on the *meso*-phenyl rings, on the length of the hydrocarbon chains, as well as on the solvent used for the deposition. Quite interestingly, when the solvent used is methanol, only non-specific globular structures of porphyrin aggregates are formed, likely by the effect of solvation. In the case of less polar and non-protic chloroform or toluene, various shapes, from needle-type to longer fibril-like structures, appear, whose alignment with the main axis of graphite depends on the length of the amide chain. In the case of compounds **11** and **17**, the ones with longer C_{18} chains, the strong interaction with the HOPG surface drives a close aligned geometry, whereas in the case of the **18** and **19**, featuring shorter C12 chains, the reduced degree of interaction with the surface allows for some preorganisation in solution by π–π or H-bond, subsequently interacting in a non-specific fashion with the apolar surface. However, the chiroptical features of the described solid-state assemblies have not been reported.

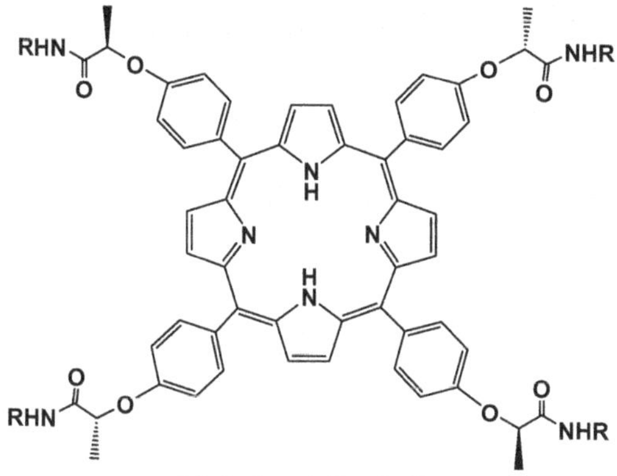

17) 3-phenyl; R = $C_{18}H_{37}$
18) 3-phenyl; R = $C_{12}H_{25}$
19) 4-phenyl; R = $C_{12}H_{25}$

Fig. 26 Molecular structure of compounds **17–19**

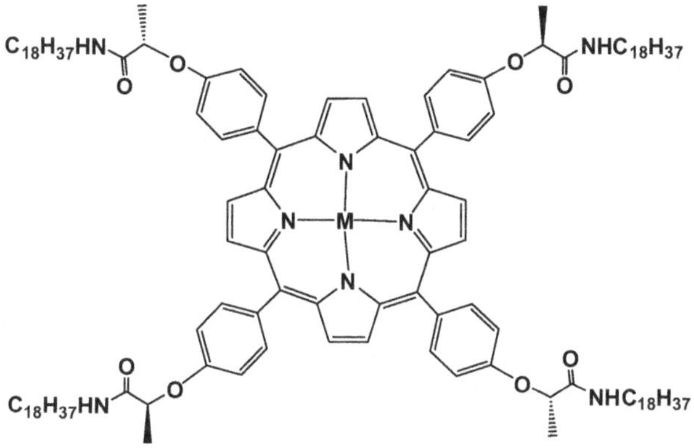

20) M = Co; 21) M = Pd
22) M = Cu; 23) M = Zn

Fig. 27 Molecular structure of the chiral metallo-derivatives **20–23**

The same authors further investigated the role of core-coordinated metal ions on the self-assembly properties of some metal derivative of the above described compounds (Fig. 27) [87]

The presence of the central metal ion affects the nature of the aggregates (H vs. J aggregation), the asymmetry of the assembled structures and the gelation properties of the solvent. The studies have been carried out in methyl-cyclohexane, at a concentration of 10 μM. In particular, for the case of Co(II) derivative **20**, CD spectroscopy studies reveal the formation of both H- and J-type structures, with positive Cotton effect, distributed in a random relative amount, depending on the temperature of the solution (25 to −10°C range). Pd(II) derivative **21** shows more

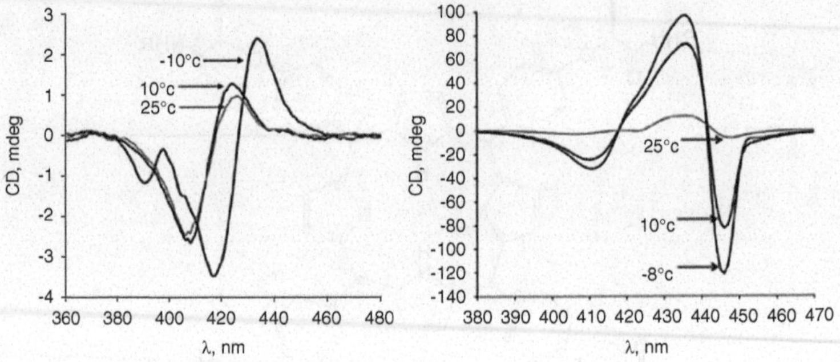

Fig. 28 Series of CD of spectra at different temperatures in methylcyclohexane for derivative **20** (*left plot*), and **23** (*right plot*) (adapted from [87])

selective H-aggregation mode, with weak positive coupled features, being the intensities of the bands in the range of few mdeg, at 5 μM concentration. A very weak uncoupled dichroic band in showed instead for the Cu(II) derivative **22**. Quite interesting differences have been reported for the case of the coordinated Zn(II) derivative **23**. In this latter case the CD spectra of the aggregated species show much more intense and negative coupled profiles (up to 100 mdeg), that increase on cooling the temperature of the solution (25 to −8°C) (Fig. 28).

Moreover, the CD bands, as well as the corresponding absorptions in the UV–vis spectra, are remarkably red-shifted, indicating the formation of J-like structures. This finding has been ascribed to the formation of specific amidic C=O to Zn^{2+} coordination, which drive the formation of offset arrangement platforms. Evidences for this hypothesis have been obtained by careful inspection of the corresponding IR spectra.

Further, the propensity of the title compounds to induce a gel phase of the hydrocarbon medium has also been exploited. The results show once more the effect of the nature of the central metal ion on the critical gel concentration and on the overall mesoscopic morphology on the dry phases (SEM), nicely in line with the studies carried out in solution. Temperature variable experiments indicate an enhancement of the chirality of the system in the range 50–80°C, probably by mending of structural defects. Above this temperature, the VCD bands disappear, as a consequence of the gel–sol transition of the material.

The fundamental importance of the supramolecular chirality of porphyrins, prompted research groups to carry on studies devoted to the full understanding of the intimate nature of the mechanisms involved in the amplification of chirality. In this respect, elegant works on the "Sergeants-and-Soldiers" and "Majority-Rules" effects on the chiral aggregation of porphyrin derivatives have been published by Meijer and co-workers. The sergeant-and-soldiers principle implies the control of the arrangement of a large number of cooperative achiral units (the soldiers) by a small number of chiral units (the sergeants). Conversely, the majority-rules effects arise in a scalemic mixture of chiral elements, in which a slight excess of one enantiomer leads to a strong bias toward the helical sense preferred by the chiral species present in a higher concentration (for earlier studies on covalent chiral polymerisation see [88]).

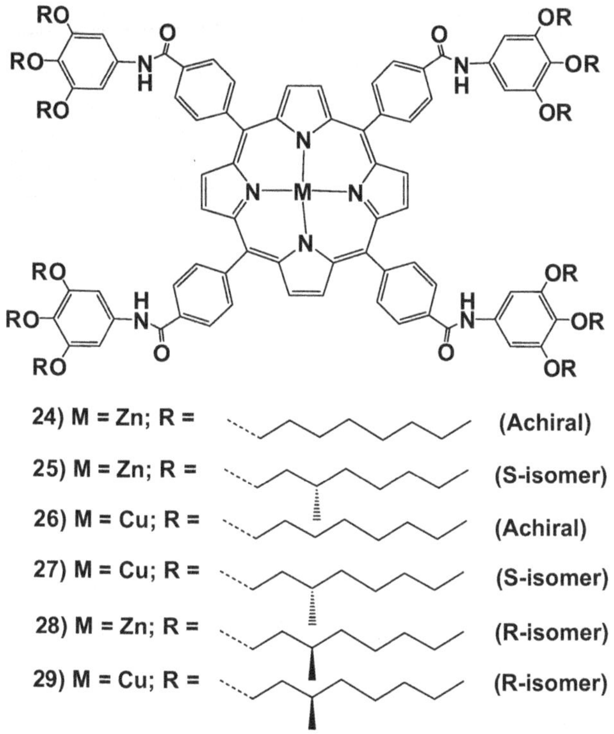

Fig. 29 Molecular structure of compounds **24–29**

In "diluted majority-rules" achiral soldiers are added to a system composed by chiral aggregates [89]. The studies carried out by Meijer concern the co-aggregation of several porphyrin metal derivatives (Cu(II) or Zn(II)) bearing chiral (*S*-configuration) or achiral chains on their phenyl *meso*-positions (Fig. 29).

Related systems have been proven to be of interest for the construction of assemblies with energy transfer properties [90]. Preliminary studies on **25** showed the attitude of this derivative to give highly chiral self-assembled structures in methyl-cyclohexane [91]. The blue shift of the Soret band indicates the achievement of stacked H-type arrangement, due to H-bonds between amide side groups. The assemblies are disrupted by heating (Fig. 30) as showed by the strong decrease of the CD intensity, but rebuilding of the aggregates occurs on cooling, indicating a highly cooperative process with high entropy release (this has been calculated by applying a temperature-dependent nucleation-elongation model: [92]). Addition of an excess of pyridine caused the depolymerisation of the assemblies, as a result of the axial ligation of the Lewis base to the Zn(II) core ion. This reflects itself on the disappearance of the CD features, and on dramatic morphological changes of solid samples on HOPG (Fig. 30). The extent of disassembly, as well as the nature of the most abundant species in solution, depend of the molar ratio of the nitrogen base, and occur upon scavenging the free monomers in equilibrium with the aggregate.

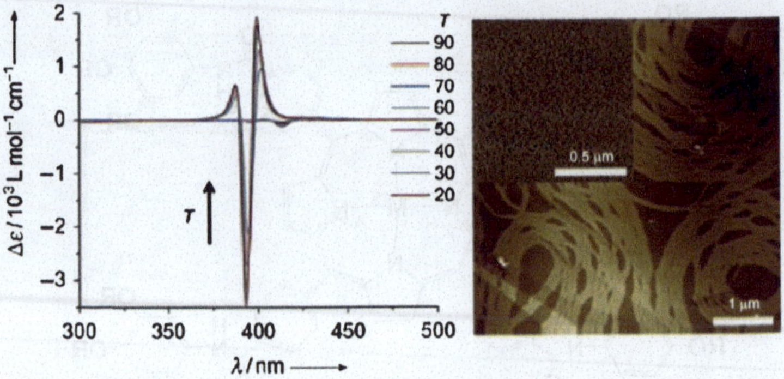

Fig. 30 (*Left*) Temperature-dependent CD spectra of **25** in methylcyclohexane. (*Right*) AFM images of **25** cast onto HOPG (methylcyclohexane solution). In the *inset* is reported the same solution with the addition of an excess of pyridine (adapted from [91])

This derivative has been shown to feature interesting sergeant-soldier effect in the co-polymerisation with related Cu(II) derivatives [93]. The chiral **25** strongly amplifies the helicity of the co-aggregate with the achiral copper derivative **26**. Same results have been obtained in the co-assembly of chiral **27** with the achiral **24**. The effect strongly increases with the **27** proportions. Modelling of the data indicated a large helix "Reversal Penalty" (RP, i.e. the loss of free energy on inversion of the aggregate supramolecular configuration) of about 15 kJ mol^{-1}, with respect to a lower amount of "Mismatch Penalty" (MMP, i.e. the loss of energy on inclusion of a chiral monomer into chiral assemblies of non-preferred screw sense) of about 1 kJ mol^{-1}.

The above **25/26** system, composed by a chiral Zn-sergeant co-assembled with achiral non-coordinating units in a 1:9 molar ratio, revealed remarkable chirality memory effect. This effect is based on a temporary kinetic inertness of a stereochemical configuration of assembled achiral monomers, obtained upon removal or substitution of the chiral effector (sergeant) [94]. In this case the approach consisted on removal of the chiral sergeant **25** by ligation to a quinuclidine (QND) base. Noteworthy, by addition of 10^4 molar excess of QND to the coaggregate (at μM concentration), only a small decrease of the intensities of the CD bands is shown, indicating the occurrence of the memory effect of the achiral **26** coaggregate upon removal of the chiral counterparts. This effect is quite strong and lasts for months.

Variable temperature experiments indicated that heating the solution at above ca. 70°C (the elongation temperature of the non-covalent polymer) caused the loss of chirality, which could not be restored by cooling down the system at room temperature. Interestingly, a partial recovery of the CD features is obtained by a temperature cycling to 60°C, at which some imprinted chiral oligomeric **26** seeds would be still present. Conversely, a full restore of the chirality has been observed in the same experiments for the coaggregate in the absence of added QND. A scheme for the possible equilibria involved is given in Fig. 31.

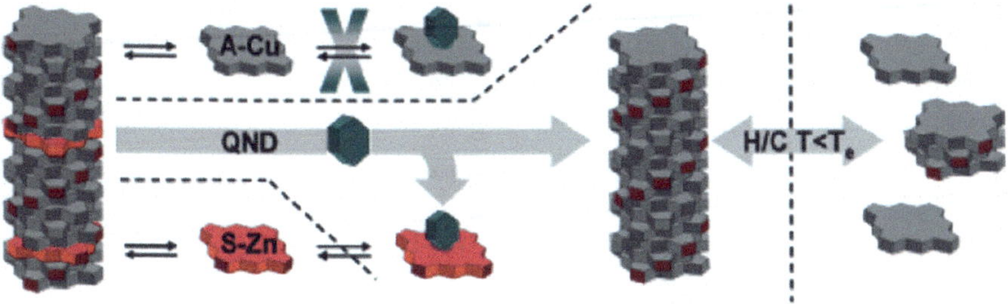

Fig. 31 Schematic picture of selective depolymerisation with retention of chirality, and temperature induced switching of the chiral memory (A-Cu corresponds to achiral **26**; S-Zn stands for chiral **25** of Fig. 29) (adapted from [93])

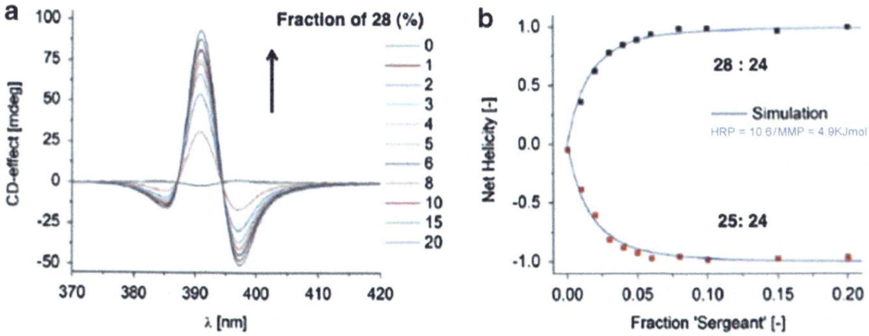

Fig. 32 Sergeant-and-Soldiers experiments of **25** (*S*-enantiomer) and **28** (*R*-enantiomer) on achiral **24**. (**a**) CD spectra of **28/24** system and corresponding normalised helicity vs. fraction of chiral sergeant (**b**) (adapted from [95])

The work was further extended to the chiral enantiomers with *R*-configuration at the carbon, namely **28** (*R*-Zn derivative) and **29** (*R*-Cu derivative) (Fig. 29), focusing on the kinetic stability of the assemblies [95].

The CD spectra of **28/24** system at μM concentration, show the increase of the bands upon increasing the concentration of the homometallic chiral dopant (sergeant), up to ca. 10% molar ratio. The non-linear effect is reported graphically in Fig. 32. The results are mirror-imaged on using the **28** enantiomer. The data are modelled [96] to give the free energy penalties HRP and MMP, whose magnitudes are in the order of the results obtained for the formerly reported mixed metal experiments, indicating a negligible effect of the nature of the coordinated metal ion on the chiral amplification.

Same sergeant-soldiers and majority-rules experiments carried out on **25** (*S*-enantiomer) and **28** (*R*-enantiomer) gave consistent HRP and MMP amounts, likely due to the presence of intermolecular hydrogen bonding, and by the structural mismatch by the methyl groups of the chiral centres. Moreover, the saturation effect is reached at ca. 50% ee. All these findings indicate the preference for "narcissistic self-sorting" onto homochiral aggregates, and the frame is further

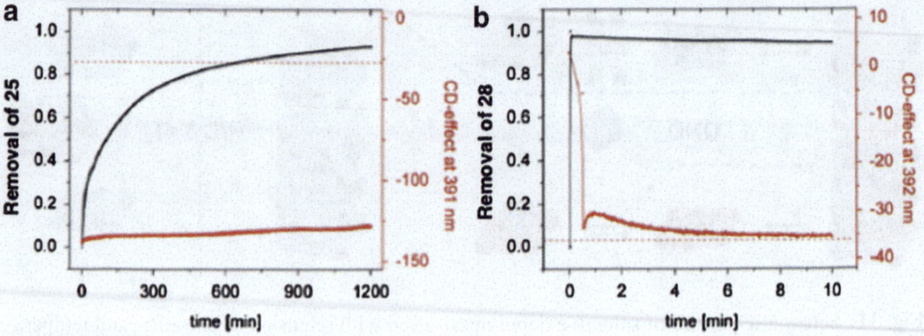

Fig. 33 (**a**) Kinetic profile of selective QND extraction of **25** from **25/26** (1:9) coaggregates by QND. (**b**) Same profile by extraction of **28** from **28/27** (1:1) coaggregates (adapted from ref [95])

confirmed by temperature variable cycles. Majority-rules experiments in the case of **25** and **29** at different ees, gave quite high amount of energy penalties of ca. 15 kJ mol^{-1} and 11 kJ mol^{-1} for HRP and MMP respectively, indicating an highly unfavourable co-assembly. Again, temperature variable experiments confirmed the formation of self-sorting homochiral conglomerates, with that formed by **25** are more stable than the ones composed by **29** counterparts.

The experiments of "dilution" of scalemic mixtures of the former species with achiral **26**, i.e. in a "diluted-majority-rules" regime, indicated the formation of mixed aggregates, with still high HRP but a substantially lower amount of MMP (ca. 10 kJ mol^{-1} and 3 kJ mol^{-1}, respectively), indicating a reduced mismatching effect by inclusion of achiral monomers. The selective scavenging process of zinc monomers from coaggregates by QND in mixed-metal diluted majority-rules (**25/29**) and sergeant-soldiers systems (**28/26**) reveals interesting properties.

The kinetic investigations of the ligation phenomenon (UV–vis and CD means) reveals a very slow extraction step of the Zn-monomer **25**, once embedded in the sergeant-soldier regime copolymer **25/26** (Fig. 33a).

Parallel CD studies revealed substantially unchanged chirality of the remaining **26** backbone, showing lasting memory effect. On the contrary, on "narcissistic" majority-rules case, i.e. **28/27** at 0% ee of the species (i.e. silent CD state), in which the self-sorted homochiral species are the most abundant in solution, a very fast depolymerisation of the **28** aggregates occurs, leaving the chiral **27** domains unchanged, which feature the expected negative Cotton effect (Fig. 33b). These overall findings can be interpreted on the basis of a shielding protective effect exerted by the **26** aggregates backbone toward the trapping of **25** by QND.

Moreover, analogous scavenging experiments carried out on the "diluted majority-rules" case of **25/29/26** (12.5:7.5:80; ee 25% at 20% **25** sergeant) in which the Zn derivative dominates the overall chirality of the heteroassemblies, revealed another interesting property of the system. Also in this case the selective extraction with QND occurs slowly (backbone protection), but the memory effect vanishes within 1 day. This is explained by the fact that the remaining "frustrated" **29/26** aggregates, being in the "wrong" handedness formerly induced by the **25** effector, slowly evolve through entropy-driven atropisomerisation equilibrium.

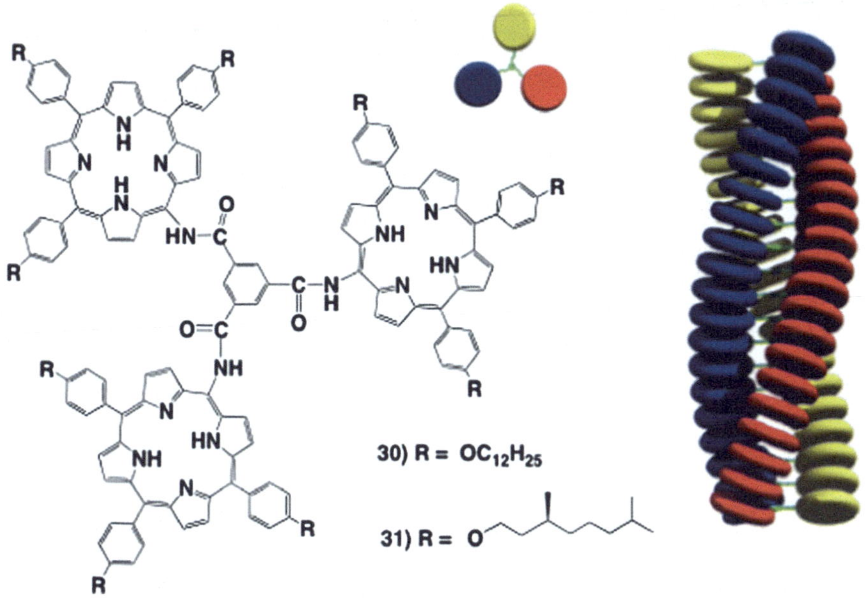

Fig. 34 Molecular structure and schematic drawing of self-assembled derivatives **30** and **31** (adapted from [98])

In concomitance of these systematic studies, the group of Elemans and Nolte exploited the aggregation behaviour of several benzenetricarboxyamide-porphyrin trimers (Fig. 34). These disk-like building blocks, owing to a programmed π–π, hydrogen bonds and dispersion forces motifs, are able to self-assemble into huge ordered porphyrin stacks featuring interesting properties.

An earlier report showed the ability of the achiral trimer (**30**) to aggregate into columnar stacks, driven by hydrogen bonds between the amide groups and π–π interactions among the aromatic platforms [97]. A drop casted chloroform solution of these species onto mica showed, on dewetting, the formation of highly regular macroscopic surfaces of squared millimetre sizes, patterned by columnar stacks (5 nm wide) at distance of ca. 1 μm. These stacks are reasonably composed by millions of trimeric porphyrin building blocks, opening interesting perspectives for new lithographic techniques. The chiral derivative **31** showed the formation in solution of chiral species, due to the presence of the molecular information stored on the side appended groups [98]. This behaviour depends on the nature of the solvent. In chloroform the prevalent phase of the porphyrin derivatives is monomeric (sharp Soret bands), whereas in hydrocarbon media such as hexane or cyclohexane, UV–vis spectra showed highly coupled blue-shifted Soret bands, indicating the formation of H-type stacked aggregates (Fig. 34). The corresponding CD spectra showed a quite complicated behaviour, as a result of the convolution of three Cotton effects.

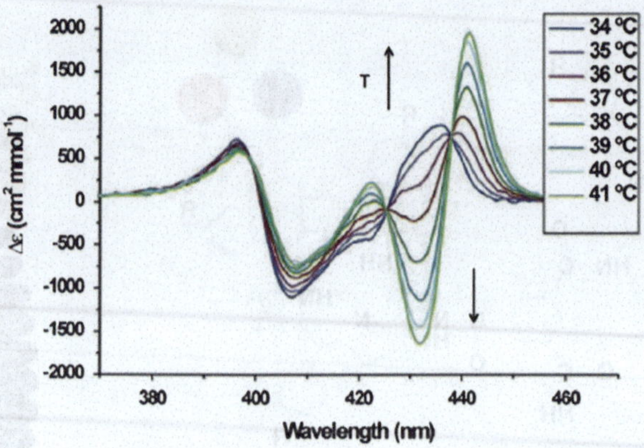

Fig. 35 CD spectra of self-assembled **31** in cyclohexane at different temperatures (adapted from [98])

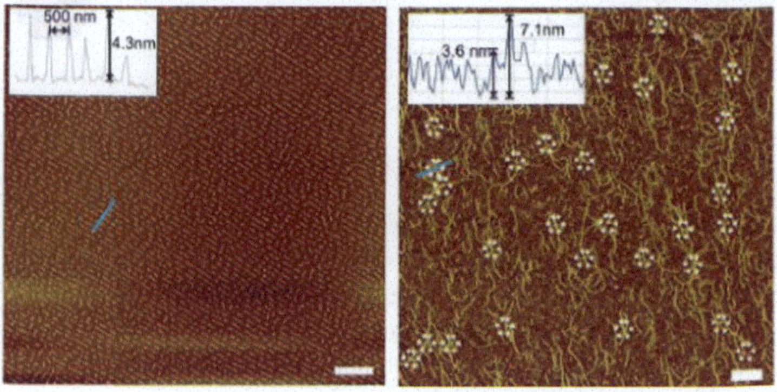

Fig. 36 AFM images of patterned surface after drop-casting of **31** in $CHCl_3$ (**a**; scale bar 1 μM) and hexane (**b**; scale bar 2 μM). The *dashed circles* highlight the crossing point of the fibres. The corresponding profiles are outlined in the *insets* (adapted from [98])

The spectral patterns change on increasing the temperature, giving an indication of some loosening of the building blocks on heating (Fig. 35). Combined Dynamic and Static light scattering studies indicate the presence in solution of thick rods of about 300 nm in length and ca. 5 nm of diameter, in agreement with the calculated dimension of the building blocks. Very interestingly, the drop casting-dewetting protocol formerly followed, results also in this case to the formation of highly ordered patterned mica surfaces. However, the overall 2D morphology depends strongly on the solvent employed, i.e. on the molecular state of the building blocks in solution (Fig. 36). In the case of more dispersing chlorinated solvent, highly ordered columnar structures are formed on the surface, as a result of the dewetting of the monomeric solution. Conversely, in the case of hydrocarbon media as hexane, the 2D morphology is the result of the deposition of the already-formed columnar stacks, in a random and superposed fashion, due to their limited mobility. The sizes and structures are quite dispersed, showing an average length of 535 ⊥ 50 nm, with the majority of them possessing a length of 100–350 nm.

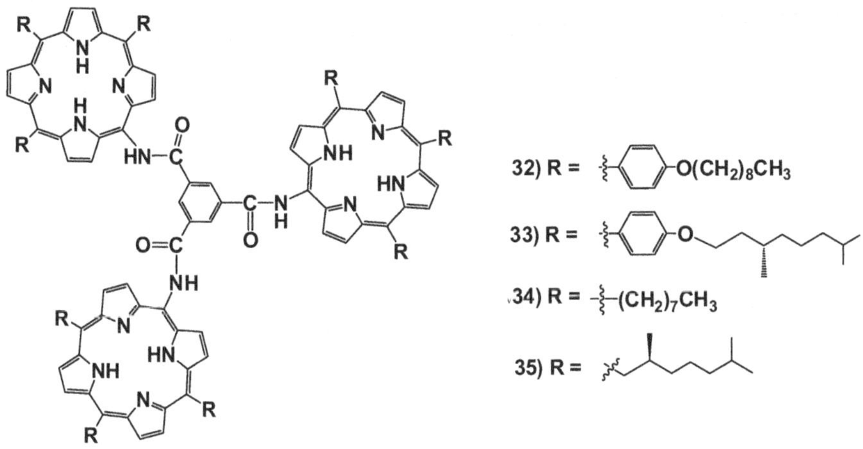

Fig. 37 Molecular structure of chiral derivatives **32–35**

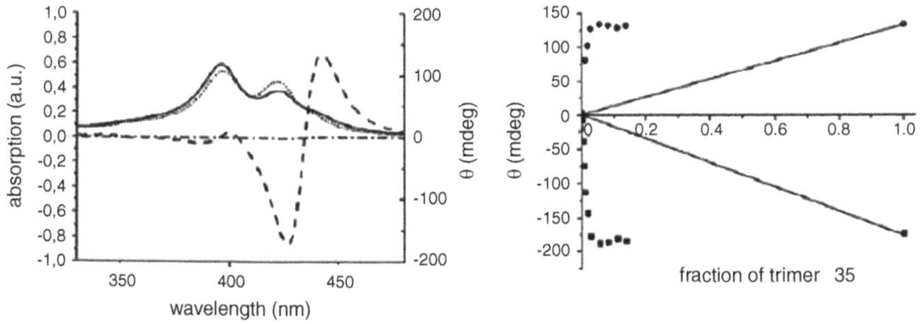

Fig. 38 (*Left*) Normalised UV–vis spectra of **34** (*dotted*) and **35** (*solid*) and respective CD spectra of **34** (*dashed-dotted*) and **35** (*dashed*) in heptane. (*Right*) Plot of CD intensity of aggregates of **34** (428 and 444 nm) showing chiral amplification by added **35** in toluene (adapted from [99])

The studies in solution were extended to other related trimeric derivatives, in order to assess the effect of the nature of the solvent, the concentration of the building blocks and the mechanisms of the self-assembly event by addressed sergeant-and-soldier studies (Fig. 37) [99].

Quite surprisingly, the co-solution of **32** and **33** in hexane, at various ratios of micromolar concentrations did not give evidences for chiral amplification, as a likely consequence of formation of homoassociated species, with kinetic inertness toward the exchange process of chiral and achiral components. Same findings are obtained in the case of the less sterically hindered **34** and **35**. Interestingly, these trimers show reversible aggregation in toluene (a competing π–π solvent), which is enhanced at higher concentration. Expectedly, aggregates of derivative **35** show CD effect, with a preponderance of a positive, red-shifted feature (Fig. 38). In this solvent, sergeant-soldier experiments between **34** and **35** performed in aggregative

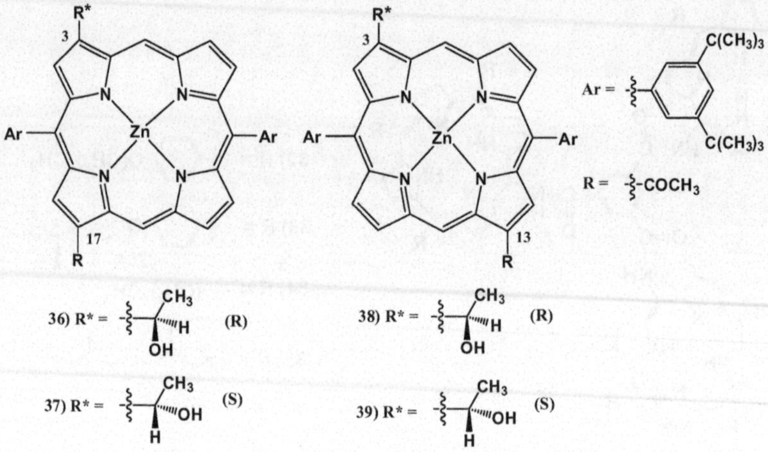

Fig. 39 Molecular structure of *BChls* mimics **36–39** used in the studies

conditions gave evident amplification of chirality, as showed by the non-linearity of the relative CD intensities vs. concentration of **35**. The fact that the maximum of the ellipticity is reached at a very early stage (2.5% of **35**) indicates that the chiral sergeant can be successfully incorporated into achiral stacks of **34** (the soldiers), with a highly efficient transfer of the molecular information.

Finally, the inertness of the aggregate in heptane that prevents successful chiral amplification has been demonstrated by an experiment in which a toluene solution of chiral sergeant-soldier aggregates was slowly evaporated to give a solid film of material. The re-dissolution of the solid in heptane, at initial concentration, resulted in a highly chiral solution of the aggregates, whose CD bands showed comparable intensities to those featured in the more favourable toluene medium, indicating the occurrence of a memory effect of the structures.

The group of Balaban and colleagues reported on intensive studies on the self-assembly of natural, semisynthetic, and synthetic chlorosomal bacteriochlorophylls (*BChls*), and on the application for the construction of artificial light harvesting antennae systems. Earlier important results have been recently reviewed elsewhere [100–104]. In this section, their more recent papers with strict connection and focusing on the formation of chiral aggregates of porphyrin derivatives will be presented. The self-assembly properties of several stereoisomers of synthetic *BChls* mimics (Fig. 39), differing in both the regiochemistry of a carboxylate group (13 vs. 17 position) and the stereochemistry of chiral hydroxyalkyl groups in 3-position, have been explored by spectroscopic means and by scanning tunneling electron microscopy (STEM) [105].

The macrocycles are equipped with recognition groups with specific molecular information (the self-assembly algorithm) for an optimal molecular recognition process close to that governing the formation of chlorosome pigments.

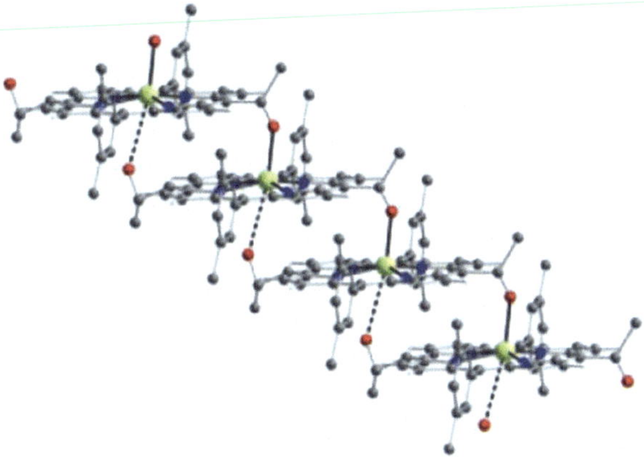

Fig. 40 X-ray structure of the *BChls* mimic showing the long range carbonyl-zinc interactions (adapted from [101])

In all of the cases, the species aggregate in heptane featuring clearly red-shifted spectra, indicating the presence of J-like assemblies. Surprisingly, the aggregation is more pronounced for the racemates with respect to the respective homochiral solutions. The supramolecular species are disrupted in the presence of stoichiometric amount of coordinating methanol, revealing the key role of the coordination to central metal ion in the molecular recognition process (Fig. 40).

Former studies indicate a columnar packing of the platforms mainly due to π–π stacking between the aromatic platforms and by coordination of the hydroxyethyl groups to the metal centre. Another weak electrostatic interaction of the C=O group of the acetyl residues can be also implied, as showed by X-ray diffraction [101]. Furthermore, UV–vis spectra indicate that the assembly is more favoured in the case of **38** and **39** regioisomers, possessing the proper substitution pattern on the periphery of the porphyrin (i.e. a collinear arrangement of the "northern" 3-hydroxyalkyl-metal ion-"southern"13-carboxy substituents), present in chlorosomal natural macrocycles. In particular, the *R*-enantiomer **38** gave more extended aggregates than its *S* counterpart **39**. The CD spectra of the homochiral aggregates, for both the regioisomers, feature intense and complex Cotton effects, and reflects the behaviour manifested by UV–vis means. The relative intensities, depending on the extent of aggregation, are ca. twofold higher in the case of more prone derivatives **38** and **39**. The sign and the shapes of the spectra, although of different intensities, are mirrored for each pair of enantiomers and resemble those of the monomeric chiral species. This indicates that the sense of the chirality is dictated by the asymmetry of the hydroxyl groups. Again, the CD spectra of **38** are more intense than those of the *S* counterpart **39**, paralleling their differences in aggregation propensity. In the case of the racemates, silent CD spectra are obtained in all cases (Fig. 41).

STEM images of either racemate or enantiopure assemblies of **36** and **37**, showed the formation of large nanorods due to the collinear arrangement of the porphyrin derivatives. These structures are composed by a linear arrangement of long aligned fibrillar fine subunits. The morphology is similar to that earlier observed for the corresponding regioisomers [101] despite the non-optimal

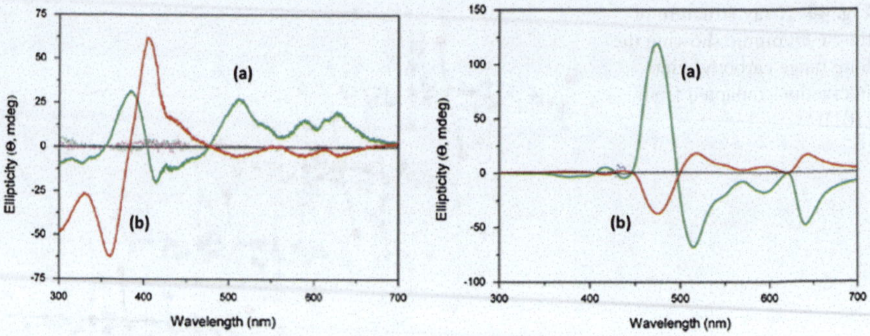

Fig. 41 (*Left*) CD spectra of *BChls* mimics in heptane: (**a**) **36** (*R*-isomer); (**b**) **37** (*S*-isomer). The background *dotted traces* refers to the solutions in the presence of methanol. (*Right*) Analogous CD spectra in heptane of (**a**) **38** (*R*-isomer); (**b**) **39** (*S*-isomer) (adapted from [105])

regiochemistry of the substituents in the former building block. In the case of racemates, the assemblies form longer structures, according to the spectroscopic evidences in solution.

The systems presented, and in particular derivatives **38** and **39**, have been lately proven to show interesting properties for the development of efficient photon-to-current conversion devices [106].

3.3 Spontaneous Symmetry Breaking at Air–Liquid Interfaces

This section deals with the formation of chiral porphyrin films at air–water interface by Langmuir-Blodgett (LB) or Langmuir-Schafer (LS) techniques [107]. Spontaneous symmetry breaking in the formation of chiral assemblies of porphyrin is a very interesting issue. Among others, an important issue proper of these techniques is that achiral amphiphilic compounds can be organised to form chiral assemblies upon interfacial mirror symmetry breaking.

Some relevant works addressed to the formation of chiral porphyrin films have been published by Liu et al., who entailed the study of the film-forming behaviour of a series of achiral tetraphenylporphyrin derivatives, with either lipophilic or hydrophilic substituents (Fig. 42) [108]. Both the compounds, although without long lipophilic "sticky" chains, showed π–A compression isotherms in line with the formation of condensed monolayers. The limiting values of *area per molecule* are 0.6 nm^2 $molecule^{-1}$ and 0.25 nm^2 $molecule^{-1}$ for the methoxy and the hydroxy derivatives, respectively, indicating the formation of disordered unstable monolayer of porphyrin rings (Fig. 42).

The transferred films on quartz substrates by LS technique showed a bathochromic shift for both of the samples, indicating a side-to-side interaction of the chromophores. Importantly, CD studies revealed that the LS films featured

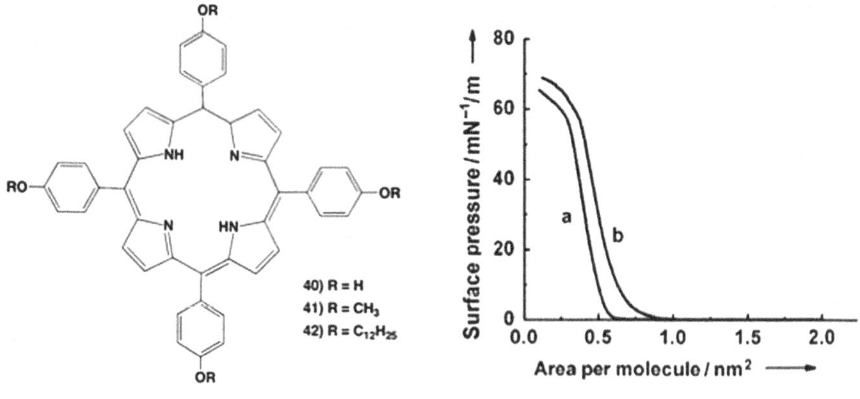

Fig. 42 Molecular structure of amphiphilic porphyrin derivatives **40–42**, and π–A isotherms (20°C) for the spreading films of **40** (**a**) and **41** (**b**) (adapted from [108])

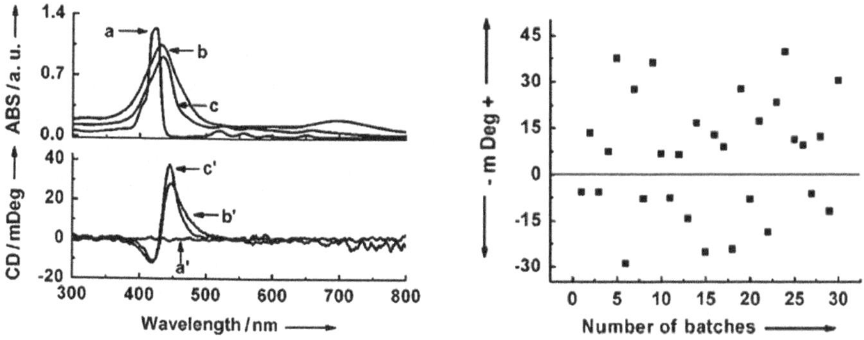

Fig. 43 (*Left*) UV–vis (*top*) and CD (*bottom*) spectra of **40** in $CHCl_3$ (*a* and *a'*), LS films (20 layers, *b* and *b'*), and LS films after annealing (*c* and *c'*). (*Right*) Statistical distribution of CD intensity of LS films (ca. 444 nm) after annealing obtained from 30 independent batches (adapted from [108])

supramolecular chirality, as indicated by the presence of coupled Cotton effects, more intense in the case of the hydroxy derivative **40**. Interestingly, the annealing of the films (heating under vacuum) resulted in a further red shift of the Soret bands, and an increase of the CD intensities, more evident in the case of the methoxy-substituted macrocycles **41**. These findings suggest that some chiral nucleation centres are formed during the filming step, and their chirality is transferred to the whole system, by a sergeant-soldiers effect.

The amplification observed during the annealing process indicates a favourable thermal rearrangement of the macrocycles (Fig. 43). The CD intensities, and more importantly, the sign of the Cotton effect of the solid-state materials are dependent on the batch of the experiments, indicating that the observed phenomena, i.e. the formation of right-handed or left-handed structures, are stochastic in nature (Fig. 43).

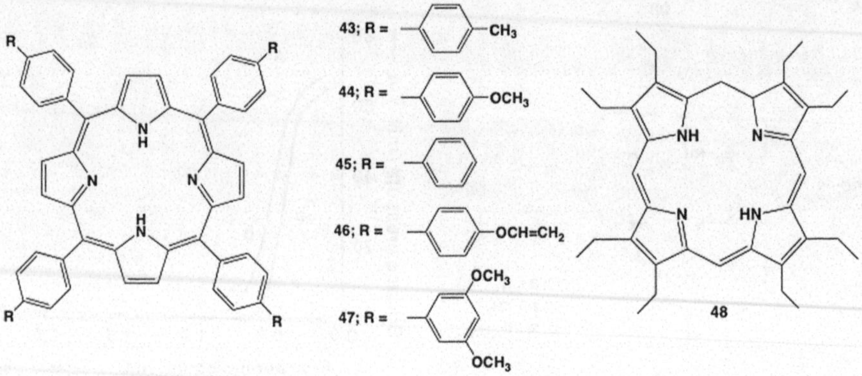

Fig. 44 Molecular structure of amphiphilic derivatives **43–48**

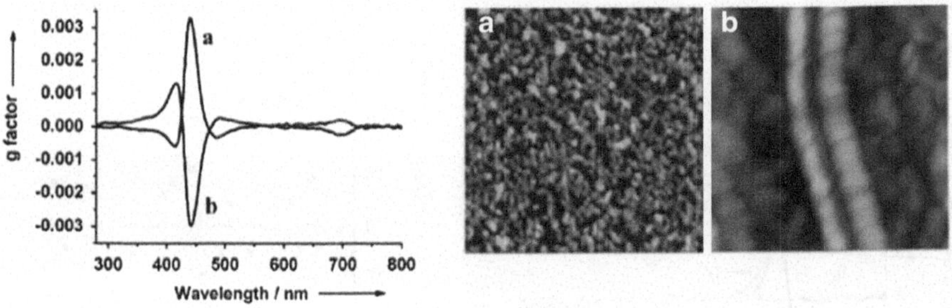

Fig. 45 Normalised CD spectra for LS films of **43** on HCl sub-phase (independent batches *a*, and *b*), and AFM images for LS films monolayer of **48** cast on pure water (**a**) and HCl sub-phase (**b**) (adapted from [109])

The work was extended to film formation of a larger series of derivatives, on aqueous hydrochloric acid surfaces (Fig. 44) [109].

In acidic media the macrocycles are in the protonated form (inner core nitrogen atoms), and this should promote some electrostatic repulsion among the aromatic platforms, resulting in wider area per molecule values, so allowing the formation of optimal solid state films.

The corresponding LS films showed a general red shift of the UV–vis spectral patterns for the *meso*-aryl-substituted porphyrin derivatives **43–47**, indicating the formation of J-aggregate. However, in the case of the octaethyl derivative **48** an H-type arrangement can be envisaged from the hypochromicity featured by the electronic spectra. All the solid substrates showed intense CD bisignated bands in the Soret regions of the spectra of the diprotonated species, indicating the formation of architectures with supramolecular chirality. It must be pointed out that also in these cases the signs of the Cotton effects depend stochastically on the batches of the deposition. AFM studies reveal that, differently from the samples collected on neutral aqueous sub-phase that are composed by irregular structures, all the films cast on acidic media showed the presence of nanorods with definite left-handed or right-handed helicity, depending on the sampled batch (Fig. 45).

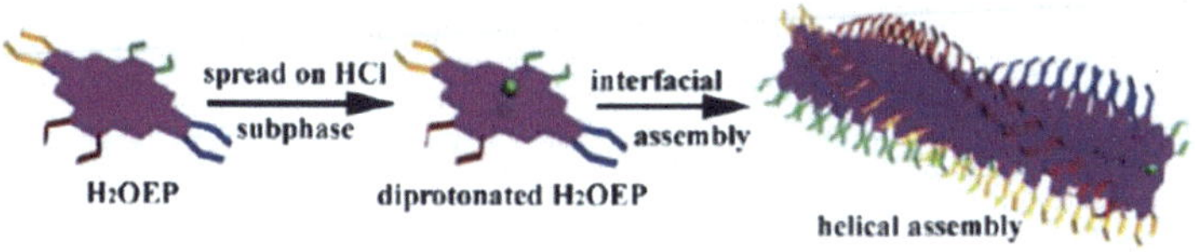

Fig. 46 Idealised drawing for the formation of ordered assemblies of **48** on HCl sub-phase (reproduced with permission from [109])

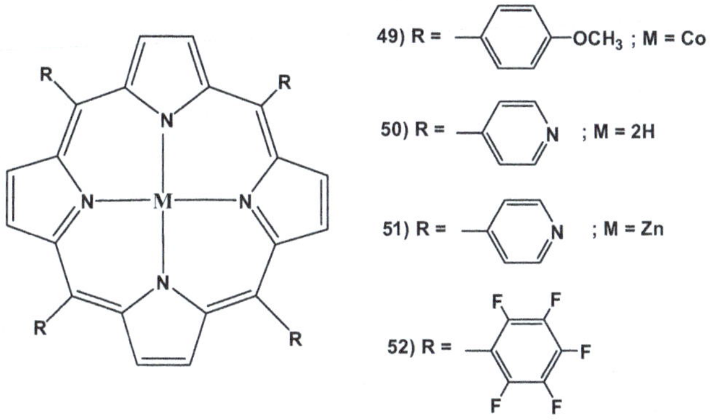

Fig. 47 Molecular structures of derivatives of different lipophilicity **49–52**

A reasonable explanation for the observed phenomena is that the protonation of the inner core, endows some amphiphilic character to the structures. The positive charges contrast, by both electrostatic repulsion and steric hindrance (presence of chloride counterions and saddling of the rings), the π–π interactions allowing for a better mutual disposition of the platforms. An idealised drawing is reported in Fig. 46.

The effect of the nature of the counterion, i.e. of the nature of the acid (HX; X = Cl; Br; I; NO_3) used for the acidification of the aqueous subphases, has been studied in the case of the dimethoxyphenyl derivative **47** [110] (for reports on the effect of the counteranion on the aggregation of porphyrin derivatives see: [111, 112]. Differently from HCl, in the case of acids with large anions such as I^- and NO_3^-, the LS films showed no supramolecular chirality, whereas in the case of Br^-, an ion of intermediate radius, only weak uncoupled CD bands are featured. Interestingly, the chirality of the films can be restored in all of the cases, when a porphyrin solution is cast over the acidic sub-phases containing a strong excess of dissolved NaCl. All these findings emphasised the crucial role of the counterion for determining the correct spacing, and partial charge neutralisation, for the achievement of an optimal distance of the porphyrins.

The effect of the amphiphilicity of the macrocycles has been evaluated in the series of derivatives showed in Fig. 47 [113]. Besides formerly studied compounds, new substrates, either with or without coordinated metal centres, have been considered.

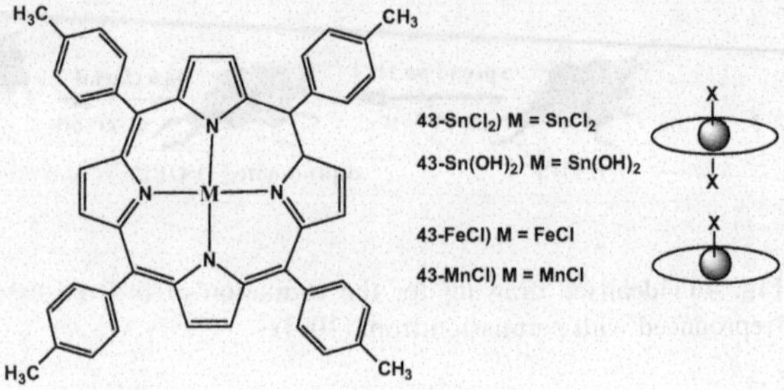

Fig. 48 Molecular structure of metalloderivatives of **43**

The important finding of the studies is that only the hydrophilic (or "less lipophilic") compounds feature supramolecular chirality of the films upon annealing procedures, whereas the more lipophilic ones, i.e. the pentafluorophenyl, the octaethyl, tolyl, and the bare thetraphenyl derivatives, do not form chiral assemblies neither after thermal treatment. These findings highlight the important role played by the hydrophilicity/lipophilicity character of the porphyrin platforms for the achievement of interface proper assembly.

The effect of the nature of the metal centre was further investigated in the case of the Sn(IV), Mn(III) and Fe(III) derivatives, which possess different coordination state (Fig. 48) [114].

The films were spread onto aqueous HCl subphases, and the casted films treated by thermal annealing. The Sn(IV) derivatives **43-SnCl$_2$** and **43-Sn(OH)$_2$**, possessing hexacoordinated metal centres, formed chiral films, whose structures have been showed by AFM to be comprised of fibrous nanostructures of average length of about 1 μm. UV–vis spectra suggest the formation of J-type assemblies. The signs of the CD features depend once more on the batch used for the filming step. Conversely, in the case of the pentacoordinated Fe(III) and Mn(III) counterparts **43-FeCl** and **43-MnCl** respectively, only inhomogeneous structures are formed, with CD silent spectra. In these latter cases UV–vis spectra show the concomitant formation of both J and H aggregates, indicating the lack of specificity in the molecular recognition.

The effect of the implementation of intermolecular hydrogen-bonding algorithm was discussed by comparing two mono-substituted macrocycles, bearing a carboxylic acid residue, and a methylester moiety (Fig. 49) [115]. The possibility to establish such non-covalent interactions, in the case of the acid residue of **53**, allows for the formation of ordered self-assembled structures that show, after casting on surface and subsequent thermal treatment, intense supramolecular chirality.

The fact that the same experiment, performed on alkaline subphase (aqueous NaOH), did not result in the formation of chiral supramolecular assemblies strongly corroborate the given hypothesis. The methylester derivative **54** did not show any supramolecular chirality in all of the experiments undertaken. Finally, in a latter work, the authors showed that the overall chirality of the filmed structures could be mechanically controlled by the direction of the closure of the barriers of the LB apparatus, during the formation of the condensed film at the water–air interfaces

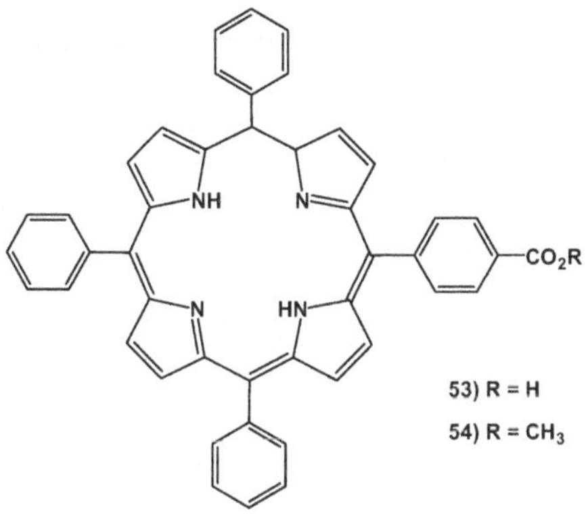

Fig. 49 Molecular structure of the carboxy-functionalised derivatives **53** and **54**

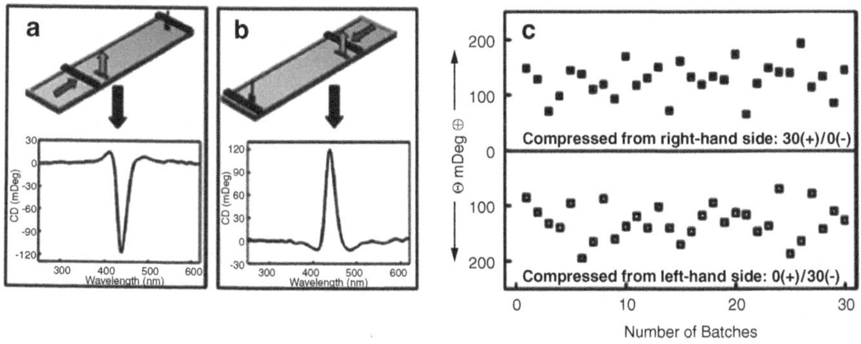

Fig. 50 Scheme of the apparatus used for the directional compression. (*Left*) *Top panels*: *left hand side* (**a**) and *right hand side* direction (**b**); *bottom panels*: CD spectra of the corresponding LB films. (*Right*) Statistical distribution of the CD features (438 nm) showing the high reproducibility of the sample handedness upon directional compression (**c**) (reproduced with permission from [116])

[116]. In other words, whereas random chirality occurs in a standard LB protocol, the left- or right-handedness of the macroscopic assemblies can be selectively induced by a choice of a "left" or "right" direction of the motion of the barriers, as depicted in Fig. 50, for compound **43** onto acidic aqueous sub-phase. This interesting important effect, that can shed further light on the symmetry breaking phenomena that occur at interfaces, is reminiscent to that found in the selection of macroscopic chirality by directional stirring, a topic that will be examined in deeper details in the next section.

3.4 *Spontaneous Symmetry Breaking by Vortex Stirring*

The induction of supramolecular chirality on achiral building blocks by chiral vortexes is a very important field of studies, thanks to its strong implication in the symmetry breaking and amplification in the evolution of prebiotic systems [117]. Seminal works

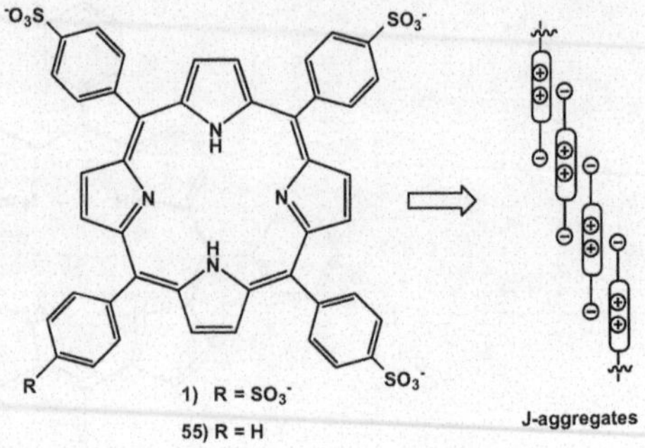

Fig. 51 Molecular structures of water-soluble derivatives and idealised drawings of J-aggregates in acidic media

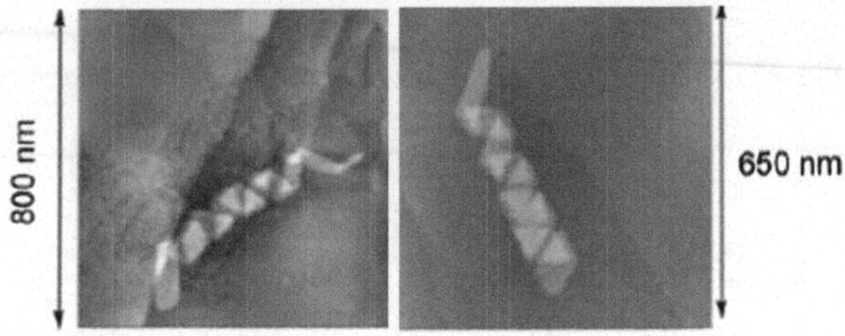

Fig. 52 AFM images of aggregates of **55**, after 6 h of directional vortex-stirring of an acidic solutions of **55**, showing long-range folding of the J-aggregates (adapted from [121])

in the subfield of water-soluble porphyrin derivatives have been published by the group of Ribò and by others [52, 118, 119] who showed the formation of chiral aggregates of zwitterionic porphyrin derivatives by a simple effect of the stirring of the solutions (for an interesting overview on this topic see [120]).

The aggregation of derivatives **1** and **55** in acidic aqueous solutions resulted in the formation of large structures in the form of tapes of micrometres in length, in which the platforms are held in a J conformation, by a fine combination of electrostatic and dispersion forces (π–π interactions), as depicted in Fig. 51.

Remarkably, in the case of the tris-sulfophenyl derivative **55**, these tapes evolved with time from 2D to 3D folded helical ribbons [121]. Key aspect of these systems is that a long-order helical sense can be selectively tuned in vortex-stirred solutions, by the effect of hydrodynamic gradient of laminar flow at the walls of the container. AFM topographies shows highlight of these effects (Fig. 52). Physicochemical studies aimed at to disentangle artefacts effects on the above CD measurements have been reported, showing unambiguously the contribution of the selective macroscopic folding to the chiroptical properties of the system [122, 123] (for a critical analysis of the effect of artefacts as Linear Dichroism (LD) or Linear Birefringence (LB) in the CD spectra see: [124]). To make things more complicated, the described former experiments point out that the signs of *screwness* of the aggregates depend also on the means used for obtaining the species, i.e. rotary evaporation or magnetic stirring of the solution, as well as on the shape of the flasks used (square vs. cylindrical sections).

These findings are interpreted by the occurrence of specific gradients of flows (laminar along the walls, ascending or descending in the middle of the solutions) in the solution containers. To get more insights on these difficulties, non-conventional

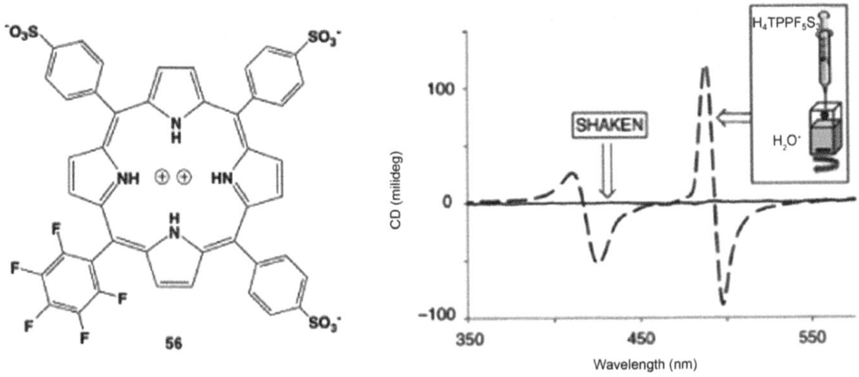

Fig. 53 Molecular structure of fluoroderivative **56**, and CD spectra of the corresponding aggregates obtained in shaken and stirred solution (adapted from [129])

optical techniques, as two-modulator generalised ellipsometry (2-MGE), were employed [125]. The results indicated that the main contribution to the chirality of the aggregates is due to the chirality of the flows, and that chirality of the flows depends on the actual size and shapes of the stirred containers. The contribution arising from the physical orientation of the formed nanostructures can be neglected.

A recent work devoted to the full comprehension of the effect of experimental artefacts (fibres alignment as a consequence of mechanical stirring) on the aggregation of **1** has been published by Purrello and coworkers [126, 127]. Their experimental evidences lent to exclude that the chiroptical properties of these J-aggregates significantly depend on the presence of trace contaminants, but rather on the formation of inherently chiral structures initially present in racemic ratio, whose final scalemic distribution depends on the direction of the stirring. The aggregates spontaneously layers onto the cuvette surfaces, showing the same helicity of the most abundant non-covalent stereoisomer in solution, which is, strictly speaking, trapped by the cuvette walls. Same results have been obtained by stirring a solution of preformed chiral J-aggregates, either in "wrong" or in "right" sense, with respect to the initial helicity of the self-assembled species. The sense of the swirling dominates the final chirality of the aggregates that can be "frozen" onto the cuvette walls. Additional experiments carried out in the presence of chiral templates corroborate the above findings.

However, the study of the aggregation of a more lipophilic pentafluorophenyl derivative (Fig. 53), which as a consequence, features a slower kinetic of aggregation, gave opposite evidences [128, 129]. It allowed to point out that the experimental protocol pursued for the preparation of the solutions, such as vortexed vs. stagnant, affects the nature of the J-aggregates that are formed due to the onset of different mechanisms of growth. This is due to a selection effect that acts in the bifurcation event that controls the primary-nucleation step. In particular, the aggregates formed in a "shaken" solution are CD silent, whereas the ones obtained by stirring show the typical pattern showed in Fig. 53, featuring three

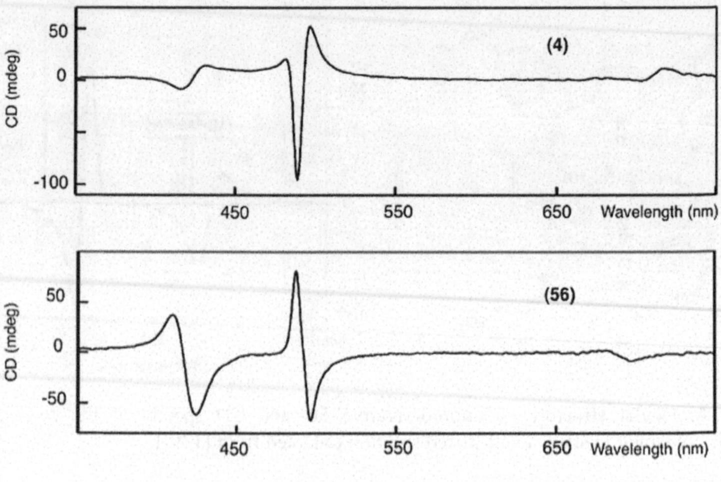

Fig. 54 CD spectra of aggregates of **4** (*top*) and **56** (*bottom*) in the presence of trace chiral contaminants of water (adapted from [129])

excitonic bands at ca. 420, 490, and 700 nm (it must be reminded that the two high-energy transitions do not correspond to different H- or J-aggregates, but are the consequence of different space direction of the exciton vectors of the macrocycles held in J-conformation. The different intensities depend on the oscillatory strengths of the transitions, as well as on the stage of hierarchical assembly (numbers of units forming the whole non-covalent polymer); see [118]. The discussions of results previously obtained for similar cases would be re-elaborated taking into account this point of view. See for example: [130]).

This indicates that the J-aggregates formed under shaking correspond to a racemic mixture, and those formed by stirring correspond to a scalemic one.

Surprisingly, differently from the previous cases discussed, in the experimental conditions used that promote slow aggregation kinetics, the final chirality does not depend on the direction of the stirring.

The chiroptical effect is then ascribed to the unavoidable presence of adventitious chiral contaminants of the water. Moreover, in all of the cases examined, tetra- and trisulfonato derivatives show all positive coupled features, whereas in the latter fluorinated species these appear to have negative signs (Fig. 54). This has been inferred to the formation of aggregates of different sizes and morphology. The effect of the physical nature of the pentafluorophenyl moiety (electronic and steric hindrance) should be neglected. Supports to this hypothesis have been given by the fact that aggregation carried out with high excess of "chiral contaminants" such as D- or L-tartaric acid that, overwhelming the initial trace contaminants, dictates the signs of the CD features also in swirled solutions. Also in this case the fluoro derivative **56** shows CD spectra with opposite signs with respect to the other two counterparts. The overcoming effect of a chiral dopant over that of the mechanical stirring has been formerly showed in an experiment of J-aggregate formation of **1** in a centrifugal-liquid-membrane cell (CLMC) at toluene/water interface [131].

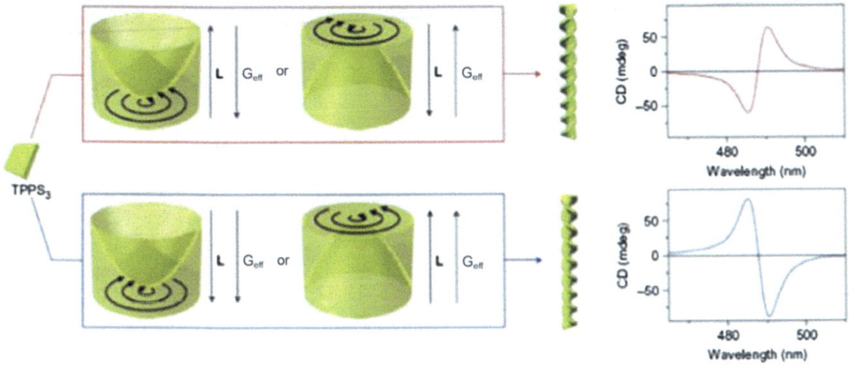

Fig. 55 Schematic drawing illustrating the correlation between the observed chirality of the assemblies and the applied physical forces (reproduced with permission from [132])

An astonishing experiment that could shed further light on this important issue has been given very recently by Monsù Scolaro and colleagues [132]. In their experiment, the aggregation of tris(sulfonate)phenylporphyrin (**55**) in acidic condition has been carried out under the combined effect of stirring, that generates an angular momentum (L) and gravity field, modulated by an applied magnetic field (magnetic levitation force, G_{eff}). The aggregation is triggered by increasing the ionic strength of the aqueous solution, to give the formation of rather small self-assembled species ($\leq$0.1 μM).

In the particular instrumental apparatus employed, the reacting solutions experience different rotational (L) and gravity strengths (G_{eff}) that can select the supramolecular chirality of the generated assemblies (Fig. 55). The notable results obtained are that a negative CD spectrum (i.e. a mutual CCW arrangement of the porphyrin platforms) is obtained in the case of a parallel alignment of both L and G_{eff} vectors, independently from the direction of stirring, whereas in the case of an antiparallel arrangement of the physical forces, a positive Cotton effect is produced, indicating a CW arrangement of the platforms. Remarkably, in the absence of applied magnetic field ($G = 0$), i.e. in conditions of "normal gravity" effect, no correlation between the direction of stirring and the handedness of the supramolecular specie is found, indicating that this factor is essential for a correct alignment of the molecules. The alignment forces are determined by the strength of the applied magnetic field, resulting from the anisotropy of the magnetic susceptibility of the porphyrins. Importantly, the direction of the applied magnetic field does not influence the supramolecular chirality, indicating a negligible magnetochiral effect (for an important recent work on this issue see: [133]).

All the results showed in this section bring to the conclusion that in this kind of hierarchical self-assembly, the precise control of the final asymmetry of the assemblies could be made possible by finely tuning the multiplicity of thermodynamic, physical and kinetic factors involved.

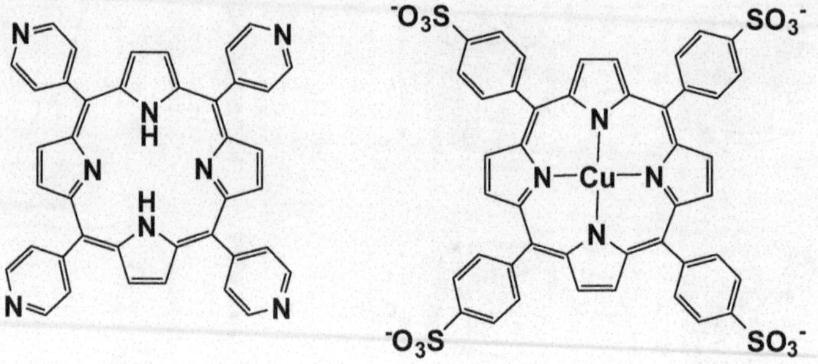

Fig. 56 Molecular structures of derivatives **57** and **58** used in the chiral template assembly

3.5 Templated Self-Assembly

This section will deal to new results achieved in well-known field of templated formation of chiral porphyrin suprastructures from achiral building blocks. Several seminal papers have published in the past on this topic by Robert Pasternack and colleagues, devoted to the comprehension of the mechanistic and structural factors that strongly influence the self-assembly processes, such as the structure of the tetrapyrrolic macrocycles, that of the chiral natural or synthetic templates, the reaction conditions and memory effects of the imprinted chirality (for some examples of important earlier contributions to the field see: [134–138]).

A recent example in which the imprinted chirality of porphyrin aggregates can be cyclically stored, erased and restored has been reported by Purrello [139]. The examined "supramolecular memory system" is constituted by a complex of an anionic **58** and a pH sensitive neutral derivative **57** (Fig. 56).

In acidic conditions the pyridine groups are protonated, steering the electrostatic interactions among the macrocycles, whereas the anionic structure remains in anionic form, owing to the high acidity of the sulfonic moieties. The presence of the inert central metal atom prevents protonation of the inner core of the porphyrin. The adducts can be templated onto chiral non-covalent L- or D-phenylalanine polymer (Fig. 56) to form chiral suprastrucures, whose helicity is dictated by the helical sense of the aminoacidic polymer. Thanks to their kinetic inertness, the supramolecular configuration of the assemblies is trapped in a local intermediate-free energy minimum, and it is retained after the removal of the template [140].

The chiral assemblies are rather inert, also remaining in their configuration after prolonged standing or by heating. The memory can be then "erased" by deprotonation of the *meso*-pyridine groups, at high pH, where the pyridine groups are in the neutral forms. In such a case, the corresponding CD spectra are silent. Remarkably, further acidification of the solution causes the reversible reassembly of the platforms, and the appearance of the CD features, indicating a restoration of the supramolecular chirality, due to the presence of residual traces (at sub-spectroscopic level) of chiral porphyrin oligo-assemblies (seeds) that acts as a highly efficient chiral template (Fig. 57).

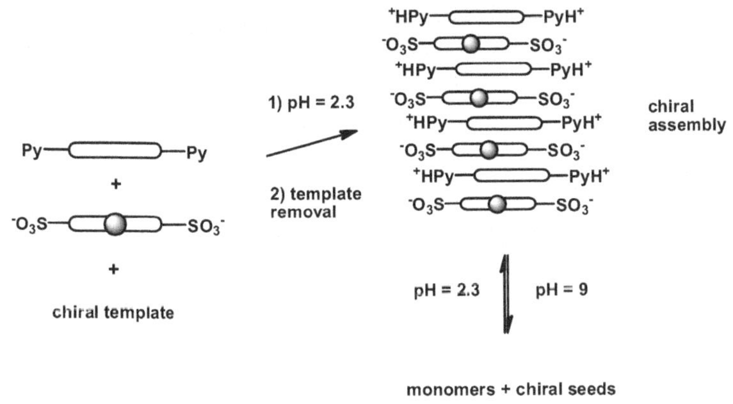

Fig. 57 Schematic drawing of the template memory effect for the chiral assembly of **57** and **58** in the presence of phenylalanine

Mechanistic insights on the template self-assembly processes have been obtained in the case of analogues derivatives, indicating the occurrence of a complex reaction path, involving fast equilibria of hetero-assembly of porphyrins onto the chiral non-covalent polymer. The rates are about two orders of magnitude higher than that observed in the absence of templates [141].

These studies have been extended to other intriguing systems consisting in the adducts of porphyrin **1** with chiral complexes of ruthenium, namely Λ and Δ enantiomers of $[Rh(1{,}2\text{-phenanthroline})_3]^{2+}$. Ruthenium complexes are great importance, being widely employed in the construction of Dye-Sensitised Solar Cells (DSSC) [142]. The studies reported indicated that the chirality of the resulting aggregates, templated by the presence of the Ru(II) enantiomers, not only can be erased and restored by the change of the pH (protonation/deprotonation of the porphyrin inner nitrogens), but also showed high inertness upon reversal of configuration by the presence of a strong excess of the "wrong" Ru-dopant isomer. Further developments involving the interaction with chirally functionalised calixarenes and chiral Ru complexes have been reported [143]. Different strategies for the construction of chiral hetero-assemblies were pursued, allowing for a fine tuning of the stoichiometry, sequence, and bi- or tri-dimensionality of the final supramolecular architectures.

As showed in the examples reported, the electrostatic interactions play a key role in the self-assembly of the porphyrin units. This algorithm can be directly implemented into the molecular frame (covalently linked cationic or anionic groups), or reversibly turned on by pH changes of the solutions (protonation/deprotonation of nitrogen atoms). It has been shown alternatively, that the protonation of **1** can be "photo-triggered" in DMSO/ethanol media [144] (for a previous report on this effect in a chlorinated solvent see: [145]). As a consequence, the formation of chiral J-aggregates is fostered, upon symmetry breaking effect. The handedness of the suprastructures occurs randomly, probably as a consequence of the formation of fractal-type species [146], but can be selectively tuned in the presence of an excess of chiral effectors as D- or L-lysine.

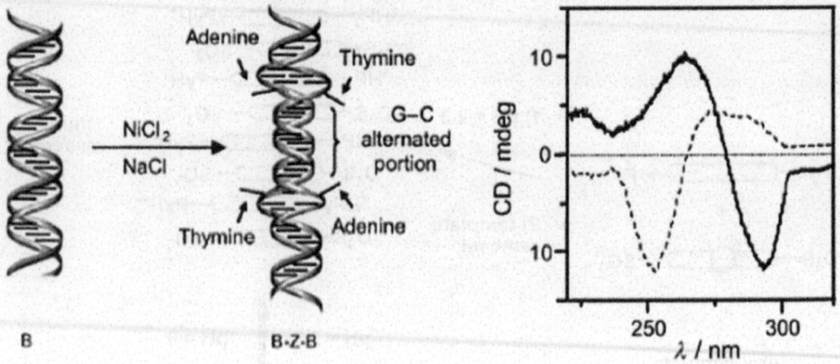

Fig. 58 Schematic presentation of B to B–Z–B DNA transition and CD spectra of B- (*dashed*) and Z-DNA structures (*continuous*) (adapted from [150])

Same philosophy resides on the spontaneous symmetry breaking, reported by Liu, in the formation of porphyrin assemblies in the presence of ionic liquids and chiral inducers [147]. The effect of the ionic liquid (IL) is to promote an optimal interaction among the aromatic porphyrin platforms **1**, resulting in a selective formation of porphyrin assemblies whose chirality is steered by the presence of D- or L-lysine. The effect of the nature of the added IL, and of its concentration has been also discussed, indicating a prevalence of effects in the case of IL with shorter alkyl chains.

Templated aggregation of porphyrin derivatives has been demonstrated to be a powerful means for unraveling the actual structure of natural or synthetic biopolymer. This is due to their unique UV–vis and CD spectroscopic properties, that are not only markedly affected by the "microenvironment", but are also featured in spectroscopic windows usually free from the usual absorption bands of the biopolymers. Along this line, the investigation of the various conformations of DNAs has been successfully carried out by using water-soluble porphyrin as conformational reporters [148] (analogous studies using porphyrin monomers have also been published: [149]). A cationic porphyrin derivative has been successfully employed, for example, for the specific recognition of left-handed Z-DNA tracts embedded in the B–Z–B sequences [150]. Earlier works by same authors showed that related porphyrins were able to unambiguously recognise B or Z forms of DNA or related synthetic sequences [137, 151]. Z-transition, promoted by the interaction of the B-form with cationic species (e.g. Na^+, Ni^{2+}, conjugated acid forms of polycationic amines) causes exposure of the guanine nitrogen N7 [152].

The conformational change results in a clear inversion of the CD spectra on the UV-region (Fig. 58) and allows for the coordination of a Zn-porphyrin, which would be employed as a chirality reporter in the visible region of the spectra, by Induced Circular Dichroism (ICD) effect.

The conformational change results in a clear inversion of the CD spectra on the UV-region (Fig. 58) and allows for the coordination of a Zn-porphyrin, which would be employed as a chirality reporter in the visible region of the spectra, by ICD effect. In the experiment, several B- and B–Z–B sequences (DNA models)

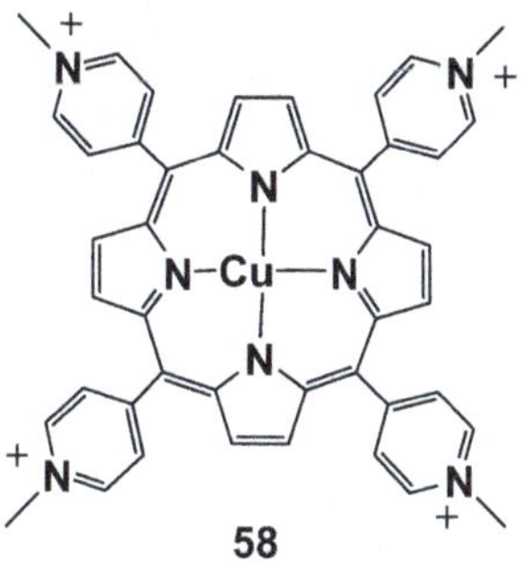

58

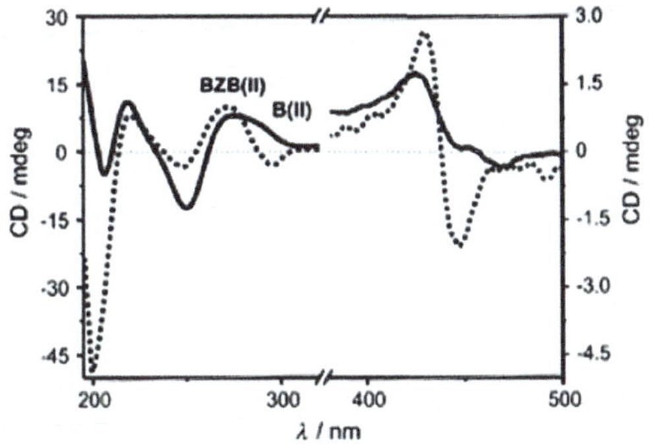

Fig. 59 Molecular structure of derivative **58** and CD variations for the B/**58** and BZB/**58** complexes in the presence of $NiCl_2$ (B/Z ratio 32:26) (adapted from [150])

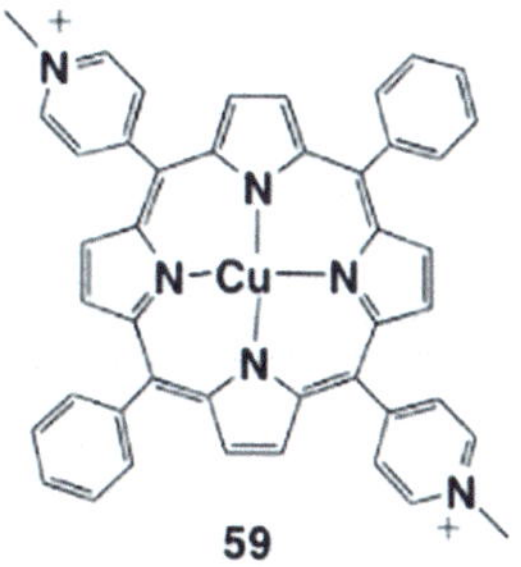

59

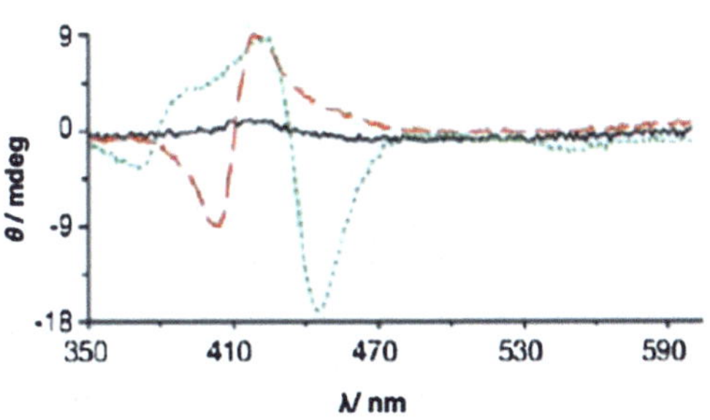

Fig. 60 (*Left*) Molecular structure of porphyrin **59**. (*Right*) CD traces of **59** in the absence of **PLG** (*continuous line*); in the presence of **PLG** as α-helix (pH 4.7, *dashed line*), and as random-coil conformations (pH 7.2, *dotted line*) (adapted from [153])

have been proven, with decreasing ratio of the Z/B ratios. The B to Z transition was promoted by the addition of proper amounts of cationic species, and in all of the cases the conformational changes could be effectively signalled by the ICD effect on the tertacationic porphyrin **58** at μM concentration (Fig. 59).

This philosophy has been extended also in the conformational studies of synthetic polymers, as model proteins. An interesting work has been reported by Monsù Scolaro and Pasternack [153], who studied the interaction of a related cationic porphyrin **59**, with a chiral poly-L-glutammate (**PLG**), as a model protein scaffold (Fig. 60). The macrocycle used in the work, possessing only two pyridyl moieties, presents increased binding ability with respect to its tetra-substituted homologue **58**, toward **PLG**. The compound **59**, in water at μM concentration, is in the form of oligomeric achiral species, as indicated by the absence of bands in the corresponding CD spectra.

The presence of **PLG** (17 kDA; 100 μM Glu residues) causes dramatic changes of the CD features, depending on the adopted conformation of the peptide, which is driven by the pH of the solution. In particular, at pH 4.7 where the polymer adopts a

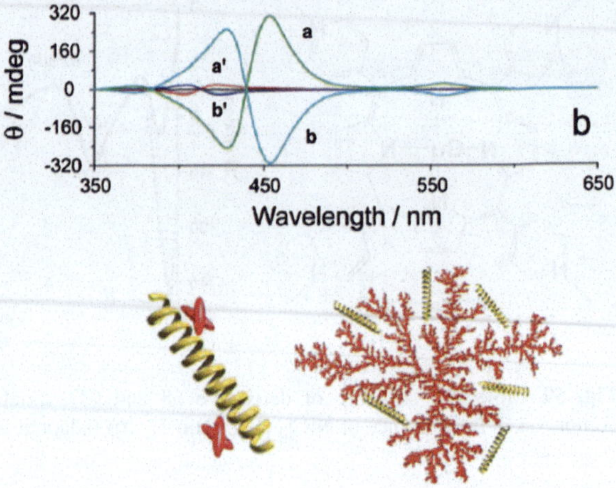

Fig. 61 (*Top*) CD spectra of **59** in the presence of L-**PGA** (*a*) and D-**PGA** (*b*) according to protocol 2, and L-**PGA** (*a′*) and D-**PGA** (*b′*) according to protocol 1. (*Bottom*) Idealised drawing for the interaction of **PGA** with oligomeric forms (*left*) and fractal structures of **59** (*right*) (adapted from [154])

α-helix conformation, a positive bisignated CD signal appears, indicating the interaction of the porphyrin with **PLG** in the form of somewhat larger aggregates, as corroborated by concomitant UV–vis and RLS studies. Quite surprisingly, an increase of the pH to 7.2, at which the polymer is in random-coil conformation, resulted in a rather complex CD signals, with negative sign, that reflects a left-handed coil conformation. UV–vis and RLS spectroscopy revealed in this case the formation of extended porphyrin aggregates, that can act as "antenna" for the conformational change of the polymer backbone. The sensitivity of the probe has also been proven in the case of shorter **PLG** polymers, less prone to adopt a α-helix motif.

Further studies demonstrated that the asymmetry of the polymer could be efficiently transferred and amplified to large fractal random aggregates of **59** [154]. These species can be obtained in water solutions at high ionic strength (ca. 0.1 M NaCl) (the effect of ionic species is to shield the charges on the porphyrin frame, and consequently turning on the aggregation by π–π interactions and hydrophobic effect: [155]). It should be reminded that aggregate clusters, owing to their peculiar structure, are "inherently chiral" objects. Nonetheless, their chaotic mechanism of growth renders a large-scale system (a solution) a "quasi-racemic" mixture [156].

Their mode of interaction with α-helical **PGA**, the protonated form of PLG polymer at pH 4.2, depends strongly on the protocol used for the preparation of the solutions. For example, the addition of NaCl as a last component to a preformed solution of **59** and **PGA** (protocol 1) resulted in a bisignated positive Cotton effect of the CD spectra. Phase inversion is obtained in the presence of D-**PGA**, as templating substrate. Both UV–vis and RLS spectra indicated the presence of small porphyrin oligomeric or dimeric species bound to the helical polypeptide. Remarkably, the addition of **PGA** to a solution of **59** as fractal clusters (protocol 2) resulted in a high amplification of the CD intensities, indicating an efficient transfer of the molecular information from the polymer to the whole aggregates. Also in this case the CD features are mirrored in the presence of the D-**PGA** enantiomer (Fig. 61).

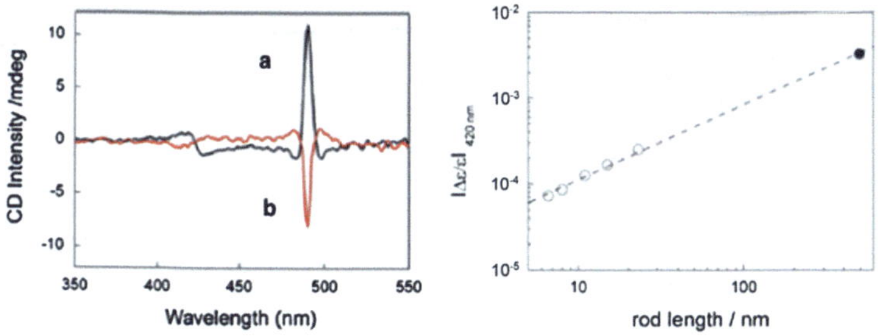

Fig. 62 (*Left*) CD spectra of aggregates of **1** in AOT microemulsion in the presence of L- or D-tartaric acid (trace *a*, and *b*, respectively). (*Right*) Double logarithmic plot of the dissymmetry *g*-factor vs. size of aggregates in AOT microemulsions (*open circles*), or in bulk solution (*full circle*) (adapted from [157])

The effect is proportional to the concentration of the dopant, showing a steep initial increase, followed by a slower step that reaches a plateau at 100 μM of **PGA**. This would indicate strong primary electrostatic interactions of the components, followed by a looser secondary interaction by hydrophobic effect, once the charges of the polymer are saturated. Further kinetic experiments suggested also that (1) the addition of **PGA** during the formation of the aggregates quenches the growth of the self-assembled structures, and (2) the efficiency of the chirality transfer is proportional to the size of the clusters.

The "scaling of chirality" in the aggregation of **1** has also been studied in microemulsions of water/sodium bis(2-ethylhexyl)sulfo succinate(AOT)/decane [157]. In these conditions AOT forms reversed micelles. The sizes of the pool of the reversed micelles, that can be varied by tuning the hexane/water ratio, controls the size and the coherence length of the porphyrin J-aggregates, resulting in the control of the overall asymmetry of the structures. The acidic environment for fostering the J-aggregation of **1** is given by enantiopure tartaric acid, which acts also as a chiral template. A CD spectra show the typical signature of the chiral J-aggregates, mirrored on using opposite acid enantiomers (Fig. 62).

Remarkably, the results showed an impressive correlation between the size of the structures (obtained by Dynamic Light Scattering experiments) and the dissymmetry *g*-factor ($\Delta\varepsilon/\varepsilon$) of the aggregates. This also implies that these species feature a quite ordered linear or helicoidal rod-like morphology, with extended electronic conjugation likely along all over their length. Finally, the fact that a good correlation is obtained, also including the size of fractal clusters obtained independently in bulk conditions (closed circle in Fig. 62), indicates that these latter species are composed by rod-like aggregates as repeating base-units.

Finally, intriguing aspects have been pointed out by kinetic investigations on the chiral aggregation of derivative **1** templated by tartaric acid. The results obtained showed an unexpected dependence of the rates of reaction, and of the anisotropy of the final aggregates, on the enantiomer employed as a chiral inducer [158]. At low porphyrin concentration (3 μM) in the presence of an excess of L-tartaric acid (100 mM), the aggregation follows a stretched exponential decay, which is

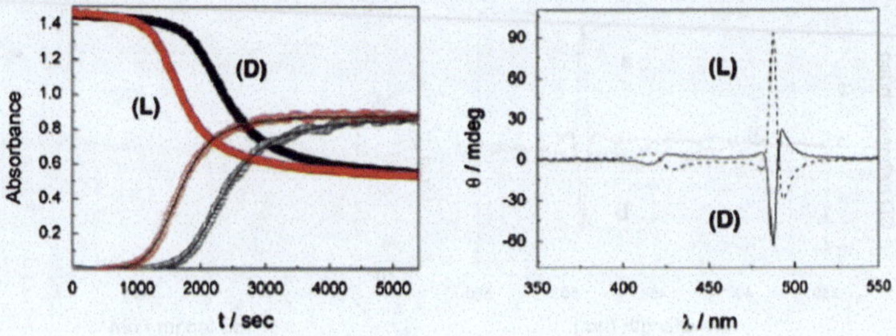

Fig. 63 (*Left*) Kinetic traces (at 434 and 491 nm) for the aggregation of **1** (3 μM) in the presence of HCl (0.5 M) and L- or D-tartaric acid (100 mM). (*Right*) Corresponding final CD spectra (adapted from [158])

complete within few hours. Conversely, in the presence of the D-enantiomer as a chiral effector, the reaction is complete within a month, with a lower dissymmetry factor of the formed aggregates. Same effect has been found in the case of aggregation fostered by HCl (0.5 M). In this case an expected autocatalytic kinetic decay is followed. Again, the presence of L- or D-tartaric acid affects both the reaction rates and the ellipticity of the aggregates (Fig. 63), indicating that these species interact with the self-assembling species presumably by hydrogen bond to the sulfonate moieties. On increasing the porphyrin concentration (20 μM), the kinetic discrimination is lost, and the transfer of the chiral information is drastically reduced for both of the enantiomers, indicating the formation of highly reactive and poorly discriminating intermediates. No differences in rates or in CD activity have been found by performing the reaction in the presence of the *meso*-form of tartaric acid. These unexpected results evidence once more the subtle balance of factors that governs the transfer of molecular information within complex systems that may strongly depend on the reaction conditions.

3.6 Asymmetric Induction by Chiral Media

The effect of the nature of the reaction media, of paramount importance for the fate of chemical reactions, cannot be overlooked. A recent work by Meijer emphasized the role of the chiral solvent in the induction of preferential handedness in supramolecular stacks of chiral organic species, giving insights into the mechanistic aspects of the phenomenon [159].

Although many of the examples reported so far evidence the strong effect of the bulk properties of the reaction media, such as polarity, pH, ionic strength and forth, the specific effects of chiral solvation on the self-assembly of chiral porphyrin-based supramolecular species is a field still almost unexplored. The effect of the (achiral) solvent on the chiral induction of an enantiopure aminoacid derivative upon binding to a bis-Zn(II)bisporphyrin system has been formerly reported by Borovkov and Inoue [160]. Some studies have been carried out in chiral phases such as gels.

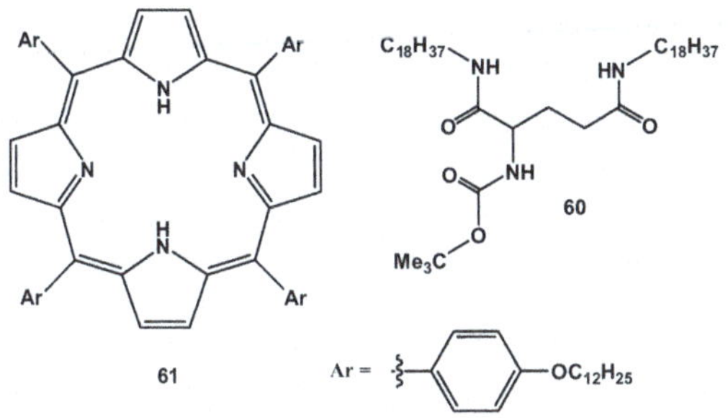

Fig. 64 Molecular structure of the gelator **60** and of the porphyrin **64**

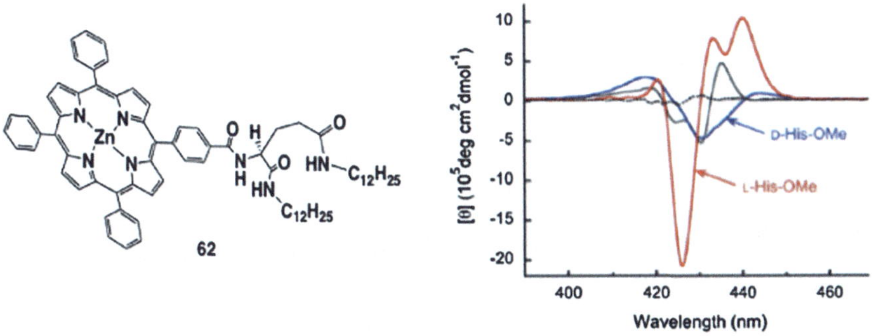

Fig. 65 Molecular structure of the receptor **62** (50 μM) and CD plot variations upon ligation with different enantiomers of histidine methylester. The inner traces of lower intensity refer to **62** in cyclohexane (gel phase; *solid line*) and $CHCl_3$ (*dotted line*) nnnnnnadapted from [162])

Liu and coworkers reported, for example, on the formation of chiral porphyrin assemblies in chiral gel, formed by DMSO and a low molecular weight chiral gelators, with transfer of chirality to the whole condensed phase [161]. The organic gelators are the *N,N′*-bis(octadecyl)-L-Boc-glutamic diamide and its D-isomer (**60**). These components gelate DMSO to give chiral phases, as indicated by the appearance of entangled helical structures of the dried xerogels (Fig. 64). The doping with the porphyrin derivatives **40**, **44** or **61**, resulted in the formation of co-gel, with the porphyrin in the form of aggregated species. CD spectroscopic studies showed that only in the case of the macrocycle possessing the longer alkyl chains, the chirality of the gel is transferred to the porphyrin systems, as indicated by the excitonically coupled bands in the Soret region. Moreover, the sign of the CD bands are in close relation with the chirality of the gelators, being positive in the case of the L-isomer and negative for the D-counterpart. On the other hand, no induction of chirality occurs in toluene gel phase, where the porphyrins are dispersed in monomeric form.

The group of Ihara reported very recently on the enantiomeric recognition of porphyrin assemblies forming chiral molecular gel in specific solvents [162]. The porphyrin used is an achiral tetraphenylporphyrin (**62**) bearing a L-glutammide (a "g" moiety) that promote a chiral staking, so forming a "secondary chiral" host (Fig. 65). The authors formerly demonstrated that a chiral Zn(II)porphyrin assembly featured enantiomeric recognition properties through axial coordination [163]. The compound aggregates in apolar solvent to give extended chiral structures showing intense CD Cotton effects. In more polar chloroform the intensities are drastically reduced, likely caused by deaggregation phenomena. The presence of the "g-moiety" infers to the macrocycle lyotropic behaviour promoting, at high concentration, gel phase transition of specific solvents as cyclohexane or toluene.

Remarkably, the addition of amino acid methyl esters that can ligate to the porphyrin in gel phase, causes dramatic variation of the CD spectral features in terms of different signs and different intensities, depending on the nature of the ligand. Quantitative effect of the selectivity in the enantiomeric recognition was obtained by fluorescence spectroscopy, showing moderate selectivity toward L-enantiomers, with the highest effect for D- and L-Histidine ($K_{(L)}/K_{(D)} = 3.74$).

4 Conclusion

An overview of the aspects of the chirality, from Life sciences to technological and practical applications has been offered. The most employed strategies for the achievement of chiral suprastructures of porphyrin derivatives has been discussed, with aim to stimulate the growth in this fascinating field of research.

References

1. Lehn J-M (1995) Supramolecular chemistry. Concepts and perspectives. VCH, Weinheim
2. Steiner T (2002) The hydrogen bond in the solid state. Angew Chem Int Ed 41:48
3. Prins LJ, Reinhoudt DN, Timmerman P (2001) Noncovalent synthesis using hydrogen bonding. Angew Chem Int Ed 40:2382
4. Fujita M (1998) Metal-directed self-assembly of two- and three-dimensional synthetic receptors. Chem Soc Rev 27:417
5. Pitt MA, Johnson DW (2007) Main group supramolecular chemistry. Chem Soc Rev 9:1441
6. Schottel BL, Chifotides HT, Dumbar KR (2008) Anion-π interactions. Chem Soc Rev 37:68
7. Metrangolo P, Meyer F, Pilati T, Resnati G, Terraneo G (2008) Halogen bonding in supramolecular chemistry. Angew Chem Int Ed 47:6114
8. Beale TM, Chudzinski MG, Sarwar MG, Taylor MS (2013) Halogen bonding in solution: thermodynamics and applications. Chem Soc Rev 42:1667
9. Dougherty DA (2013) The cation-π interaction. Acc Chem Res 46:885
10. Crego-Calama M, Reinhoudt DN (eds) (2006) Supramolecular chirality. Topics in current chemistry. Springer, New York, p 265
11. Deisenhofer J, Michel H (1989) The photosynthetic reaction center from the purple bacterium *Rhodopseudomonas viridis*. Angew Chem Int Ed Engl 28:829
12. Kendrew JC, Bodo G, Dintzis HM, Parrish RG, Wychoff HW, Phillips DC (1958) A three-dimensional model of the Myoglobin molecule obtained by X-ray. Nature 181:662
13. Perutz MF, Rossmann MG, Cullis AF, Muirhead H, North ACT (1960) Structure of Hæmoglobin: a three-dimensional Fourier synthesis at 5.5-Å. Resolution, obtained by X-ray analysis. Nature 185:416

14. Poulos TL, Finzel BC, Howard AJ (1986) Crystal structure of substrate-free Pseudomonas putida cytochrome P-450. Biochemistry 25:5314
15. Kadish KM, Smith KM, Guilard R (eds) (2010) Handbook of porphyrin science. Synthesis and coordination chemistry, vol 2. World Scientific, Singapore
16. Paolesse R, Monti D, Nardis S, Di Natale C (2010) Porphyrin-based chemical sensors. In: Kadish KM, Smith KM, Guilard R (eds) Handbook of porphyrin science, vol. 12, Chapter 54. World Scientific, Singapore, p. 121
17. de Visser SP, Nam W (2010) High-valent iron-oxo porphyrins in oxygenation reactions. In: Kadish KM, Smith KM, Guilard R (eds) Handbook of porphyrin science, vol. 10, Chapter 44. World Scientific, Singapore, p. 85
18. Fukuzumi S (2010) Artificial photosynthetic systems composed of porphyrins and Phthalocyanines. In: Kadish KM, Smith KM, Guilard R (eds) Handbook of porphyrin science, vol. 10, Chapter 46. World Scientific, Singapore, p. 183
19. Ethirajan M, Patel NJ, Pandey RK (2010) Porphyrin-based multifunctional agents for tumor-imaging and photodynamic therapy of tumors. In: Kadish KM, Smith KM, Guilard R (eds) Handbook of porphyrin science, vol. 4, Chapter 19. World Scientific, Singapore, p. 249
20. Kelvin WT (1894) J. Oxford Univ. Junior Scientific Club 18:1
21. Mislow K (2002) Stereochemical terminology and its discontents. Chirality 14:126
22. Lectures derived to *Société Chimique de Paris*; January 20 and February 3, 1860
23. Bentley R (2006) The nose as a stereochemist. Chem Rev 106:4099
24. Strong M (1999) Regulations and policies affecting FDA-regulated products. Food Drug Law J 54:3969
25. Klabunovsky E (2002) Short definitions of Life. In: Pàlyi G, Zucchi C, Caglioti L (eds) Fundamentals of life. Elsevier, New York
26. Kondepudi DK, Kaufman RJ, Singh N (1990) Chiral symmetry breaking in sodium chlorate crystallization. Science 250:975
27. Kondepudi DK, Asakura K (2001) Chiral autocatalysis, spontaneous symmetry breaking, and stochastic behavior. Acc Chem Res 34:946
28. Thiemann W (1974) Disproporptionation of enantiomers by precipitation. J Mol Evol 4:85
29. Thiemann W, Darge W (1974) Experimental attempts for the study of the origin of optical activity on earth. Orig Life Evol Biosphere 5:263
30. Lee TD, Young CN (1957) Experimental test of parity conservation in beta-decay. Phys Rev 105:1413
31. Goldhaber ML, Grodsins L, Sunyar A (1958) Helicity of neutrinos. Phys Rev 109:1015
32. Weinberg S (1967) A model of leptons. Phys Rev Lett 19:1264
33. Rein DW (1974) Some remarks on parity violating effects of intramolecular interactions. J Mol Evol 4:15
34. Woods CS, Bennett SC, Cho D, Masterson BP, Robert JL, Tanner CE, Wieman CE (1997) Measure of parity nonconservation and an anapole moment in cesium. Science 275:1759
35. Hegström R, Rein D, Sandars P (1980) Calculation of the parity nonconserving energy difference between mirror-image molecules. J Chem Phys 73:2329
36. Quack M (2002) How important is parity violation for molecular and biomolecular chirality? Angew Chem Int Ed 41:4618
37. Quack M, Stohner J (2001) Molecular chirality and the fundamental symmetry of physics: influence of parity violation on rotovibrational frequencies and thermodynamic properties. Chirality 13:745
38. Schwerdtfeger P, Gierlich J, Bollwein T (2003) Large parity-violation effects in heavy-metal-containing chiral compounds. Angew Chem Int Ed 42:1293
39. Schwerdtfeger P, Bast R (2004) Large parity violation effects in the vibrational spectrum of organometallic compounds. J Am Chem Soc 126:1652
40. Meierhenrich UJ, Nakon L, Alcaraz C, Bradehöft JH, Hoffmann SV, Barbier B, Brock A (2005) Asymmetric vacuum UV photolysis of the amino acid leucine in the solid state. Angew Chem Int Ed 44:5630

41. Bonner WA (1991) The origin and amplification of biomolecular chirality. Orig Life Evol Biosph 21:59
42. Keszthelyi L (1995) Origin of the homochirality of biomolecules. Quart Rev Biophys 28:473
43. Bailey J, Chrysostomou A, Hough JH, Gledhill TM, Mc Call A, Clark S, Ménard F, Tamura M (1998) Circular polarization in star – formation regions: implications for biomolecular homochirality. Science 281:672
44. Bailey J, Chrysostomou A, Hough JH, Gledhill TM, Mc Call A, Clark S, Ménard F, Tamura M (2001) Astronomical sources of circularly polarised light and the origin of homochirality. Orig Life Evol Biosph 31:167
45. Engel MH, Macko SA (1997) Isotopic evidences for extraterrestrial non-racemic amino acids in the Murchison meteorite. Nature 389:265
46. Kooper G, Kimmich N, Belisle W, Sarinana J, Brabham K, Garrell L (2001) Carbonaceous meteorites as a source of sugar-related organic compounds for the early Earth. Nature 414:879
47. Engel MH, Mecko SA, Silfer JA (1990) Carbon isotope composition of individual amino acids in the Murchison meteorite. Nature 348:47
48. Cronin JR, Pizzarello S (1999) Amino acid enantiomer excess in meteorites: origin and significance. Adv Space Res 23:293
49. Cronin JR, Pizzarello S (1997) Enantiomeric excess in meteoritic amino acids. Science 275:951
50. Ehrenfrend P, Glavin DP, Botta O, Cooper G, Bada JL (2001) Extraterrestrial amino acids in Orgueil and Ivuna: Tracing the parent body of Cl type carbonaceous chondrites. Proc Natl Acad Sci USA 98:2138
51. Nelson DR, Trendall AF, Altermann W (1999) Chronological correlation between Pilbara and Kaapvaal cratons. Precambrian Res 97:165
52. Ribò JM, Crusats J, Sagués F, Claret J, Rubires R (2001) Chiral sign induction by vortices during the formation of mesophases in stirred solutions. Science 292:2063
53. Frank FC (1953) On spontaneous asymmetric synthesis. Biochim Biophys Acta 11:459
54. Kondepudi DK, Prigogine I (1998) Modern thermodynamics. From heat engines to dissipative structures. Wiley, New York
55. Soai K, Shibata T, Morioka H, Choji K (1995) Asymmetric autocatalysis and amplification of enantiomeric excess of a chiral molecule. Nature 378:767
56. Soai K, Shibata T, Sato I (2000) Enantioselective automultiplication of chiral molecules by asymmetric autocatalysis. Acc Chem Res 33:382
57. Lavabre D, Micheau J-C, Islas JR, Buhse T (2008) Kinetic insight into specific features of the autocatalytic Soai reaction. Top Curr Chem 284:67
58. Blackmond DG (2004) Proc Natl Acad Sci USA 101:5732
59. Schiaffino L, Ercolani G (2008) Unravelling the mechanism of the Soai asymmetric autocatalytic reaction by First-Principle-Calculations: induction and amplification of chirality by self-assembly of hexamolecular complexes. Angew Chem Int Ed 47:6832
60. Sato I, Sugia R, Matsueda Y, Furumura Y, Soai K (2004) Asymmetric synthesis utilizing circularly polarized light mediated by the photoequilibrium of chiral olefins in conjunction with asymmetric autocatalysis. Angew Chem Int Ed 43:4490
61. Soai K, Kawasaki T, Sato I (2006) In: Rappoport Z, Marek I (eds) The chemistry of organozinc compounds, part 2. Wiley, Chichester, p. 380
62. El-Hachemi Z, Mancini G, Ribò JM, Sorrenti A (2008) Role of hydrophobic effect in the transfer of chirality from molecules to complex systems: from chiral surfactants to porphyrin/surfactants aggregates. J Am Chem Soc 130:15176
63. Rubires R, Farrera J-A, Ribò JM (2001) Stirring effects on the spontaneous formation of chirality in the homoassociation of diprotonated meso-tetraphenylsulfonato porphyrins. Chem Eur J 7:436
64. El-Hachemi Z, Escudero C, Arteaga O, Canillas A, Crusats J, Mancini G, Purrello R, Sorrenti A, D'Urso A, Ribò JM (2009) Chiral sign selection of the J-aggregates of diprotonated tetrakis-(4-sulfonatophenyl)porphyrin by traces of unidentified chiral contaminants present in the ultra-pure water used as solvent. Chirality 21:408

65. Andrade SM, Teixeira R, Costa SM, Sobral AJFN (2008) Self-aggregation of free base porphyrins in aqueous solution and in DMPC vesicles. Biophys Chem 133:1
66. Andrade SM, Costa SMB (2006) Spectroscopic studies of water-soluble porphyrins with protein encapsulated in bis(2-ethylhexyl)sulfosuccinate (AOT) reverse micelles: aggregation versus complexation. Chem Eur J 12:1046
67. Andrade SM, Costa SMB (2011) Tetrakis(4-sulfonatophenyl)porphyrin fluorescence as reporter of human serum albumin structural changes induced by guanidine hydrochloride. J Photochem Photobiol A Chem 217:125
68. Zhao L, Ma R, Li J, Li Y, An Y, Shi L (2008) J- and H-aggregates of 5,10,15,20,-(4-sulfonatophenyl)-porphyrin and interconversion in PEG-*b*-P4VP micelles. Biomacromol 9:2601
69. Zhao L, Wang X, Ma R, An Y, Shi L (2009) Chiral micelles of achiral TPPS and diblock copolymer induced by amino acids. Macromolecules 42:6253
70. Zhao L, Xiang R, Ma R, Wang X, An Y, Shi L (2011) Chiral conversion and memory of TPPS J-aggregates in complex micelles: PEG-*b*-PDMAEMA/TPPS. Langmuir 7:1154
71. Monti D, Venanzi M, Mancini G, Di Natale C, Paolesse R (2005) Supramolecular chirality control by solvent changes. Solvodichroic effect on chiral porphyrin aggregation. Chem Commun 2471
72. Monti D, De Rossi M, Sorrenti A, Laguzzi G, Gatto E, Stefanelli M, Venenzi M, Luvidi L, Mancini G, Paolesse R (2010) Supramolecular chirality in solvent-promoted aggregation of amphiphilic porphyrin derivatives: kinetic studies and comparison between solution behaviour and solid-state morphology by AFM topography. Chem Eur J 16:880
73. Pasternack RF, Fleming C, Herring S, Collings PJ, dePaula J, DeCastro G, Gibbs EJ (2000) Aggregation kinetics of extended porphyrin and cyanine dye assemblies. Biophys J 79:550
74. Monti D, Venanzi M, Stefanelli M, Sorrenti A, Mancini G, Di Natale C, Paolesse R (2007) Chiral amplification of chiral porphyrin derivatives by template heteroaggeration. J Am Chem Soc 129:6688
75. Stepanek P, Dukh M, Saman D, Moravcova J, Kniezo L, Monti D, Venanzi M, Mancini G, Drasar P (2007) Synthesis and solvent-driven self-aggregation studies of meso-"C-glycoside"-porphyrin derivatives. Org Biomol Chem 5:960
76. Monti D, Venanzi M, Gatto E, Mancini G, Sorrenti A, Stepanek P, Drasar P (2008) Study of the supramolecular chiral assembly of meso-"C-glucoside"-porphyrin derivatives in aqueous media. New J Chem 32:2127
77. Zelenka K, Trnka T, Tislerova I, Monti D, Cinti S, Naitana ML, Schiaffino L, Venanzi M, Laguzzi G, Luvidi L, Mancini G, Novakova Z, Simak O, Wimmer Z, Drasar P (2011) Spectroscopic, morphological, and mechanistic investigation of the solvent-promoted aggregation of porphyrins modified in meso-positions by glucosylated steroids. Chem Eur J 17:13743
78. Lettieri R, Monti D, Zelenka K, Trnka T, Drasar P, Venanzi M (2012) Glucosylated steroid-porphyrins as new tools for nanotechnology applications. New J Chem 36:1246
79. Dukh M, Saman D, Lang K, Pouzar V, Cerny I, Drasar P, Kral V (2007) Steroid-porphyrin conjugate for saccharide sensing in protic media. Org Biomol Chem 1:3548
80. Sakakibara S, Nakatsubo F, French AD, Rosenau T (2012) Chiroptical properties of an alternatingly functionalized cellotriose bearing two porphyrin groups. Chem Commun 48:7672
81. Lv F, He X, Lu L, Wu L, Liu T (2011) A novel water-soluble near-infrared glucose-conjugated porphyrin: synthesis, properties and its optical imaging effect. J Porphyrins Phthalocyanines 15:218
82. Iavicoli P, Xu H, Feldborg LN, Linares M, Paradinos M, Stafström S, Ocal C, Niet-Ortega B, Casado J, Lopez Navarrete JT, Lazzaroni R, De Feyter S, Amabilino DB (2010) Tuning the supramolecular chirality of one- and two-dimensional aggregates with the number of stereogenic centers in the component porphyrins. J Am Chem Soc 132:9350
83. Linares M, Iavicoli P, Psychogyiopoulou K, Beljonne D, De Feyter S, Amabilino DB, Lazzaroni R (2008) Chiral expression at the solid–liquid interfaces: a joint experimental and theoretical study of the self-assembly of chiral porphyrins on graphite. Langmuir 24:9566

84. Ma S, Cao X, Mak M, Sadik A, Walkner C, Freedman TB, Lednev IK, Dukor RK, Nafie LA (2007) Vibrational circular dichroism shows unusual sensitivity to protein fibril formation and development in solution. J Am Chem Soc 129:12364
85. Nieto-Ortega B, Ramìrez FJ, Amabilino DA, Linares M, Beljonne D, Lòpez Navarrete JT, Casado J (2012) Electronic and vibrational circular dichroism spectroscopies for the understanding of chiral organization in porphyrin aggregates. Chem Commun 48:9147
86. Iavicoli P, Simòn-Sorbed M, Amabilino DB (2009) Surface morphology of chiral porphyrins as a function of constitution and amphiphilic nature. New J Chem 33:358
87. Feldborg NL, Saletra WJ, Iavicoli P, Amabilino DB (2011) Central metal ion determined self-assembly of intrinsically chiral porphyrins. J Porphyrins Phthalocyanines 15:996
88. Green MM, Peterson NC, Sato T, Teramoto A, Cook R, Lifson S (1995) A helical polymer with a cooperative response to chiral information. Science 268:1860
89. van Gestel J, Palmans ARA, Titulaer B, Vekemans JAJM, Meijer EW (2005) "Majority-rules" operative in chiral columnar stacks of C_3-symmetrical molecules. J Am Chem Soc 127:5490
90. Hoeben FJM, Wolffs M, Zhong J, De Feyter S, Leclère PELG, Schenning APHJ, Meijer EW (2007) Influence of supramolecular organization on energy transfer properties in chiral oligo (*p*-phenylene vinylene) porphyrin assemblies. J Am Chem Soc 129:9819
91. Helmich F, Lee CC, Nieuwenhuizen MML, Gielen JC, Christianen PCM, Larsen A, Fytas G, Leclère PELG, Schenning APJH, Meijer EW (2010) Dilution-induced self-assembly of porphyrin aggregates: a consequence of coupled equilibria. Angew Chem Int Ed 49:3939
92. Smoulders MMJ, Schenning APHJ, Meijer EW (2008) Insights into the mechanisms of cooperative self-assembly: the "sergeants-and-soldiers" principle of chiral and achiral C_3-symmetrical discotic triamides. J Am Chem Soc 130:606
93. Helmich F, Lee CC, Schenning APHJ, Meijer EW (2010) Chiral memory via chiral amplification and selective depolymerisation of porphyrin aggregates. J Am Chem Soc 132:16753
94. Purrello R (2003) Supramolecular chemistry: lasting chiral memory. Nat Mater 2:216
95. Helmich F, Smulders MMJ, Lee CC, Schenning APHJ, Meijer EW (2011) Effect of stereogenic centers on the self-sorting, depolymerisation, and atropisomerisation kinetics of porphyrin-based aggregates. J Am Chem Soc 133:12238
96. Smoulders MMJ, Filot IAW, Leenders JMA, van der Scoot P, Palmans ARA, Schenning APHJ, Meijer EW (2010) Tuning the extent of chiral amplification by temperature in a dynamic supramolecular polymer. J Am Chem Soc 132:611
97. van Hameren R, Schön P, van Buul AM, Hoogboom J, Lazarenko SV, Gerritsen JW, Engelkamp H, Christianen PCM, Heus HA, Maan JC, Rasing T, Speller S, Rowan AE, Elemans JAAW, Nolte RJM (2006) Macroscopic hierarchical surface patterning of porphyrin trimers via self-assembly and dewetting. Science 314:1433
98. van Hameren R, van Buul AM, Castriciano MA, Villari V, Micali N, Schön P, Speller S, Monsù Scolaro L, Rowan AE, Elemans JAAW, Nolte RJM (2008) Supramolecular porphyrin polymers in solution and at the solid–liquid interface. Nano Lett 8:253
99. Veling N, van Hameren R, van Buul AM, Rowan AE, Nolte RJM, Elemans JAAW (2012) Solvent-dependent amplification of chirality in assemblies of porphyrin trimers based on benzene tricarboamide. Chem Commun 48:4371
100. Balaban TS (2005) Tailoring porphyrins and chlorins for self-assembly in biomimetic artificial antenna systems. Acc Chem Res 38:612
101. Balaban TS, Linke-Scaetzel M, Bhise AD, Vanthuyne N, Roussel C, Anson CO, Buth G, Eichhöfer A, Foster K, Garab G, Glieman H, Goddard R, Javorfi T, Powell AK, Rösner H, Schimmel T (2005) Structural characterization of artificial self-assembling porphyrin that mimic the natural chlorosomal bacteriochlorophylls c, d, and e. Chem Eur J 11:2268
102. Balaban TS, Berova N, Drain CM, Hauschild R, Huang X, Kalt H, Lebedkin S, Lehn J-M, Nifaitis F, Pescitelli G, Prokhorenko VI, Riedel G, Smeureanu G, Zeller J (2007) Syntheses and energy transfer in multiporphyrinic arrays self-assembled with idrogen-bonding recognition groups and comparison with covalent steroidal models. Chem Eur J 13:8411

103. Huijer A, Merek PL, Savenije TJ, Siebbeles LDA, Scherer T, Hauschild R, Szmytkowski J, Kalt H, Hehn H, Balaban TS (2007) Photosensitization of TiO2 and SnO2 by artificial Self-Assembling Mimics of the Natural Chlorosomal Bacteriochlorophylls. J Phys Chem C 111:11726
104. Szmytkowski J, Conradt J, Kuhn H, Reddy CM, Balaban MC, Balaban TS, Kalt H (2011) Self-assemblies of novel magnesium porphyrins mimicking natural chlorosomal bacterio-chlorophylls. J Phys Chem C 115:8832
105. Balaban TS, Bhise AD, Bringmann G, Bürk J, Chappez-Gillot C, Eichhöfer A, Fonske D, Götz DCG, Knauer M, Mizoguchi T, Mössinger D, Rösner H, Roussel C, Schraut M, Tamiaki H, Vanthuyne N (2009) Mimics of the self-assembling chlorosomal bacteriochlorophylls: region- and stereoselective synthesis and stereoanalysis of acyl(1-hydroxyalky)porphyrins. J Am Chem Soc 131:14480
106. Chappaz-Gillot C, Marek PL, Blaive BJ, Canard G, Bürck J, Garab G, Hahn H, Javorfi T, Kelemen L, Krupke R, Mössinger D, Ormos P, Reddy CM, Roussel C, Steinbach G, Szabo M, Ulrich AS, Vanthuyne N, Vijayaraghavan A, Zupcanova A, Balaban TS (2012) Anisotropic organization and microscopic manipulation of self-assembling synthetic porphyrin microrods that mimic chlorosomes: bacterial light harvesting systems. J Am Chem Soc 134:944
107. Ulman A (1991) An introduction to ultrathin organic films – from Langmuir-Blodgett to self-assembly. Academic, Boston, MA
108. Chen P, Ma X, Duan P, Liu M (2006) Chiral amplification of porphyrin assemblies exclusively constructed from achiral porphyrin derivatives. ChemPhysChem 7:2419
109. Zhang Y, Chen P, Liu M (2008) A general method for constructing optically active supramolecular assemblies from intrinsically achiral water-insoluble free-base porphyrins. Chem Eur J 14:1793
110. Zhang Y, Chen P, Ma Y, He S, Liu M (2009) Acidification and assembly of porphyrin at an interface: counterion matching, selectivity, and supramolecular chirality. ACS Appl Mater Interfaces 1:2036
111. De Luca G, Romeo A, Monsù Scolaro L (2005) Role of the counteranions in the acid-induced aggregation of isomeric tetrapyridylporphyrins in organic solvents. J Phys Chem B 109:719
112. Doan SC, Shanmughan S, Aston DE, McHale JL (2005) Counterion dependent dye aggregates: nanorods and nanorings of tetra(p-carboxyphenyl)porphyrin. J Am Chem Soc 127:5885
113. Qiu Y, Chen P, Liu M (2010) Interfacial assemblies of atypical amphiphilic porphyrins: hydrophobicity/hydrophilicity of substituents, annealing effects, and supramolecular chirality. Langmuir 26:15272
114. Yao Y, Qiu Y, Chen P, Ma Y, He S, Zheng J-H, Liu M (2010) Interfacial molecular assemblies of metalloporphyrins with two trans or one axial ligands. ChemPhysChem 11:722
115. Rong Y, Chen P, Wang D, Liu M (2012) Porphyrin assemblies through the air/water interface: effect of Hydrogen bond, thermal annealing, and amplification of supramolecular chirality. Langmuir 28:6356
116. Chen P, Ma X, Hu K, Rong Y, Liu M (2011) Left or right? The direction of compression-generated vortex-like flow selects the macroscopic chirality of interfacial molecular assemblies. Chem Eur J 17:12108
117. Avetisov V, Goldanskii V (1996) Mirror symmetry breaking at the molecular level. Proc Natl Acad Sci USA 93:11435
118. Ohno O, Kaizu Y, Kobayashi H (1993) J-aggregate formation of a water soluble porphyrin in acidic aqueous media. J Chem Phys 99:4128
119. Crusats J, Claret JM, Diez-Pérez I, El-Hachem Z, Garcia-Hortega H, Rubires R, Sagués F, Ribò JM (2003) Chiral shape and enantioselective growth of colloidal particles of self-assembled meso-tetra(phenyl and 4-sulfonatophenyl)porphyrins. Chem Commun 1588
120. Amabilino DB (2007) Nanofibre whirlpools. Nat Mater 6:924
121. Escudero C, Crusats J, Diez-Perez I, El-Hachemi Z, Ribò JM (2006) Folding of 5-phenyl-10,15,20-tris-(4-sulfophenyl)porphyrin. Angew Chem It Ed 45:8032

122. El-Hachemi Z, Arteaga O, Canillas A, Crusats J, Escudero C, Kuroda R, Harada T, Rosa M, Ribò JM (2008) On the Mechano-chiral effect of vortical flows on the dichroic spectra of 5-phenyl-10,15,20-tris-(4-sulfophenyl)porphyrin J-Aggregates. Chem Eur J 14:6438
123. Arteaga O, Escudero C, Oncins G, El-Hachemi Z, Llorens J, Crusats J, Canillas A, Ribò JM (2009) Reversible mechanical induction of optical activity in solutions of soft-matter nanophases. Chem Eur J 4:1687
124. Spada GP (2008) Alignment by the convective and vortex flow of achiral self-assembled fibers induces strong circular dichroism effects. Angew Chem Int Ed 47:636
125. Arteaga O, Canillas A, Crusats J, El-Hachemi Z, Llorens J, Sacristan E, Ribò JM (2010) Emergence of supramolecular chirality by flows. ChemPhysChem 11:3511
126. D'Urso A, Randazzo R, Lo Faro L, Purrello R (2010) Vortexes and nanoscale chirality. Angew Chem Int Ed 49:108
127. D'Urso A, Fragalà ME, Purrello R (2010) From self-assembly to noncovalent synthesis of programmable porphyrins' arrays in aqueous solution. Chem Commun 48:8165
128. Sorrenti A, El-Hachemi Z, Crusats J, Ribò JM (2011) Effects of flow-selectivity on self-assembly and auto-organization processes: an example. Chem Commun 47:9551
129. Sorrenti A, El-Hachemi Z, Arteaga O, Canillas A, Crusats J, Ribò JM (2012) Kinetic control of the supramolecular chirality of porphyrin J-aggregates. Chem Eur J 18:8820
130. Zhang L, Liu M (2009) Supramolecular chirality and chirality inversion of tetraphenylsulfonato porphyrin assemblies on optically active polylysine. J Phys Chem B 113:14015
131. Wada S, Fujiwara K, Monjushiro H, Watarai H (2007) Optical chirality of protonated tetraphenylporphyrin J-aggregate formed at the liquid-liquid interface in a centrifugal liquid membrane cell. J Phys Condens Matter 19:375105
132. Micali N, Engelkamp H, van Rhee PG, Christianen PCM, Monsù Scolaro L, Maan JC (2012) Selection of supramolecular chirality by application of rotational and magnetic forces. Nat Chem 4:201
133. Kitagawa Y, Sagawa H, Ishi K (2011) Magneto-chiral dichroism of organic compounds. Angew Chem Int Ed 50:9133
134. Pasternack RF, Goldsmith JL, Szép S, Gibbs EJ (1998) A spectroscopic and thermodynamic study of porphyrin/DNA Supramolecular assemblies. Biophys J 75:1024
135. Pasternack RF, Gibbs EJ, Bruzewicz D, Stewart D, Engstrom KS (2002) Kinetic of disassembly of a DNA-bound porphyrin supramolecular array. J Am Chem Soc 124:3533
136. Purrello R, Raudino A, Monsù Scolaro L, Micali N, Purrello R (2002) From achiral porphyrins to template-imprinted chiral aggregates and further. Self-replication of chiral memory from scratch. J Am Chem Soc 124:894
137. Balaz M, De Napoli M, Holmes AE, Mammana A, Nakanishi N, Berova N, Purrello RA (2005) Cationic zinc porphyrin as a chiroptical probe for Z-DNA. Angew Chem Int Ed 44:4006
138. Onouchi H, Miyagawa T, Morino K, Yashima E (2006) Assisted formation of chiral porphyrin homoaggregates by an induced helical poly(phenylacetylene) template and their chiral memory. Angew Chem Int Ed 45:2381
139. Mammana A, D'Urso A, Lauceri R, Purrello R (2007) Switching off and on the supramolecular chiral memory in porphyrin assemblies. J Am Chem Soc 129:8062
140. Matassa R, Carbone M, Lauceri R, Purrello R, Caminiti R (2007) Supramolecular structure of extrinsically chiral porphyrin hetero-assemblies and achiral analogues. Adv Mater 19:3961
141. Lauceri R, Fasciglione GF, D'Urso A, Marini S, Purrello R, Coletta M (2008) Kinetic investigation of porphyrin interaction with chiral templates reveals unexpected features of the induction and self-propagation mechanism of chiral memory. J Am Chem Soc 130:10476
142. Grätzel M (2009) Recent advances in sensitized mesoscopic solar cells. Acc Chem Res 42:1788
143. D'Urso A, Nicotra PF, Centonze G, Fragalà ME, Gattuso G, Notti A, Pappalardo A, Pappalardo S, Parisi MF, Purrello R (2012) Induction of chirality in porphyrin-(bis)calixarene assemblies: a mixed covalent-non-covalent vs a fully non-covalent approach. Chem Commun 48:4046

144. Jiang S, Zhang L, Liu M (2009) Photo-triggered J-aggregation and chiral symmetry breaking of an anionic porphyrin (TPPS) in mixed organic solvent. Chem Commun 6252
145. Monsù Scolaro L, Romeo A, Castriciano M, De Luca G, Patanè S, Micali N (2003) Porphyrin deposition induced by UV-irradiation. J Am Chem Soc 125:2040
146. Monsù Scolaro L, Romeo A, Castriciano MA, Micali N (2005) Unusual properties of porphyrin fractal J-aggregates. Chem Commun:3018
147. Zhang L, Tian Y, Liu M (2011) Ionic liquid induced spontaneous symmetry breaking: emergence of predominant handedness during the self-assembly of tetrakis(4-sulfonatopphenyl) porphyrin(TPPS) with achiral ionic liquid. Phys Chem Chem Phys 13:17205
148. Pescitelli G, Di Bari L, Berova N (2011) Conformational aspects in the studies of organic compounds by electronic circular Dichroism. Chem Soc Rev 40:4603
149. Balaz M, Bitsch-Jensen K, Mammana A, Ellestad GA, Nakanishi K, Berova N (2007) Porphyrin as spectroscopic sensors for conformational studies of DNA. Pure Appl Chem 79:801
150. D'Urso A, Holmes AE, Berova N, Balaz M, Purrello R (2011) Z-DNA recognition in B-Z-B sequences by a cationic zinc porphyrin. Chem Asian J 6:3104
151. D'Urso A, Mammana A, Balaz M, Holmes AE, Berova N, Lauceri R, Purrello R (2009) Interactions of tetraanionic porphyrin with DNA: from a Z-DNA sensor to a versatile supramolecular device. J Am Chem Soc 131:2046
152. Ha SC, Lowenhaupt K, Rich A, Kim YG, Kim KK (2005) Crystal structure of a junction between B-DNA and Z-DNA reveals two extruded bases. Nature 437:1183
153. De Luca G, Romeo A, Monsù Scolaro L, Pasternack RF (2010) Conformations of a model protein revealed by an aggregating CuII porphyrin: sensing the difference. Chem Commun 46:389
154. Occhiuto I, De Luca G, Villari V, Romeo A, Micali N, Pasternack RF, Scolaro LM (2011) Supramolecular chirality transfer to large random aggregates of porphyrins. Chem Commun 47:6045
155. Pasternack RF, Bustamante C, Collings PJ, Giannetto A, Gibbs EJ (1993) Porphyrin assemblies on DNA as studied by resonance light-scattering techniques. J Am Chem Soc 115:5393
156. Katzenelson O, Hel-Or HZ, Anvir D (1996) Chirality of large random supramolecular structures. Chem Eur J 2:174
157. Castriciano MA, Romeo A, De Luca G, Villari V, Scolaro LM, Micali N (2011) Scaling the chirality in porphyrin J-nanoaggregates. J Am Chem Soc 133:765
158. Castriciano MA, Romeo A, Zagami R, Micali N, Monsù Scolaro L (2012) Kinetic effects of tartaric acid on the growth of chiral J-aggregates of tetrakis(4-sulfonatophenyl)porphyrin. Chem Commun 48:4874
159. George SJ, Tomovic Z, Schenning APHJ, Meijer EW (2011) Insight into the chiral induction in supramolecular stacks through preferential chiral solvation. Chem Commun 47:3451, and references therein
160. Borovkov VV, Hembury GA, Inoue Y (2003) The origin of solvent-controlled supramolecular chirality switching in a bis(zinc porphyrin) system. Angew Chem Int Ed 42:5310
161. Li Y, Wang T, Liu M (2007) Gelating-induced supramolecular chirality of achiral porphyrins: chiroptical switch between achiral molecules and chiral assemblies. Soft Matter 3:1312
162. Jintoku H, Takafuji M, Oda R, Ihara H (2012) Enantioselective recognition by a highly ordered porphyrin-assembly on a chiral molecular gel. Chem Commun 48:4881
163. Jintoku H, Sagawa T, Sawada T, Takafuji T, Ihara H (2010) Versatile chiroptics of peptide-induced assemblies of metalloporphyrins. Org Biomol Chem 8:1344

Index

The manufacturer's authorised representative in the EU is Springer
Nature Customer Service Centre GmbH, Europaplatz 3, 69115 Heidelberg,
Germany. If you have any concerns regarding our products, please
contact ProductSafety@springernature.com

Printed and bound by CPI Group (UK) Ltd, Croydon, CR0 4YY
30/09/2020
02225591-0011